A NEW PLANTATION SOUTH: Land, Labor, and Federal Favor in Twentieth-Century Arkansas

Carter G. Woodson Institute Series in Black Studies

Armstead Robinson—*General Editor*

Michael Plunkett
Afro-American Sources in Virginia: A Guide to Manuscripts

Sally Belfrage
Freedom Summer

Armstead L. Robinson and Patricia Sullivan, eds.
New Directions in Civil Rights Studies

Leroy Vail and Landeg White
Power and the Praise Poem: Southern African Voices in History

Robert A. Pratt
The Color of Their Skin: Education and Race in Richmond, Virginia, 1954–89

Ira Berlin and Philip D. Morgan, eds.
Cultivation and Culture: Labor and the Shaping of Slave Life in the Americas

Gerald Horne
Fire This Time: The Watts Uprising and the 1960s

Sam C. Nolutshungu
Limits of Anarchy: Intervention and State Formation in Chad

Jeannie M. Whayne
A New Plantation South: Land, Labor, and Federal Favor in Twentieth-Century Arkansas

A NEW PLANTATION SOUTH: Land, Labor, and Federal Favor in Twentieth-Century Arkansas

JEANNIE M. WHAYNE

University Press of Virginia
Charlottesville & London

THE UNIVERSITY PRESS OF VIRGINIA

Copyright © 1996 by the Rector and Visitors
of the University of Virginia

First published 1996

♾ The paper used in this publication meets the minimum requirements of the American National Standard for Information Sciences—Permanence of Paper for Printed Library Materials, ANSI Z39.48-1984.

Library of Congress Cataloging-in-Publication Data

Whayne, Jeannie M.
    A new plantation south : land, labor, and federal favor in twentieth-century Arkansas / Jeannie M. Whayne.
        p.   cm. — (Carter G. Woodson Institute series in Black studies)
    Includes bibliographical references and index.
    ISBN 0-8139-1655-0 (cloth : alk. paper)
    1. Plantations—Arkansas—History. 2. Agricultural laborers—Arkansas—History. 3. Arkansas—Economic conditions. 4. New Deal, 1933–1939. I. Whayne, Jeannie M. II. Title. III. Series.
HD1471.U52A887 1996
338.1'09767—dc20   95-42606
                                                                        CIP

Printed in the United States of America

To Armstead Robinson

CONTENTS

Acknowledgments	xiii
Introduction	1
1. "A Wild and Sickly Country"	9
2. From Railroad Camp to Plantation Town	21
3. Laboring in the "American Congo"	47
4. Sunk Lands and Lost Hopes	86
5. Different Roads to Development	113
6. Delta Blues	138
7. The New Deal and the Old Plantation	157
8. Meeting on the "Turn Row"	184
9. No Way for a Man to Live	219
Appendix: Quantitative Methods	237
Notes	245
Bibliography	277
Index	307

ILLUSTRATIONS

following page 77

1. Railroad work crew
2. High-water marks on cypress trees
3. Ritter store
4. Ritter family
5. Mules dragging a car out of the water
6. Charlie B. Greenwood
7. Eliga ("Kidd") Abbott family
8. Surveyors in the swamp
9. Cabin in the swamps
10. Moving time

following page 175

11. Sharecroppers "going home"
12. Children chopping cotton
13. Evicted from Dibble plantation
14. Ward A. Rodgers and H. L. Mitchell at Rodgers's trial
15. Norman Thomas speaks at a sharecropper meeting
16. Rev. E. B. ("Britt") McKinney, STFU vice president

17. Arkansas farmworkers at a union meeting

18. Summer church

Maps

1. Arkansas	xviii
2. Crowley's Ridge in Arkansas and Missouri	11
3. Poinsett and surrounding counties	13
4. Roads in Poinsett County, 1910–20	41
5. Counties involved in sunk lands and lakes dispute	89
6. St. Francis River basin and major rivers of Missouri and Arkansas	109

Graphs

1. Acres in cotton, soybeans, rice, and corn in Poinsett County, 1930 and 1940	170
2. Gross profit in cotton and soybeans in Poinsett County, 1924–40	170

TABLES

2.1. Occupations of adult men in Marked Tree, Arkansas, 1900–1920 — 33

2.2. Population of Marked Tree, Arkansas, 1900–1920 — 38

2.3. Land tenure of farmers who were heads of households in selected Poinsett County townships, 1900–1920 — 40

2.4. Sources of credit for lenders and borrowers of real property mortgages in selected locations, 1909–10 — 44

3.1. Nativity of all delta farmers in selected townships including owners and farm laborers, 1900–1920 — 71

3.2. Race of all delta farmers including owners and farm laborers, 1900–1920 — 73

3.3. Race and land tenure of delta residents who classified themselves as farmers and who were heads of households, 1900–1920 — 74

4.1. Occupations of adult men in Lepanto, Arkansas, 1910–20 — 98

5.1. Population of Poinsett County by section, 1900–1920 — 127

5.2. Population of delta townships in Poinsett County, 1900–1920 — 127

6.1. Land forfeiture in Poinsett County, 1920–30 — 142

6.2. Land tenure in Poinsett, Cross, Mississippi, and Crittenden counties, 1920–30 — 143

6.3.	Livestock owned by residents of Tyronza Township, 1920–33	144
7.1.	County committee for 1933 plow-up campaign	162
8.1.	Change in personal property tax filers in Tyronza Township, first quartile, 1920–34	200
8.2.	Mule ownership by race in Tyronza Township, first quartile, 1920–34	201
8.3.	Decline in personal property wealth and change in number of filers in Tyronza Township by quartile, 1920–34	203

ACKNOWLEDGMENTS

Many individuals contributed immeasurably to the completion of *New Plantation South*. Librarians and archivists, colleagues and critics, and, of course, granting institutions have provided assistance, advice, criticism, and support, for which I am extremely grateful.

My first debt is to the people of Poinsett County who spoke candidly with me about their history. Mary Ann Ritter Arnold, Virgie Waskom Dawson, Noah and Sally Hazel, Sue Chambers, Billy McClellan, Gertrud Marshall, and Merce Whayne are among those who have provided thoughtful observations and answered numerous queries. Judge Steve Ryan and his compatriots in the courthouse in Harrisburg made records available to me and were always courteous and kind. I am grateful also to Jean Thatcher of Cabot, Arkansas, and to Oscar Fendler, Michael Wilson, and Dr. Eldon Fairley of Mississippi County, Arkansas, for their many insights and suggestions.

Librarians and archivists at the University of Arkansas, University of California, San Diego, and the University of Virginia have made my research a great deal easier, and I wish to thank especially Michael Dabrishus, Andrea Cantrell, and Ellen Shipley at the University of Arkansas Special Collections Division. Their efforts to collect and maintain archival material on Arkansas is truly outstanding and greatly appreciated. I am also indebted to Dr. John Ferguson and Russell Baker at the Arkansas History Commission, Linda Pine at the Archives and Special Collections Division of the University of Arkansas at Little Rock, Tom Dillard at the Torreyson Library at the University of Central Arkansas, Barbara Rust at the Federal Regional Archives in Fort Worth, Texas, and the staffs of the Library of Congress and the National Archives in Washington, D.C.

I am deeply indebted to my many colleagues in the Department of History at the University of Arkansas who have so unselfishly given of their time and attention in reading drafts of various chapters, especially Willard Gatewood, Elliott West, and Richard Sonn. Insights inspired by discussions with various other faculty colleagues, including Randall Woods, Dan Sutherland, Anne Bailey, David Sloan, and James Chase have made their way into the book. The office staff, both past and present, are to be thanked for their assistance: Lindi Holmes, Suzanne Smith, Mary Kirkpatrick, Miriam Twyford, and Kim Chenault. I am especially grateful to Gretchen Gearhart who has been unfailingly competent and professional. Thanks are also due John Wampler who assisted very ably in the final editing.

A fellowship at the Smithsonian Institution provided an opportunity for me to focus my energies on the project and to work with a number of scholars who have contributed significantly to *New Plantation South*. At the Smithsonian, Pete Daniel, whose own work on the twentieth-century South has been an inspiration, gave freely of his time and attention. I wish also to thank several fellows at the Smithsonian for their many suggestions and insights: Stephanie McCurry, Grace Palidino, and Wayne Durrell. I am especially indebted to the CL group and to those involved in Pete's Tuesday afternoon seminar.

The Carter G. Woodson Institute for the Study of Civil Rights provided another fellowship year, allowing me, once again, to concentrate fully on the manuscript. I am grateful to Armstead Robinson and William Jackson for their support, and I wish to thank the talented group of fellows present at the institute when I arrived and to whom I owe a great debt for fellowship and intellectual stimulation: Hannah Rosen, Cynthia Blair, Mary Johnson Osirim, Daryl Scott, David Murray, Cynthia Taylor, and Darrin Smith. I am also indebted to the office staff at the institute: Gail Shirley and Mary Grace. The faculty in the history department at the University of Virginia were very generous with suggestions and criticisms, and I wish to thank especially: Edward Ayers, Paul Gaston, and Nelson Lichtenstein. Thanks are due to Richard Holway of the University Press of Virginia for his enthusiasm for the project and for understanding what I was attempting to accomplish in *New Plantation South*.

I was fortunate to have been associated with the Southern History Seminar conducted by three campuses of the University of California—

San Diego, Irvine, and Riverside—and I owe a number of individuals connected with that seminar a great debt. I am especially grateful to Steven Hahn for his willingness to read countless drafts of the manuscript. Others who have contributed significantly of their time and advice include Julie Saville, Rachel Klein, Michael Johnson, Dickson Bruce, Jonathan Wiener, David Rankin, Roger Ransom, Jerry Shenk, Michael Gorman, Jewell Spangler, and Chris Webb.

A number of other historians have read drafts or chapters of *New Plantation South* or listened to outlines of the general thesis. They have offered criticisms and suggestions which I have attempted to incorporate. Thanks are due James Cobb, Jack Temple Kirby, Robert McMath, Nan Woodruff, Alan Gallay, Peter Coclanis, and Robin D. G. Kelley. Economic historians Roger Ransom, Paul Paskoff, and Harry Scheiber as well as political scientists Jeff Ryan and William Miller provided crucial assistance with the computerized data and the arrangement of tables. I am very grateful to Mary Kennedy of the Arkansas Archaeological Survey for developing the maps and illustrations. I wish also to thank my fellow members of the Board of Trustees of the Arkansas Historical Association for their many insights and suggestions, especially John Graves and Morris Arnold. A special thank you is due to Paul McBride for assistance in acquiring some crucial data concerning homesteaders in the Arkansas delta. Thanks also to Michael Dougan for his helpful suggestions. I am grateful to Richard Buckelew, Lori Bogle, Terry Buckalew, Gary Battershell, Gary Zellar, and Tom DeBlack. Countless discussions with them over the years have helped me rework some of the ideas in *New Plantation South*.

A part of chapter 2 was published in *Agricultural History* (Winter 1992); and an earlier version of chapter 4 was published in *Forest and Conservation History* (April 1993); I am grateful to the journals for granting permission to reprint the material here. Two other publications resulted from ideas originally articulated in this book: a full-blown treatment of the prohibition movement, comparing Marked Tree and Osceola, was published in *Locus* (Spring 1995); and a fuller analysis of the segregated farm extension program in Poinsett County was explored in an article in the *Mississippi Quarterly* (Fall 1992).

My most personal thanks belong to my family whose support and patience have been astonishing. My brother, Earl Whayne, with whom I play correspondence chess, has generously kept my mind in touch with complex and shifting variables. My sister, Roberta Powell, has always eagerly read and commented on my work. My aunt, Katy Cotter, has been a great inspiration to me. My husband, John Cook, has always understood my need to focus on my work intensely and yet unfailingly sensed when a drive to our place on the lake was necessary. To him I owe my greatest thanks.

A NEW PLANTATION SOUTH: Land, Labor, and Federal Favor in Twentieth-Century Arkansas

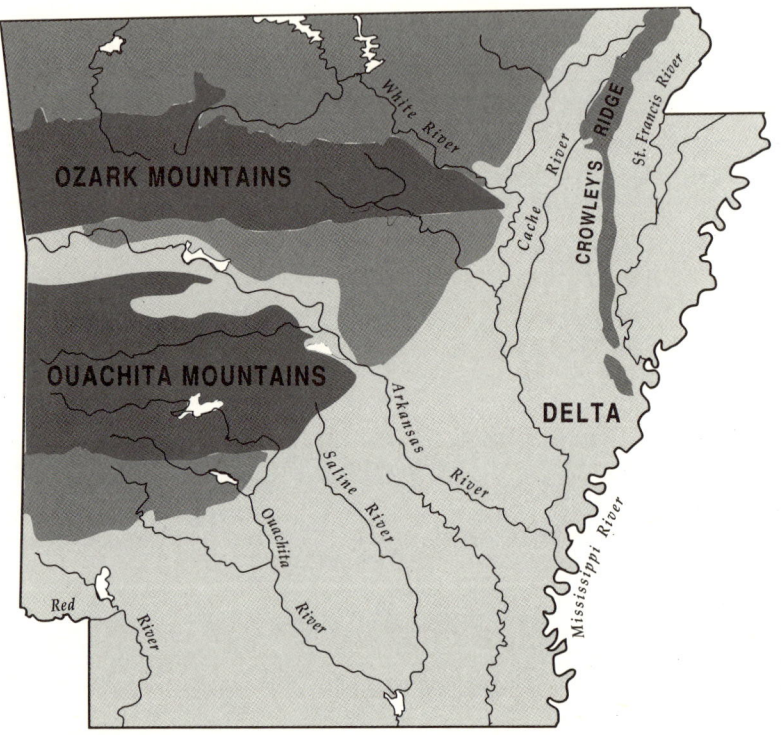

Map 1. Arkansas

INTRODUCTION

At the end of the nineteenth century, the northeastern Arkansas delta included one of the last frontiers in the United States (map 1). Majestic stands of virgin timber covered hundreds of thousands of acres, but the presence of a swamp made this timber virtually inaccessible to the lumber industry. Beginning in the 1880s railroad men penetrated the swamp to lay track and devoted considerable energy and expense to erecting earthwork roadbeds above the floodplain. Once the railroads were in place and enterprises like the Chapman and Dewey Lumber Company set up operations in remote places like Poinsett County, the business of clearing the swamps began in earnest. With the felling of the old growth timber and the appearance of clearings, drainage enterprises began, and crops were planted. It became readily apparent that this final frontier included some of the most fertile soil remaining uncultivated in the country. Thus began the exploitation of the region's last resource: the land itself.

The men who pushed into the northeastern Arkansas delta beginning in the late nineteenth century included both rich and poor, black and white, southerners and midwesterners. Most of them lacked capital and were from the South. Attracted by the railroads, the timber industry, and the sawmills, they wandered into and out of the area. A few opened mercantile establishments and put down roots in the raw young communities that sprang up in the backwoods. While others moved from Arkansas to Missouri to southern Illinois to Kentucky to Tennessee to Mississippi and back to Arkansas in an endless and often fruitless search for steady work, those who remained in place began to build roads, dig drainage ditches, establish plantations, and vie for political power.

Even as they all struggled to tame a wild and watery domain, they began to do battle with one another over the direction that the area's development would take and over the very nature of the society they would create. While small businessmen in towns like Marked Tree insisted on plantation agriculture and its satellites—cotton, sharecropping, the crop lien, and the commissary—other men clamored for the opportunity to homestead a few hundred acres and carve out an independent existence for themselves. A union of homesteaders sprang up in the delta to do battle with the emerging delta elite, and the ensuing encounters took place in the courtroom, in the legislature, and in the dead of night.

The businessmen-planters of the delta, who became a kind of plantation elite, were really middle-class businessmen who had a great deal in common with their counterparts in an older area of the county, Crowley's Ridge, where the county seat was located. Crowley's Ridge was sparsely settled, largely because it was hilly and relatively infertile, but a powerful courthouse clique existed there and the businessmen-planters in the delta would find it necessary to engage in a protracted struggle with them over control of the county government. Sometimes their political differences were temporarily put aside to fight a common foe. Two issues that united them were prohibition and segregation. In Arkansas, as elsewhere in the South, prohibition became intimately linked to segregation. Some middle-class whites in the delta imbibed the "progressive" impulse sweeping across the South in the early twentieth century and insisted on replicating the segregated communities they were familiar with elsewhere. Others in the delta violated Jim Crow attitudes and mingled in legal and illegal saloons, provoking a struggle over prohibition which was linked with segregation and pitted some middle-class merchants against the workers. A few merchants, however, founded their fortunes by providing "working-men's clubs" and struggled alongside the sawmill workers against local progressives determined to close the saloons. The fluid nature of race relations within one segment of the population—young, unmarried, male mill hands—suggests that despite both de facto and de jure segregation elsewhere, something else was possible in a place where prevailing social constraints had yet to be imposed. That some of the delta merchants were unconcerned about racial intermingling in their establishments may say more about their attention to the bottom line. The saloon business was quite lucrative, and what linked the delta merchants on both sides of the prohibition issue was their farming interests. Whether they owned one hundred or five thousand acres, they supported the founding of plantations.[1]

By the end of the second decade of the twentieth century, both the plantation and prohibition had prevailed. The remaining homesteaders struggled on, and bootleggers became folk heroes. But the 1920s confronted businessmen-planters, small farmers, and sawmill workers alike with even greater challenges. A serious crisis in the farm economy compounded by two natural disasters, the flood of 1927 and drought of 1930–31, resulted in massive tax defaults. Great expectations during the World War I boom had led to expensive drainage enterprises that proved to be impossible to pay for during the disastrous 1920s. In the next decade federal government programs rescued the plantation economy and a final battle over the course of development ensued, pitting tenants and sharecroppers against plantation owners. With the help of the federal government, the planters triumphed once again and presided over the transformation of the plantation after World War II.

With the exception of a few significant studies, the spread of the plantation into the Old Southwest (Arkansas, Missouri, and Texas) has received very little attention in historical literature. A few historians, however, have included an analysis of Arkansas's plantation system in their broader studies of the South. Using a wide lens to view the South as a whole or to analyze a discrete subregion, these historians have focused on federal policies and socioeconomic changes.[2] In doing so, they magnified the roles played by the federal and state governments and the big planters. While political and economic structures must be incorporated into any analysis of the twentieth-century South, this focus can obscure both the dynamic interaction of groups on the local level and the role they played in shaping events. Those most responsible for the conceptual framework that dominates current historiography on the modernization of twentieth-century southern agriculture include historians Jack Temple Kirby, Pete Daniel, and James Cobb and economist Gavin Wright. Kirby suggests that modernization was orchestrated by "a determined corps of men and women and their government." Pete Daniel argues that this "transformation [was] forced, over time, by mechanization and government policy." James Cobb, meanwhile, looks to even larger forces at work: the global marketplace. In his study of the

Yazoo Mississippi delta, Cobb focuses on "the close and consistent interaction with prevailing national and international economic influences and trends" and the tendency of white leaders "to promote the Delta's economic expansion and secure its effective integration into the world and national economy." Gavin Wright, finally, focuses on the "separate regional labor market" that characterized the South and its "interaction with national and international economies."[3]

The organizing principles of these studies necessarily force Kirby, Daniel, Cobb, and Wright to understand the South from the top down. While it is essential to appreciate the larger forces that helped shape conditions in the region, the individual tends to get lost in the analysis, and defeated alternatives to the course of development tend to be forgotten or dismissed. This is not to imply that Kirby, Daniel, Cobb, or Wright ignore the individual or fail to address some of the defeated alternatives, but their studies do not focus on what took place on the local level. It is there that one can more fully understand the role of the individual and the power and pathos of defeated alternatives.

Planters were by no means of one mind on the issue of mechanization, and the federal government was inconsistent in its policies. The contest over modernization commenced on the local level, and it was the actors in the remote towns and farms who joined battle and inherited the consequences. The farmers and small-town entrepreneurs-turned-businessmen-planters of Poinsett County, for example, actively participated in—and struggled fiercely with one another over—the modernization of southern agriculture. *New Plantation South* provides a corrective to the top-down perspective. At the same time, by focusing on a particularly unique area—an Arkansas delta county in the heart of a new plantation region—it fills a gap in an otherwise rich scholarship. While historians James Cobb and Robert Brandfon have focused on the rise of the plantation system in the Yazoo Mississippi delta in the late nineteenth century and Nan Woodruff has examined race relations in the Mississippi and Arkansas deltas in the early twentieth century, no one has closely examined the emergence of the twentieth-century plantation system in northeastern Arkansas.[4] Because the greatest challenge to New Deal agricultural programs emerged in northeastern Arkansas, indeed, in Poinsett County, such an examination is all the more rewarding. In launching their challenge, tenants and sharecroppers in the county were questioning planter authority and defying the federal government. The activities of their Southern Tenant Farmers' Union (STFU) gained an international audience, and although the tenants and sharecroppers who mounted the challenge eventually lost the battle, some of their organizers became key players in other unions and persisted in their efforts to represent the interests of landless farmers. These men were not passive bystanders when the transformation underwritten by the New Deal intensified during World War II and threatened a massive displacement of tenants and sharecroppers in the postwar era. Written from the perspective of the county that gave birth to the most significant agricultural protest movement in the twentieth century, this study attempts to provide an inside view of the larger issues and to shift the focus away from federal policies and socioeconomic changes to the small farmers trying to make a living off the land and the entrepreneurs who turned lumber, land, and labor into commodities.

Conflict between contending factions helped shape not only the rise of the neoplantation system in the post–World War II era but also the emergence of the plantation system, which came to exist in Poinsett County in the early twentieth century.[5] The entrepreneurs who fashioned a plantation economy out of a virtual wilderness in the county were small-town businessmen and were themselves new to the area and new to plantation agriculture. While engaging in the internal improvements necessary to develop the area, they also launched a protracted struggle for control over the county government. Tenancy and sharecropping suited their purposes because neither required their undivided attention; they could focus on their business enterprises and their political battles while tenants and sharecroppers, often more familiar than they with cotton and mules, brought their land under cultivation. The businessmen-planters could draw on no local traditional power base to rationalize their authority. Their unfamiliarity with the plantation system and their limited success in the political wars with elites in other parts of the county further eroded their authority. The tenants and sharecroppers who settled in Poinsett County understood that the planters had an uncertain hold on power. This weakness encouraged greater resistance to the demands of planters and helps to explain the rise of the STFU.

Sharecropping arose in the years immediately following the Civil War as a compromise between what planters wanted—landless laborers—and what freedman wanted—land ownership. Planters accepted sharecropping because freedmen, who could not secure land for themselves, refused to work for wages under the close supervision of planters. The idea of partnership with freedmen rankled southern planters, but their labor needs forced them to consider the appearance of such an alliance a necessary evil. While black sharecropping arose from compromise between freedmen and planters, white tenancy evolved from the economic crises of the late nineteenth century. White tenants, many of them drawing on their own experiences as small farm owners, assumed even more authority over their own lives than did black sharecroppers.[6]

In the first decades of the twentieth century, when the new planters of Poinsett County desperately needed labor, they accepted the fiction of partnership with sharecroppers and tenants in order to bring the land into cultivation. Historian Harold Woodman has described the tenancy and sharecropping system which took shape in the late nineteenth and early twentieth centuries and has emphasized the laws enacted to control landless farmers of both races. In his *New South, New Law* (expanding on a series of articles), Woodman explains the legal basis of tenancy and sharecropping and identifies the way in which planters narrowed the options for both blacks and whites who found themselves enmeshed in the plantation system.[7] Woodman emphasizes how the law distinguished between tenants and sharecroppers in a manner which especially disadvantaged sharecroppers. While tenants technically owned the crop and thus had some legal standing in court if a dispute between tenant and planter arose, the law viewed sharecroppers as laborers. However, as sociologist Edward Royce suggests, the commissary system, whereby tenants became inextricably indebted to planters, eroded what little advantage tenants had. According to Royce, "However clearly established in law, the distinction between sharecropping and tenancy tended to break down in practice; this distinction, furthermore, was not clearly recognized or acted upon by landlords and laborers."[8]

A certain ambiguity regarding the nature of the relationship between tenants and sharecroppers, on the one hand, and the planters, on the other, contributed to the rapid adoption of the plantation system. Rarely did the parties sign formal contracts specifying the precise arrangement they were entering into, whether a tenancy or a sharecropping arrangement; even when they did sign contracts, the tenure relationship was not always clearly defined. Woodman suggests that even the courts sometimes had "trouble deciding whether a particular contract or oral agreement created a landlord-tenant or a landlord-cropper relationship."[9] The ambiguity allowed a perceptual difference to complicate the relationship. As time passed, planters chose to view sharecroppers as laborers; sharecroppers and tenants regarded themselves as farmers in partnership with planters. Landless farmers resisted the efforts of state legislatures all over the South to clarify the relationships by enacting restrictive laws. When blacks confronted planters openly, however, they often paid with their lives. Mobility became the most pervasive form of resistance. The movement of tenants and sharecroppers from plantation to plantation, seeking "a little bit better break," enabled landless farmers to avoid complete domination and maintain some control over their lives.[10] Like E. P. Thompson's English working class, tenants and sharecroppers were unwilling to accept the low status conferred upon them by planters.[11] This very mobility, however, limited their political participation because they often could not establish the residency requirements necessary for taking advantage of the franchise. And, as Gavin Wright suggests, the movement of these tenants and sharecroppers (in what Wright refers to as a regional labor market) from other states into Arkansas enabled the plantation to expand and survive.

The differing perspectives held by the planters and the landless farmers who worked their plantations were headed for a collision in the 1930s. Planters instructed the federal government on how to interpret the relationship, and the federal government chose to take the legal definitions seriously when implementing New Deal programs. Its strict adherence to these definitions created problems for both tenants and sharecroppers and their landlords and ultimately contributed to the adoption of wage labor. In a sense the system had survived because ambiguity produced a certain flexibility, and that flexibility was lost when the federal government defined the status of sharecroppers and tenants. The definitions had added weight because the New Deal programs funneled cash to the planters, thus augmenting their power. In the face of this new challenge to their independence, tenants and sharecroppers asserted what they considered their rights and precipitated an embarrassing controversy which, ironically, forced an even closer relationship between the planters and the federal government.

But before joining battle with the planters, tenants and sharecroppers had to overcome a major impediment to solidarity. They had to drop the illusion that they were in partnership with planters, and black and white tenants and sharecroppers had to recognize their common interests. This recognition had long eluded them because "class is a relationship, and not a thing," and relationships even in the most stable of circumstances are mutable, ever-changing.[12] In the tenancy and sharecropping system, where ambiguity was accentuated, these relationships were particularly unclear. Added to the protean character of class relations was the tortured nature of race relations, which made it nearly impossible for white and black tenants and sharecroppers to realize their similarities and become class conscious enough to bury their differences and confront the planter class.

That most of the tenants and sharecroppers in Poinsett County did not labor in gangs for large centralized operations like the Delta and Pine Land Company of Mississippi was one of the greatest inhibitors to the development of class consciousness. Instead, they worked off on their own on 20- to 40-acre farms and enjoyed a sense of independence and detachment. This possibility of independence may well have been one of the factors that attracted landless farmers to the area. The arrival of a St. Louis financier in Poinsett County in the early 1930s precipitated the crisis leading to the formation of the STFU. Treating his plantation like a business enterprise, Hiram Norcross violated the customs governing the system operating in Poinsett County. His actions led directly to the formation of the interracial STFU. The loosely administered plantation operations of Poinsett County had forestalled the recognition of class interests between blacks and whites, but Norcross personified "capital," cavalierly treating tenants and sharecroppers as "labor." Thus class consciousness finally coalesced during the 1930s when planter actions and New Deal programs sharply defined class relations, and black and white tenants and sharecroppers took one final stand against proletarianization. They lost the battle and melted into the cities in search of jobs outside the farming sector. Their status as independent, even though landless, farmers was lost, and the experiment in interracial cooperation was abandoned.[13]

1

"A WILD AND SICKLY COUNTRY"

As ethnologist Edwin Palmer made his way through the densely forested northeastern Arkansas swamps in 1882 looking for Indian mounds to catalog, he passed into Poinsett County and encountered mad bees but "neither bird nor animal on the road."[1] A recent flood topping at eighteen feet had swept away everything in its path. Palmer marveled at the high-water marks on the trees and noticed that "men in their boats" had severed treetops during flood stage "to mark their return road." The desolation was only temporary, for the Arkansas delta had long been known as a hunter's paradise with an abundance of "deer, turkey, bear, bison, panthers and wolves" as well as a vast variety of waterfowl and other birds.[2] Indeed, some men supplemented their incomes by selling wildcat and wolf scalps to county governments eager to reduce the population of those predators.[3] Because it was one of the last counties in the state to be settled, a traveler entering "the wilds of Poinsett County . . . could frequently see a wolf, [a] bear, a deer, a raccoon, a catamount, a polecat, a bobcat,"[4] and the notoriously dangerous wild razorback hogs which still roamed the backwoods. Malaria-carrying mosquitoes and other insects pestered and sickened people in the swamps, and countless diaries, reminiscences, and letters describe the chills, ague, fevers, and other maladies associated with life there.[5]

But as desolate and remote, as lifeless and seemingly impenetrable as Palmer found the swamp, he also encountered cotton on high ground that topped at six feet.[6] For decades men had passed through or around this Arkansas delta swamp looking for land that could be more easily cultivated. Some of those who chose to settle in the swamp found life and farming distinctly uncertain. In 1879, for example, twenty-five men planted ninety-three acres of cotton. Six of them harvested only one-tenth of a bale to an acre and almost certainly were victims of high water. Most of the rest enjoyed a yield of a bale to the acre.[7] They may have marketed those bales by floating them down the St. Francis River to Helena, or they may have taken a roughly hewn road to Osceola, in neighboring Mississippi County, the road that Palmer took into Poinsett County on his trip.

It would have been unlikely for them to have gone west toward Harrisburg, the county seat of Poinsett County, for the swamp and dense forest were most daunting in that direction. When Palmer made a trip to Harrisburg in the spring of 1882, he no doubt traveled along Crowley's Ridge, avoiding the swamp altogether. He found nothing impressive in Harrisburg and complained of insects in his drinking water and wretched beds at the Harrisburg Hotel. The badly cooked and poorly served food disappointed him, and his inability to find a shoe mender put him in a poor humor. He described a "poor brick courthouse" that housed a doctor's office, a printing shop, a post office, and an unfinished jail in such rough condition that prisoners had to be sent elsewhere.[8] Saloons were outlawed, but Palmer discovered that "Kansas eggs" filled with whiskey could be had for ten cents apiece or a dollar a dozen, and although he apparently did not join them, he wrote disparagingly of a "Saturday crowd" of drinking men frequenting the inner room of a grocery store on weekdays.[9]

Harrisburg, designated the seat of the most sparsely populated county in the state in 1856, was not officially incorporated until the year following Palmer's visit. Poinsett County was dwarfed by its neighbor counties and isolated by a transportation system crippled by geographical happenstance and the limited resources of the state government. Neither the county nor the town appeared very promising in 1882.[10] Considered in 1830 one of the rising towns in the territory of Arkansas, Harrisburg never realized its promise.[11] Conditions in the late nineteenth century thwarted its economic growth and oriented development away from Harrisburg and the people who settled the surrounding countryside on the ridge. The vision of the early settlers was restricted to that relatively narrow strip of land known as Crowley's Ridge which stretched 150 miles from New Madrid, Missouri, to Helena, Arkansas, and bisected Poinsett County, where it rose 250 feet

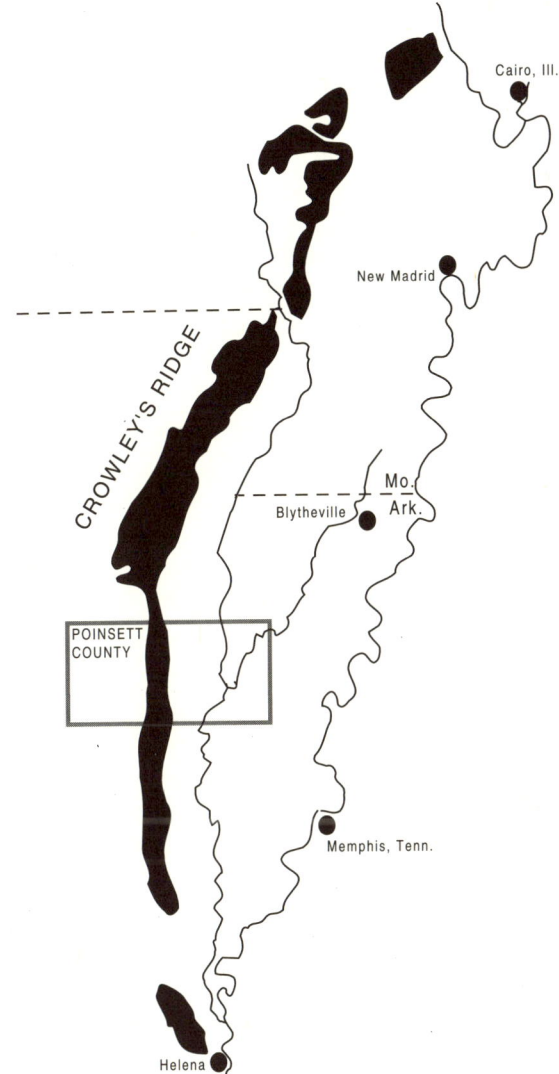

Map 2. Crowley's Ridge in Arkansas and Missouri

and measured from three to five miles wide (map 2). The existence of Crowley's Ridge effectively created three separate geographical regions with their own distinct soil types, leading in part, to the development of three different social and economic structures in the county.

To the west of Crowley's Ridge was an inhospitable prairie covered with a thick brush stretching up from the Arkansas Valley region and ending just above Poinsett County. The few intrepid souls who settled there in the mid nineteenth century made their living "hunting, fishing, bee keeping, and raising cattle," and if they had need of items they could not themselves produce, they had to travel twenty-four miles west to Newport in Jackson County, "a three days journey by ox wagon."[12] While the lands west of the ridge hardly beckoned to prospective settlers, to the east was a densely wooded and mostly uninhabitable swamp. Described as a "wild and sickly country" by an immigrant ex-slave from North Carolina in 1889, the Arkansas delta stretched irregularly along the Mississippi River and was situated in the St. Francis River floodplain. The northeastern portion in particular lay under water for much of the year.[13] The notorious swamp very nearly prohibited travel east of the ridge in Poinsett County; only the most daring or foolhardy would have attempted to reach Harrisburg by passing through it.[14]

But if Edwin Palmer had traveled far enough into the Poinsett County swamp, he would have reached a primitive railroad camp at a spot where the Little River and the St. Francis River come together (map 3). Far rougher than the town of Harrisburg, the railroad camp in the early twentieth century became a thriving sawmill town named Marked Tree from which a new planter elite emerged. The lumber industry boomed, the population exploded, and entrepreneurs drained and cultivated the delta. A few individuals bought up most of the cutover land, introduced the plantation system, and concentrated on the cultivation of cotton.

The farmers in Bolivar Township, where Harrisburg was located, did not initiate development of the delta and continued to concentrate their attention on farming enterprises along Crowley's Ridge. In the last two decades of the nineteenth century, they focused on the problems plaguing most Arkansas farmers: the drop in farm commodity prices, the rise in the number of mortgages, and the increase in tenancy,

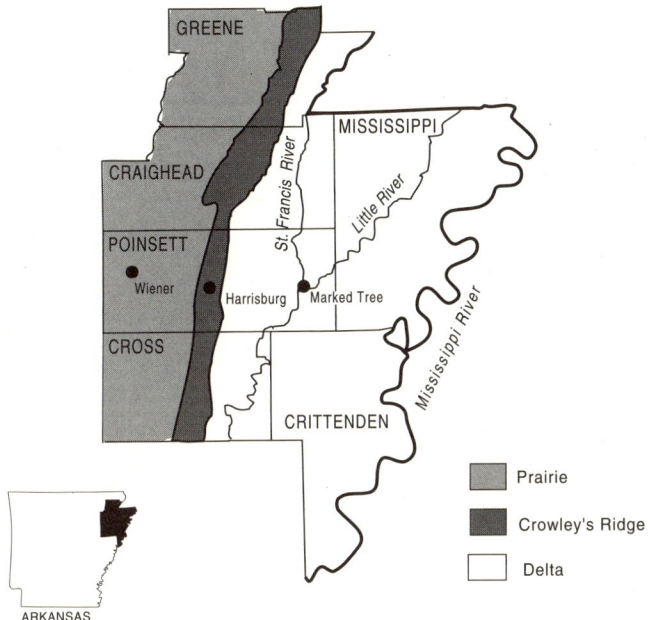

Map 3. Poinsett and surrounding counties

which all combined to limit the horizons of the farmers around Harrisburg.[15] Even had they had the capital to expand, they would hardly have turned their attention to the malaria-infested swamplands east of the ridge. And even had they foreseen the creation of an extensive system of levees and drainage ditches in the early twentieth century, which would uncover some of the richest delta remaining in the South, they lacked the requisite political connections to acquire title to any of the swampland lying within Poinsett County. That land belonged to the state, and its disposition was administered by a commission which served the interests of a small clique of planters and businessmen. Thus, when it came to developing the delta, the farmers along Crowley's Ridge in Poinsett County had neither the political clout nor the economic wherewithal necessary to play a meaningful role.[16]

The vulnerability of the politicians of Crowley's Ridge to the designs of businessmen-planters had long been a subject of concern to the Harrisburg courthouse clique. In fact, the challenge of the new Poinsett

County delta planters in the twentieth century was not without precedent. During the Civil War, for example, the Harrisburg establishment had been unable to defend itself against the designs of wealthy planter David Cross, who had a plantation in the only part of Poinsett County's delta that was habitable in the years before the major drainage enterprises of the twentieth century. Cross held thousands of acres and had eighty slaves by 1860, and he possibly resented the political dominance exerted by those on Crowley's Ridge.[17] They had settled there a mere ten to fifteen years before his arrival, and the placement of the county seat in remote Harrisburg must have been an inconvenience to him. Because of the primitive transportation network, it would have taken Cross and the other planters in the county's delta days to reach the courthouse, and if Palmer's description of the Harrisburg in 1882 is any indication, the town had little to recommend itself beyond the possibility of transacting official business.[18]

The fall of Crowley's Ridge to Union forces early during the Civil War disrupted the functioning of county government, and David Cross decided to lead a delegation through Union lines to the Confederate governor in Little Rock and petition to have a portion of Poinsett and two other counties appropriated for the creation of Cross County. He owned eighty-six thousand acres in these three counties and had served with the Confederate governor as a delegate to the secession convention in South Carolina just two years before. Benjamin Harris, the county's state senator and a founder of Harrisburg, had organized a regiment and was in the field at the time, so that he could not raise any objections.[19] The governor granted Cross's request, leaving Poinsett one of the poorest and least populated counties in the state.[20]

The partitioning of Poinsett County stimulated controversy, however, and ill feeling resurfaced during the local elections two years later. When Alexander M. Winn ran for reelection as constable of Searcy Township, which had been part of Poinsett County before the 1862 division, he was opposed by Grandisom M. Sharp, who accused him of hostility to the formation of the new county. Winn defended himself by arguing "that he had been elected to represent Poinsett County and had done his duty and . . . that the population of St. Francis was equally distributed over the county and she would not be materially affected by the loss of part of her territory. But the new county would take from

Poinsett the bulk of her territory and wealth and leave her one of the least populous and poorest counties in the state."[21]

By the time Benjamin Harris was paroled from the Confederate army in 1865, the creation of Cross County was a fait accompli. One year later, during Presidential Reconstruction, the state of Arkansas officially sanctioned the establishment of Cross County and other such actions of the Confederate governor. Benjamin Harris became the county judge (1866–68) of Poinsett County, until Radical Reconstruction displaced him, but he presided over a county of considerably reduced circumstances. Poinsett still had approximately two hundred thousand acres in swamplands, but a geographic depression known as the sunk lands in the heart of the post 1862 Poinsett County delta created virtual lakes there. The delta Cross appropriated was at a slightly higher elevation and therefore much more desirable. The delta remaining in Poinsett County was in the hands of a state swamplands commission with which the Crowley's Ridge farmers had little influence.[22]

New businessmen-planters emerged in the delta in the early twentieth century, drained the swamplands of Poinsett County, and eventually mounted a fresh challenge to the yeomen of Crowley's Ridge. The conflict that was to arise in the twentieth century was created, in part, by events of the late nineteenth century. Even as Edwin Palmer made his way along Crowley's Ridge in 1882, the isolation of the Crowley's Ridge farmers and the neglect of the rest of Poinsett County was about to end.

Central to the incorporation of the county into the state economy was the development of an efficient railroad system, but those who fashioned the railroad network in the state during the late nineteenth century had little interest in the consolidation of Poinsett County. They wanted access to the raw materials Arkansas had to offer, and their attention was focused on the rather difficult task of financing the enterprise, for the construction of railroads in the state was not easily accomplished.

An ill-fated partnership of planters in the delta and business interests centered in Little Rock had actively courted the railroad industry as early as the Reconstruction period but failed to bring railroads to the state. In 1872 Arkansas defaulted on its railroad bonds. David Cross, the prominent planter of newly formed Cross County, was involved in one such scheme and sold forty thousand acres of land in 1871 to cover the losses incurred by the railroad he tried to organize in southeastern Arkansas. The partnership of railroad interests and planters got a new boost from the emergence of the lumber industry in the 1880s when Jay Gould, interested in Arkansas's forests, helped to finance the construction of two thousand miles of railroad in the state.[23]

The coming of the railroad to Poinsett County served the interests of the farmers on Crowley's Ridge, but it also set in motion the forces that presented a challenge of major proportions in the early twentieth century.[24] The Kansas City, Fort Scott and Gulf Railroad encouraged the founding of the first settlement in the Poinsett County delta, Marked Tree. Although the railroad proved to be of paramount importance to the development of the delta, it failed to draw the population of the county's three disparate areas together. Different railroad interests built three north/south lines through Poinsett County in the 1880s: the St. Louis Southwestern Railroad ran through the prairie in 1881; the St. Louis, Iron Mountain and Southern ran along the ridge in the same year; and the Kansas City, Fort Scott and Gulf Railroad opened up the delta in 1882.

In the prairie west of Crowley's Ridge, Weiner was founded and named for the engineer who oversaw the construction of the St. Louis Southwestern Railroad in 1881. Almost immediately, settlers began to build homes and stores near the station and established the Weiner Post Office in 1884. Small timber camps sprang up in the nearby woods where men fashioned crossties, hauled them by oxcart, and stacked them along the railroad right-of-way for inspection by railroad officials. Settlers during the 1870s, like the John P. Phillips family of Macon, Georgia, came by ox wagon. Most of the new settlers, understandably, came by rail.[25] Despite its growth after the coming of the St. Louis Southwestern Railroad, the prairie remained underpopulated, economically unstable, and politically dependent upon the ridge. Yet differences existed between those who settled the prairie and the yeomen on the ridge: prairie farmers were much more likely to be northerners than southerners, they eventually turned to the cultivation of rice rather than orchard crops, and they relied on wage labor to harvest their crops.[26] The prairie farmers, however, often looked to the ridge for supplies and credit, and it would take decades for the thinly populated prairie to develop an economic infrastructure independent of the ridge.

The St. Louis, Iron Mountain and Southern Railroad, meanwhile, which ran through Harrisburg in 1881, was in large part responsible for the tripling of Harrisburg's population between 1880 and 1900. The railroad brought in new blood and an infusion of capital, expanded Bolivar Township's participation in the market, not only for agricultural goods but for forest products, and encouraged a greater orientation to the outside market. Along with this new orientation away from a localized, semisubsistence farming economy came a significant rise in farm mortgages and farm tenancy on the ridge.[27] While the St. Louis, Iron Mountain and Southern Railroad linked Bolivar Township with the railroad from Kansas City to Memphis at Jonesboro, north of Harrisburg, and with the railroad from Little Rock to Memphis at Forrest City, south of Harrisburg, no direct line joined Harrisburg to Marked Tree, the young community in Poinsett County's delta.[28] Thus, the railroad also rendered it unnecessary to improve the transportation system between the ridge and the delta.

The Kansas City, Fort Scott and Gulf Railroad oriented the county's delta toward Memphis rather than the county seat of Harrisburg. The kind of society that emerged in the delta, moreover, differed sharply from that evolving in the county's prairie. Unlike the prairie where the land deserved the cheap price the midwestern farmers paid for it, the delta, once drained and readied for cultivation, was rich and promising. Businessmen rather than small farmers became the dominant landowners, and they amassed small estates rather than the 160-acre homesteads so common in the prairie.[29] Rather than family farms, the businessmen of the Poinsett County delta established plantations. They looked to the plantation economy in the counties contiguous to Poinsett's delta and employed tenants and sharecroppers rather than wage laborers. The presence of many southerners in the population of newcomers who flocked to the county's delta eased the transition to a plantation economy, for unlike the prairie which attracted midwestern farmers, chiefly from Illinois and Indiana, the delta attracted southern migrants, principally from Mississippi and Tennessee.[30] By the dawn of the new century, the Poinsett County delta stood poised for a phenomenal transformation which would overshadow the prairie and the ridge.

Once the Kansas City, Fort Scott and Gulf Railroad established its work camp at the spot between the Little River and the St Francis River in 1882, hunters and trappers moved into houses facing the St. Francis

River and took advantage of the company's mail service.[31] Small sawmills began to spring up and one of them, the Oliver Davis Mill, attracted a young Iowa man of German parentage who was destined to become the single most powerful man in the Poinsett County delta. Ernest Ritter, whose uncle was part owner of the Davis mill, came to Marked Tree in 1886 to work in the mill. But he labored there only a short time, for he seriously injured his foot in a work-related accident and in 1887 had to return to Iowa, where it took him nearly a year to recover. Young Ritter might have been discouraged by his adventure in Arkansas, but he must have seen the opportunity awaiting an enterprising man. He returned in 1889, with his bride, Anna Hirschman Ritter, and began to homestead 160 acres. At the same time he opened a mercantile business which would flourish in the years to come and serve as the foundation of his growing prosperity.[32]

Even as young Ritter and his wife were moving to their homestead in August 1889, the railroad work camp and depot at Marked Tree attracted the attention of the Chapman and Dewey Lumber Company. Chapman and Dewey purchased the mill site in 1890 from another Iowa man, a Mr. Harding, who had moved his family to Marked Tree in the spring of 1890, watched them sicken with malaria in the summer, and became disgusted enough with conditions to sell out by the fall. Chapman and Dewey, which had invested in more than one hundred thousand acres of land spread over several delta counties in the early 1880s, chose Marked Tree as a base from which to operate.[33] In 1897 the town incorporated, and by 1900 its population stood at 352. Of the 181 adult males in Marked Tree in 1900, over half of them worked in the mill or the timber industry.[34] Another 10.1 percent were involved in service work that largely supported the mill. This stands in stark contrast to Harrisburg, where only 18 percent of the 156 adult males labored in sawmills.[35]

While 181 (51.6 percent) men resided in Marked Tree, only 85 women (24.2 percent) made their homes there. And both men and women were far more likely to live in boarding establishments than they were in single-family homes. Indeed, life in this rough sawmill town must have been very like what Mary Hamilton describes in her recently published autobiography, *Trials of the Earth*. First brought to the "wild Arkansas country" by her father when she was still a teenager,

Mary later followed her husband to sawmill towns all over the Arkansas and Missouri deltas. She frequently ran boarding establishments catering to the men who worked in the mills, sometimes housing and cooking for one hundred men or more. She was witness to the harsh realities of the sawmiller's life: disease, accidents, and loneliness followed them from sawmill towns to lumber camps, and economic security almost always eluded them. Like other sawmillers who were always on the lookout for a better situation, the Hamiltons moved about frequently and buried four of their children along the way. Recording floods, tornadoes, disease, accidents, and the butchery of poorly trained medical practitioners, Mary portrays the delta as a dangerous and fickle place.[36]

Certainly, natural disasters and manmade scourges plagued those who settled in Marked Tree. The dirt streets were frequently muddied from the heavy rains and overflows, and the poorly constructed screenless shacks offered little shelter against the elements and no protection from the malaria-carrying mosquitoes. The largest boarding establishment in Marked Tree in 1900 was operated by George Harris who housed thirty-one individuals, including five blacks who cooked and served the tables. Most of his boarders were mill hands, day laborers, or railroad workers, but he also housed seven white-collar workers. W. A. Fuller, who had purchased the Oliver Davis sawmill in 1891, resided in Harris's establishment. Except for Harris's kitchen and dining room staff, those who lived in his establishment were white, and although most of them were married, only three of the wives were present.[37]

Their wives would have found little to entertain them in Marked Tree, for the town boasted only a few saloons, one black church, and no schools in 1900. Dominated by sawmill hands and economic instability, the settlement had few comforts for families. Its greatest asset was its placement at the confluence of the St. Francis River and Little River, which gave it a unique advantage in the heavily forested swamps. Lumber interests were intensely interested in exploiting those forests, and with the advent of drainage enterprises in the first decade of the twentieth century, the businessmen of Marked Tree would soon discover another source of wealth and power: the rich delta soil that could produce high yields of cotton and corn. Once the lumbermen denuded the forests and the businessmen drained the swamps, the foundation for vast plantation enterprises was laid. The abundant game that had once made the region a sportsman's paradise gradually disappeared, and those who created the drainage districts were heedless of the indigenous plant and animal swamp life they were destroying. The future belonged to the entrepreneurs, and an astonishing transformation of the Poinsett County delta's landscape was about to get under way.

2

FROM RAILROAD CAMP TO
PLANTATION TOWN

When federal district judge William J. Driver charged in 1918 that Poinsett County was a "hot bed for bootleggers" who "are so thick . . . they have to tag one another to prevent the fraternity from soliciting among its own membership," he was striking a raw nerve in Marked Tree.[1] A certain segment of the town's elite had struggled for nearly a decade to close Marked Tree's saloons. But the sawmill hands who dominated the population continually voted for local option and petitioned to keep the saloons, their working men's clubs, open. Indeed, Marked Tree was one of the last four towns in the state to have open saloons. It took statewide prohibition, imposed in 1916, to close them permanently. The battle over the saloons reflected deep divisions within the community, and the drive on the part of some of the elite to implement prohibition exemplified their determination to supervise the social behavior of the labor force, keep black and white laborers divided, and at the same time create a town that would attract capital and new business enterprises.[2]

Marked Tree's infancy coincided with the early years of the Progressive movement, and it was no coincidence that its prohibition drive became inextricably intertwined with the politics of race and disfranchisement. In Arkansas, as elsewhere in the South, many progressives supported disfranchisement in the interest of purifying the electoral process. Although prohibition did not immediately appeal to some of Arkansas's business progressives, the issue eventually recommended itself to other progressives who viewed it as yet another moral reform. The drive to eliminate the trade in liquor became a major part of the progressive agenda in Arkansas, and Marked Tree's millworking population guaranteed that the town would become a focal point of the struggle.[3]

Arguing that the ability of some whites to manipulate the black vote threatened the integrity of the democratic process, progressives supported the implementation of the white primary in 1907 and secured the placement of a grandfather clause on the ballot in 1912. Both the grandfather clause and a prohibition statute, also on the ballot in 1912, failed to secure enough votes to become law. But during the election prohibitioners paid particular attention to the behavior of black voters, who came out in record numbers, and when they realized that they had lost in the areas having the largest black populations, they, like progressives, identified black voters as enemies to moral reform and targeted them for special treatment. Yet in places like Marked Tree, their efforts were not sufficient to close saloons, for the white millworkers proved as obstinate as blacks, and at least some of the local businessmen were eager to provide them with their workingmen's clubs.

From the time that the town was incorporated in 1897, much attention had been focused on the need to control the unruly millworking population. That this effort became entwined with progressivism is hardly surprising. But much of what occurred in those early years should not be automatically accorded to that movement. Although born in the dawn of the Progressive era, most of the town-building activity which might be labeled progressive was, in fact, typical of frontier towns anywhere at any time.[4] In the first two decades of the twentieth century, the businessmen who came to power in Marked Tree focused on the necessary task of fashioning a community in a virtual wilderness. They built schools and churches to serve the local population, and they constructed roads that snaked into the countryside. The young town council grappled with the problems associated with a phenomenal increase in population. The railroad men and fishermen who initially peopled the high ground between the St. Francis River and Little River had been joined by the men who came in to work the Chapman and Dewey lumber mills, and although the later group was as likely to be as transient as the railroad men, more of them brought their families with them.[5] Those individuals who settled there to open shops to support the expanding population of lumber workers promised even greater permanence as the town flourished and the number of women and churches, children and schools grew. The desire to attract new settlers and new businesses, meanwhile, stimulated a drive to make the town more presentable, and by 1909 the custom of allowing domestic animals to roam unattended became a subject of debate on the town council. The stilted houses in which Marked Tree residents traditionally resided were an artifact of life in the swamp, and hogs wandered freely over the deeply rutted and often muddied streets and came to rest beneath them. While some citizens complained and warned of the danger of disease, boys used the hapless hogs for target practice.[6]

Between 1900 and 1910 the town's citizens passed a limited stock law, paved or graveled the principal roads, and laid sidewalks. An abundance of single males who worked the mills continued to dominate the population, but the town council passed ordinances regulating their behavior, and they lived less frequently in boarding establishments.[7] New housing subdivisions were developed, attracting more families, churches were organized, and the segregated school system expanded to accommodate the town's white and black children.[8] The leading businessmen of Marked Tree also built roads into the countryside, began to operate plantations peopled by sharecroppers and tenants, and became their creditors. For the typical Marked Tree businessman-planter, the plantation was an extension of his business enterprise rather than a way of life. This turn to plantation farming by the town's business leaders signaled a transformation of the town itself. By 1920 it was no longer merely a sawmill settlement; it had become a plantation town.

One key player in the making of Marked Tree was local businessman Ernest Ritter, who shared center stage with the Chapman and Dewey Lumber Company. While homesteading his 160-acre tract in 1889, Ritter opened an ice plant and bought fish from the local fishermen who had been working the rivers modestly for some years. He soon began shipping the fish to the Memphis market, and by the turn of the century his fish were reaching St. Louis on a refrigerated boxcar.[9] He held the position of postmaster during the McKinley administration and served on the town council from 1897 to 1905. In the decade between 1900 and 1910, he branched into other entrepreneurial activities, building the subdivisions that housed many of the men employed in Chapman and Dewey's mills and opening plantations in the surrounding countryside. In the Poinsett County delta alone he owned 1,962 acres by 1910 and 3,185 acres by 1920. At the time of his death in 1921, he owned another 7,000 acres in contiguous counties, his estimated wealth stood at $750,000, and he was regarded as one of the wealthiest men in Arkansas.[10] Not only did he buy land and open up his own plantations; he also sold both improved plantations and unimproved acres to interested buyers. He was one of the initial petitioners for the establishment of the county's first drainage district, which drained the swamps and made plantation agriculture possible, and he was the contractor who hired the men to build the roads that provided access to the marketplace. Ritter became the single most powerful man in the Poinsett County delta, and only Chapman and Dewey exercised as much influence in the area.

E. R. and W. B. Chapman had different origins and, in the beginning, greater means than Ernest Ritter. The Chapman brothers were Kansas City investors and mortgage bankers who, together with W. C. Dewey, an old school chum and postmaster in Glasgow, Missouri, formed the Chapman and Dewey Lumber Company in 1886 and bought the Harding sawmill site in 1890.[11] By 1893 they had accumulated over one hundred thousand acres in five Arkansas counties, including over thirty thousand in Poinsett's delta, and by the turn of the century they were the region's chief employers. Although they initially demonstrated the lumber industry's usual resistance to drainage, they came to appreciate its possibilities as their own enterprises expanded to include speculation in real estate and the establishment of cotton plantations. Unlike the land owned by large lumber companies in other parts of Arkansas and elsewhere in the nation, Chapman and Dewey's land could be profitably turned to agricultural uses.[12] Although they sold more land than they eventually put into cultivation, they profited handsomely because of the soil's fertility. All that Chapman and Dewey had to do was drain their overflowed lands. By the second decade of the twentieth century, they controlled the most important drainage district in the county.[13] Once the land was recovered, the next logical step was the founding of plantations.

Ritter and Chapman and Dewey dominated the available labor in the town of Marked Tree at the same time that they implemented the tenancy and sharecropping system on their rural plantations. The class and caste structure that arose in the delta plantation area between 1900 and 1920 had its counterpart in the town of Marked Tree, where the workforce was highly mobile. There were few options for those seeking jobs in the town, and there were fewer jobs than there were job seekers. The mobility of the workforce was indicative of a period of economic chaos that began in the 1890s, eased somewhat in the first few years of the twentieth century, and then intensified with the panic of 1907. But it was also quite typical of the lumber industry itself.

The industry was so notorious that one small town in the Ouachita Mountains of northwestern Arkansas persuaded mill operator Thomas Rosborough not to open a mill there in 1906. "They did not want the transient mill workers, nor did they want blacks."[14] The men who came to work the mills of Marked Tree were precisely those the Ouachita Mountain town feared most. Seventy of the ninety-six men working the Marked Tree mills were black. None of the forty-five men listing themselves as "day laborers" and only four of the fifty-one mill workers appearing in the 1900 census remained in the town for a decade.

The promise of work, which had attracted men to the area, was not always fulfilled. Day laborers gathered at the mill entrance every morning hoping for an opportunity to stack lumber or get a day's work doing some other menial labor. Rarely did the mill hire skilled laborers (i.e., sawyers) on a daily basis. The day laborers, therefore, were those least likely to have the skills necessary to gain more permanent employment. Of the ninety-six men employed in the town's mills in 1900, forty-five worked as day laborers only. Although the black millworkers outnumbered their white counterparts, they occupied the lowest positions in the mill. Of the fifty-one men listing themselves as millworkers, thirty-one were black, but only three of them were sawyers. The other twenty-eight black men who identified themselves as millworkers were simple mill hands.[15] Of the forty-five day laborers, moreover, thirty-nine were black.[16]

Whether sawyer or day laborer, whether white or black, millworkers had no guaranteed employment. None of the millworkers avoided at least 30 days of unemployment during the year, and on average they found themselves out of work for 127 days.[17] Underemployment plagued the sawmill industry, dependent as it was upon lumbering operations throughout the nation, and it was an especial problem in the South where seasonal heavy rains made the ground too soft for logging. The swampy conditions prevailing in Poinsett County's delta rendered the difficulties especially acute. The erratic nature of the lumbering industry in the delta, and in the South generally, discouraged the workforce of the Great Lakes lumber industry from migrating to the southern forests "when the industry shifted its center of gravity there."[18] Those who chose to work the forests and sawmills in Poinsett County, therefore, were southerners, increasingly from the lower South, who may have been somewhat unfamiliar with the industry's routines but were hungry for employment.

When a job was available, the mill hand found the hours long and the work in the noisy, dust-filled sawmill grueling and dangerous. Accidents, like the one which disabled Ernest Ritter for nearly a year in 1887, were not uncommon and a man counted himself a good sawyer if he had all his fingers.[19] The men who operated the carriage upon which the individual logs were placed faced the greatest likelihood of injury. The head sawyer had the least dangerous job; he merely operated the switch which set the carriage in motion and pushed the log:

> past a band saw that sliced off one outside part of the log—a slab, with the bark on. Back came the carriage, the log was turned or set forward on the saw, then back again past the saw that . . . sliced off another slab, or board. The outside parts of the log were removed first, leaving a squared piece of timber (called a cant) that was then sawed into boards. Back and forth, back and forth went the carriage, in a few seconds the log was reduced to slabs and boards. To move the log into the best position for each cut, the sawyer and the men riding the carriage had to do rapid mental calculations.[20]

Beyond the head saw were a series of smaller saws that trimmed the slabs and boards. The men responsible for operating the carriage and turning the logs were the ones exposed to the most danger, and the necessity to work quickly accentuated the risks. A mill having two band saws—and Chapman and Dewey was such a mill—expected its men to produce one hundred thousand board feet of lumber in a day. The men who carried the logs from the river or from the tracks to the mill and the men who stacked the finished boards were engaged in less dangerous if not less strenuous labor.[21] It was this latter group that was most likely to be black and to be employed on a day-by-day basis only. The men possessing the greatest expertise, those who operated the equipment and ran the logs through the saws, were almost exclusively white and were those who were more permanently employed by the sawmills.

Whether white or black, the men who succeeded in securing positions in one of the town's mills found themselves involved in a dependency relationship resembling that in which tenants and sharecroppers were advanced supplies from the plantation commissary. Mill hands typically had an account at the company commissary and often drew upon that account during the week.[22] The essential difference between the plantation commissary and the company commissary was the time span of the indebtedness. Most mill hands could hope to "pay out" on a weekly basis while sharecroppers and tenants had to wait until harvest. Mill hands, moreover, like tenants and sharecroppers on some of the larger plantations, were sometimes paid in scrip, often called doodlums, issued by the mill and redeemable only at the company commissary. Chapman and Dewey often issued doodlums. The practice was widespread in the lumber mill industry and enabled the millowners to avoid parting with cash.[23] Even with an abundance of cheap labor available, the lumbering and agricultural industries suffered from a lack of sufficient working capital. Both were highly seasonal enterprises with months of inactivity. While agriculturalists realized their profits only upon harvest, the lumber mill owners were subject to an erratic market and the vagaries of the lumbering industry with its vulnerability to weather conditions.

While it served the purposes of the millowners to have excess labor available, it placed strains on the fledgling community that the town's civic leaders and the newly arrived women addressed in the decade between 1900 and 1910. The social fabric of Marked Tree was so thin in 1900 that E. R. Chapman's wife refused to live there and settled instead in Memphis, Tennessee, some forty miles to the southeast.[24] The town had little to recommend it to the St. Louis belle. There were a few unpainted houses built on stilts, a number of shacks along the St. Francis River in an area which came to be known as Yellow Town, and a few saloons and honky-tonks. Only two merchants operated there in 1900, and a circuit preacher rode into town on horseback every other Sunday or so to hold services in someone's home.[25]

Men who worked the mills dominated the town's population. Of the 266 adults there in 1900, 181 (68 percent) were male, and 109 (60.2 percent) of these men were unmarried. Their average age was thirty, and 105 (58 percent) boarded rather than headed their own households; 96 of them worked in the mills.[26] Only two institutional diversions served the population, and the unattached men rarely frequented one of those, the church. The saloons flourished, however, and provided a place where white and black men commingled. They drank and gambled together and outraged the middle-class sensibilities of the local elite. By the end of the decade, temperance became a major issue. In the next decade the unmarried workingmen who frequented the honky-tonks faced the combined forces of the Woman's Christian Temperance Union and the local employers. Ernest Ritter, a Bull Moose Progressive Republican, provided the WCTU with a meeting place and underwrote some of its expenses. Once the millowners saw the merit in a sober workforce, a struggle over prohibition ensued.[27]

While Marked Tree merchants were divided over the issue of prohibition, with some of them believing the existence of saloons in town was good for business, saloon owners failed to segregate their facilities despite the objection of at least some of the town's leading citizens. Marked Tree's millworkers, therefore, associated freely in a social setting outside the control of the millowners. Even so, despite this tendency of blacks and whites to frequent the same saloons in town, Marked Tree had never been a particularly safe place for blacks to settle. As early as 1891 whites dismissed from the Fuller brothers' mill during one of the typical downturns in the sawmilling business marched into the black camp and threatened black workers who had not been dismissed. The Fuller brothers had chosen to retain some of their black workers, who were paid a lower wage, while dismissing many white workers. In this case the millowners stood shoulder to shoulder with the black millworkers against the irate whites.[28]

By the early twentieth century, that old injury seemed forgotten. The high turnover rate among millworkers left little historical memory alive, and blacks and whites were fraternizing in the town's honky-tonks, as attested by letters to the editor that appeared in the *Marked Tree Gazette* throughout the decade. In 1908, for example, one white citizen complained of the rowdy honky-tonks in Marked Tree frequented by both "worthless negroes and low down white men," and a prominent local attorney complained that allowing blacks and whites to mingle in the honky-tonks might lead to amalgamation of the races. Such letter writers were especially disturbed by the presence of white and black women in these establishments. By 1911 the editor himself was decrying the intermingling and predicting amalgamation of the races.[29]

The fact that blacks and whites frequented the same saloons in Marked Tree did not serve as an inducement for blacks to settle there. The existence of a viable and self-sustaining black business community would have been a much greater attraction, and the failure of such a group to emerge in Marked Tree was another important factor discouraging black immigration to the town. In 1920 there were no black merchants operating in Marked Tree, and the one black doctor who had settled in town before 1900 had sold out and left the state by the end of the second decade of the twentieth century. Four black preachers and one black teacher served the community in 1920, but 85.7 percent of the black men in Marked Tree either worked in the mills or as unskilled or menial laborers. Only five black men (3.4 percent) were skilled workers or artisans. The remaining 7.2 percent were farmworkers. In other words, the black community had to turn to white merchants and bankers when in need of goods and capital. While the merchants were only too glad to extend credit at exorbitant interest, the bankers were most unlikely to accommodate entrepreneurial black men seeking capital to open businesses.

In the face of such limitations, blacks who wished to settle in a town chose places like nearby Osceola in Mississippi County where a flourishing black community attracted immigrants well into the twentieth century. Indeed, given the choice between Marked Tree and towns like Osceola, it is small wonder that blacks selected the latter. By 1920 blacks no longer dominated the millworking population. As early as 1910 the percentage of black men working the mills had declined from 75.5 percent to 45.4 percent. It dropped another percentage point by 1920. While some seventy-two black men still found employment in the town's mills, they occupied the least favored and most vulnerable jobs. The lack of opportunity, outside of uncertain employment in the mills, stifled black immigration to Marked Tree. In 1900 blacks made up 45.7 percent of Marked Tree's population; by 1920 they constituted only 28.2 percent. Thus, as early as 1910 the black population of Marked Tree had declined dramatically, and so blacks per se did not present the greatest obstacle to prohibition.[30]

Although the drive in Marked Tree to close saloons was aided by the statewide prohibition drive, local people battled one another over it and put their own stamp on the struggle. Indeed, the first town council, organized in 1897, married the need for town services to the necessity to control the millworkers by passing ordinances, imposing fines, and hiring out jailed offenders. It was a simple solution fashioned by a few men who occupied favored positions within the community. Whether this first town council was an elected or appointed body remains unclear, but the town council members were voted upon in 1902. Only 44 votes were cast, even though the population of the town must have exceeded 500 by that time, with perhaps 250 millworkers of voting age.[31] The fact that it was an off-year election probably reduced the number of voters, but it is likely that the transient millworkers either ignored local politics or did not fulfill the residency requirement. In any case, only a fraction of the town's population elected the individuals fashioning the local ordinances.[32]

Men destined to play important roles in the development of the town and the founding of plantations in the countryside sat on Marked Tree's first town council in 1897. They included Ernest Ritter, who founded E. Ritter and Company in 1907 and became one of the most powerful men in the delta; John Krier, the first mayor, who ran a profitable mercantile establishment and became one of the leading businessmen-planters within a decade; W. B. Miller, a business associate of Ritter's who assumed the presidency of the politically powerful St. Francis Levee Board in 1903.[33] By 1902 T. G. Staton, the manager of Chapman and Dewey Lumber Company, the largest landholder in the county, served on the council, bringing the number of aldermen to six, and by 1903 banker M. W. Hazel had replaced John Krier as mayor.[34]

From the beginning the town council members perceived a need to control the activities of the lumber mill workers. Nine of the first twelve ordinances passed by the new town council were clearly directed at the transients who worked the mills of Marked Tree. The ordinances also revealed the kinds of problems a largely single, male population presented. The city councilmen prohibited "the disturbance of the peace" and the use of violent and/or profane language. They circumscribed the carrying of guns, bowie knives, clubs, and other deadly weapons. They passed ordinances dealing with public drunkenness and gambling. They made it unlawful to keep a "bawdy house" and passed an ordinance against "indecent or lewd and lascivious behavior" which was directed against local prostitutes who operated outside the confines of a "house of ill fame or of assignation." Finally, they made it "unlawful to disturb religious services" or to "labor on the Christian sabbath." All of the ordinances were passed unanimously by the five aldermen and the mayor.[35] Many of them must have been enforced, for by September 1897 the council began plans to erect a jail, and by November of that year it sanctioned the use of prisoners in making town improvements.[36] The ordinances passed by the town council restricted the activities of the millworkers, added needed capital to the town's coffers, and provided labor to make certain improvements.[37]

Aside from helping to maintain the town's streets and sewers, convict laborers undoubtedly constructed some of the homes occupied by the families that settled there between 1900 and 1910. The lack of housing had prevented men from moving their families into Marked Tree in the early period, and the businessmen who sat upon the town council were set on a course of economic development which necessitated the provision of housing to the expanding population. The revenue generated by the enforcement of the local ordinances passed in 1897 proved to be insufficient to meet the need for expanded town services, and by 1905 the town council placed a tax upon restaurants and grocery and dry goods stores.[38] Agitation for a hog law began in 1906, and in the next year dogs were labeled a "common nuisance," and a drive to tax them ensued.[39] In the same period the local newspaper printed articles decrying weeds and filth and importuned the town's citizens to undertake a cleanup campaign to make the town more presentable to outside capitalists who might wish to settle there and bring industry and progress. The town's mayor had to give the Frisco Railroad an ultimatum in 1908 to drain unsightly and unsanitary cesspools located beside the depot.[40] In 1907 the nationwide economic panic temporarily weakened the local economy, partly explaining the high turnover in the Marked Tree mills between 1900 and 1910. The box factory closed in 1907 and did not reopen until mid-1908. It normally employed 250 men and carried a payroll of $600 a week. Similarly, a handle factory, employing 100 men, ceased operation for almost twelve months the following year.[41]

By the end of the decade many of the economic problems that had beset Marked Tree had eased. In late 1909 the handle factory reopened, and a stave factory, a hoop factory, and a shingle mill began operation, promising to bring new citizens to the town.[42] Indeed, the workforce of 1910 bore little resemblance to that of 1900. Only 4 remained of the men who worked the early mills, and the mill hands no longer suffered the underemployment that had plagued their 1900 counterparts. Twenty-five of 465 men working the mills in 1910 reported an average of 41 days unemployed in the previous twelve months compared to the 96 who had reported an average of 127 days unemployed in 1900. The transformation of the workforce reflected a dramatic transformation of the community itself. The population of the town had increased from 352 to 2,026, of whom 1,832 can be identified in the poorly preserved manuscript census for 1910. This is a significant enough sample to conclude that the percentage of millworkers in the population had increased dramatically.[43] Approximately 466 (62.1 percent) of the 751 adult males in Marked Tree in 1910 who listed their occupations were employed in the mills; in 1900 only 96 (54 percent) of 178 men who listed their occupations had been so engaged.[44] An increase in occupations supported the growth in population, but even though the number of merchants and professionals and skilled and service workers had increased, the growth in the millworking population had far outstripped it (table 2.1).

The pressure of this exploding population of millworkers helped forge the battle over prohibition, and Marked Tree attracted the attention of prohibitioners and progressives throughout the state. Two organizations operating on the state level typically established local branch organizations and sent speakers to small towns all over Arkansas in order to proselytize on the issue of prohibition. One of those, the Woman's Christian Temperance Union, was especially active in Marked Tree. The Marked Tree branch was founded in 1908 by the wives of leading businessmen, including Anna Ritter, Ernest Ritter's wife. And while there is no evidence of any local activity in Marked Tree on the part of the other major prohibition organization in the state, the Anti-Saloon League, that organization's records indicate that it was extremely concerned about the situation in Marked Tree.[45]

The Arkansas Anti-Saloon League was first organized in 1899, but in 1906 a splinter group advocating a change in strategy split off from the original state league and was soon recognized by the national organization as the official branch in Arkansas. Its members wanted to abandon the local option approach and turn instead to agitation for a statewide prohibition statute. Local option had been initiated by temperance advocates and sanctioned by state legislation in 1873. The legislation allowed voters in communities and rural areas to vote at every

Table 2.1. Occupations of adult men in Marked Tree, Arkansas, 1900–1920

	1900		1910		1920	
	No.	%	No.	%	No.	%
Millworkers						
Operators	51	28.7	271	36.1	161	36.3
Day laborers	45	25.3	195	26.0		
Merchants	2	1.1	8	1.0	17	3.8
Professionals	7	3.9	18	2.4	27	6.1
White collar	10	5.6	35	4.7	94	21.2
Skilled and artisan	13	7.3	51	6.8	34	7.7
Service	18	10.1	35	4.7	17	3.8
Misc. nonfarm	22	12.4	124	16.5	52	11.7
Agricultural						
Owners	1	.6	8	1.0	26	5.9
Other	9	5.0	6	.8	15	3.4
Total	178	100.0	751	100.0	443	99.9
Missing cases*	3		23		45	
Total	181		774		488	

Source: Federal Manuscript Census, Arkansas, Poinsett County, Schedule of Population, 1900, 1910, 1920.

*Six pages of the 1910 manuscript census were not legible because they were badly microfilmed, making the percentages for 1910 of greater importance than the absolute numbers (see Appendix).

election on whether to permit the licensing of saloons. If the voters approved local option, anyone wishing to establish a saloon would have to circulate a petition which he would present to the county judge. The judge was bound to honor that petition if a majority of the registered voters in that particular town or township had signed it.[46]

Marked Tree demonstrated precisely why the Anti-Saloon League sought to adopt a different strategy. Local option did not work to close the saloons, and some town leaders became so frustrated by 1911 that they were willing to go beyond the law in order to bring prohibition to their community. In January 1911 the town council voted to allow the town's saloons to remain open until July 8, 1911. At that time they would then be arbitrarily closed even though citizens had voted in favor of licensing saloons and had legally petitioned to keep them. On Saturday night, July 8, 1911, the saloons were closed, and even though there was a "large crowd in town," there were no reports of "disorderly conduct or drunks seen on the streets." The legal saloons remained closed for fully a year, but "blind tiger" joints operated on the outskirts of town and regularly attracted the attention of law enforcement officials. To the dismay of the local prohibitioners, the town went "very wet" in the fall 1912 elections. Although Ritter and others claimed that the election had been fraudulent, liquor was again on sale legally in Marked Tree by the first of January 1913.[47]

One of the men to decry the reopening of saloons was a local merchant who had at one time "taken a leading role in trying to keep saloons open." Now "irrevocably opposed" to open saloons, that merchant may have been S. P. Thompson, who had made his fortune through the saloon business. Saloonkeepers only rarely achieved the kind of success that Thompson eventually enjoyed, but the saloon business could underwrite a shrewd and careful businessman's rise to prominence. It was a lucrative business; hence the willingness of some local merchants in Marked Tree to defy Ritter and others over the issue. Thompson's rise can be traced from his first appearance in the historical record as a twenty-eight-year-old saloonkeeper in Marked Tree in 1900. Thompson and his younger brother Joe, who tended bar for him, had arrived in Marked Tree sometime before the turn of the century, and both lived in the same boarding establishment. By 1906 Thompson was listed in a business directory as owning his own saloon, and by 1908 a Sanborn map of Marked Tree identified his saloon and mercantile establishment as one of only three brick structures in town.[48] The 1910 census lists him as a merchant rather than as a saloonkeeper, and in 1911 he coorganized and became first president of the Farmers and Merchants Bank of Marked Tree, with $50,000 in capital stock.[49]

Although no other saloonkeeper achieved Thompson's level of success in Marked Tree, several joined him in organizing the bank. C. C. Sloane, Newt Fisher, and W. D. Shelton were saloon men, and three merchants on the board may have numbered among those merchants who believed that open saloons were good for business.[50] The editor of the local newspaper, who had waffled on the prohibition issue and had defended the "saloon men" as "good men," referred to those organizing the bank as "some of the leading businessmen and farmers in this section." According to the editor, "these men form a nucleus of wealth equal to any like number of citizens in the county." Certainly the town was booming by 1910 and may have well had room for another bank, but it is possible that their opposition to prohibition had dried up capital from the Marked Tree Bank and Trust Company. Ritter was a chief depositor in the bank and M. W. Hazel, one of the directors of E. Ritter and Company, was adamantly opposed to open saloons. Or perhaps other differences existed with Ritter which encouraged them to go their own way. Thompson, for example, had been engaged in a court duel with Ritter over a valuable parcel of town property they both claimed along the Frisco Railroad. Thompson won the suit just months before the organization of the new bank in 1911. Ritter claimed he "would carry the case to the supreme court."[51]

Whatever quarrels Thompson and Ritter had up to 1911, they shared one thing in common which led to a strange alliance between them in 1912. Both Republicans, they formed an independent ticket in the town elections with Thompson running for alderman and Ritter for mayor. Amid charges from the incumbent mayor that the independents ran a "mean and bad" campaign, the election was termed "the hottest ever pulled off in Marked Tree." Ritter and Thompson went down to a resounding defeat but did not abandon the public limelight. At a meeting of lily-white Bull Moose Republicans several months later, they determined to "eliminate negroes from the party," saying it was exclusively "a white man's affair." They later attended a meeting in Little Rock to form the Arkansas Progressive party. Like many of their fellow progressives, these small-town Bull Moosers regarded the elimination of blacks as necessary to purify politics.[52]

Together with progressive Democrats, many Republicans supported another measure on the 1912 ballot, a grandfather clause which would have further restricted black voting in general elections. Also on the ballot was a statewide prohibition statute. But voters decisively rejected both measures. The prohibition statute had been placed on the ballot through the initiative process and the efforts of the Anti-Saloon League.[53] As early as 1910 the league had taken over visible state leadership of the drive to end drinking in Arkansas, but the WCTU had the better grass-roots organization. When Anti-Saloon League officials managed to get a prohibition statute on the ballot for the 1912 election, WCTU leaders reluctantly went along. They believed it was too soon, that not enough groundwork had been done to gain passage of the legislation. They knew that in communities like Marked Tree no outright prohibition statute would pass, and they were afraid that a premature attempt to impose such a statute on the state of Arkansas would backfire.

Their fears were not entirely realized. The prohibition statute did not pass in 1912, but prohibition forces recovered rapidly from the defeat and managed to influence legislation that modified the local option law, making it more difficult for individuals to petition to license a saloon even if the community had voted in favor of local option. While certain Marked Tree men, including Ernest Ritter, declared that the election had been rife with fraud, a longtime associate of Ritter's, L. C. Going, introduced the new legislation.[54] Going, who had once served as the attorney for the Marked Tree Town Council, represented the Twenty-ninth Congressional District, which included Poinsett County, and proposed legislation modifying the petition procedure so that only white registered voters could petition the court to open a saloon.[55] The Going amendment reflected prohibition leaders' perception that black voters posed a threat to their efforts. In 1912 when the statewide prohibition statute failed, prohibition leaders had paid particular attention to the fact that counties with heavy black populations voted against them. The grandfather clause on the ballot in 1912 had stimulated a large black voter turnout, and prohibitioners subsequently identified black voters as one of their principal obstacles. Heedless of the fact that middle-class black communities supported prohibition but perhaps aware that most blacks were impoverished agricultural workers, mill hands, or railroad workers, prohibitioners supported the Going amendment. Yet in Marked Tree, the effort was still not enough to close the saloons. There remained enough white mill hands willing to sign a petition, despite the fact that the names of the petition signers now had to be published in a local newspaper. The men proved to be unconcerned about any public censure, however, and petitioned to reopen the saloons. Some of the local elite thereupon mounted a sustained campaign utilizing injunctions, suits, and intimidation. The saloons opened and closed several times in 1914 and 1915.[56]

By 1915 prohibition forces joined with progressives in Arkansas to devise a different strategy for eliminating traffic in liquor. They decided to influence legislators rather than take the issue directly to the people. The legislature passed a "bone dry law" in February 1915, and January 1, 1916, dawned a dry day in the state. Finally the saloons in Marked Tree were permanently closed. Significantly, Marked Tree was one of the last five places in Arkansas to close saloons. Blind tiger joints, bootlegging, and moonshining continued in the backwoods, and law enforcement was lax at best, but the forces of prohibition on the state level had achieved what those on the local level had been unable to accomplish.

The success of the prohibition drive not only closed the saloons but eliminated a place where black and white millworkers commingled in a social setting. But by the time prohibition was finally imposed on Marked Tree, the town was in the midst of an important transformation. The number of millworkers had declined from 466 in 1910 to 161 in 1920, reflecting a change in the town's economy, now increasingly dependent upon the rise of the plantation sector. Both the number of adult men employed in the town and the town's population decreased in the second decade of the twentieth century. The population had never been particularly stable, for it fluctuated according to the boom and bust cycle of the lumber industry. Business directories for 1906 and 1912 illustrate this dynamic. In 1906 the population was estimated at 2,500, and in 1912 it was fixed at 2,100.[57] The drop to 1,232 in 1920 corresponded with the declining fortunes of the lumber industry, notorious for denuding forests and moving on. Chapman and Dewey continued to operate its mills, for its far-flung holdings sustained it until the mid 1940s, but there were fewer smaller mills and the number of men employed in mills decreased by 70 percent between 1910 and 1920, while the population of the town dropped by only 39 percent. The number of merchants, professionals, and agriculturalists increased. Indeed, by 1920 the plantation sector in the surrounding countryside had developed sufficiently to sustain the town. The transition from sawmill settlement to plantation town was fully under way.

The town's economy had become far more complex, and despite the decline in population, there were eight times as many merchants in 1920 as there had been in 1900. The number of professionals had quadrupled, and the number of white-collar workers, who largely clerked in stores and banks, had risen from eleven in 1900 to ninety-four in 1920. The merchants who hired these clerks included Ernest Ritter and John Krier, both of whom served on the first town council. They operated booming businesses in the next two decades and, along with merchants like John B. Claunch and S. P. Thompson, provided credit to farmers in the surrounding countryside. Ritter increased his landholdings by almost one thousand acres in Poinsett County between 1910 and 1920 while Krier, Thompson, and Claunch came to own over twelve hundred acres among them. Having risen from saloonkeeper to merchant and banker, Thompson became very wealthy. The four merchants, moreover, all operated plantations in the Poinsett County delta.[58]

By 1920 Marked Tree was no longer simply a community of millworking men. While the percentage of adult males in the population had dropped from 51.6 percent in 1900 to 39.6 percent in 1920, the percentage of women had increased from 24.2 percent to 32.8 percent, and the percentage of children had risen from 24.2 percent to 27.5 percent (table 2.2). The women were credited with having "brought christianity" and having "contributed to the upbuilding of the city."[59] Although a small black church, Anderson's Chapel, had existed since 1894, it was 1908 before the first white churches were erected in Marked Tree: a Methodist Episcopal and a Baptist church. In 1914 another Baptist church was organized, and in the 1920s four more began to hold services.[60] Although Ernest Ritter's wife had moved to

Table 2.2. Population of Marked Tree, Arkansas, 1900–1920

	1900		1910		1920	
	No.	%	No.	%	No.	%
Adult males	181	51.6	774	41.6	488	39.6
Adult females	85	24.2	476	25.6	404	32.8
Children	85	24.2	612	32.9	339	27.5
	351	100.0	1,862	100.1	1,231	99.9
Missing cases*	1		151		1	
Total	352		2,013		1,232	

Source: Federal Manuscript Census, Arkansas, Poinsett County, Schedule of Population, 1900, 1910, 1920.
*Sex or age was not reported or was illegible.

Memphis as the fortunes of her husband rose and as her children required education beyond the capacity of the local school, she maintained a residence and an interest in Marked Tree. Along with other women in the town, she established organizations that made it a more comfortable place to live.[61] The women were largely responsible for the erection of the town's white churches, and they launched three ladies' organizations: a garden club, a Ladies' Improvement Association, and the St. Matthias Guild. The growing number of families created other needs, and by 1909 the town boasted two schools, including one for the growing black population. The school for white children had to be enlarged in 1908, and plans for a high school were formulated.[62] The housing shortage that had plagued the town had been addressed as early as 1907 when some "Memphis capitalists" acquired property in Marked Tree, divided it into lots, and began building dwelling houses for sale or rent. Ritter was offering lots for sale at the same time and later put some of his men to work building homes.[63]

While Marked Tree's elite concentrated on developing their town and controlling the activities of the millworkers, they also turned their attention to expanding their interests in the countryside. Trends that revealed themselves as early as 1910 were directly related to the activities of Marked Tree's businessmen. Within the first decade of the twentieth century, the Poinsett County delta adopted certain essential elements of the southern plantation system. Businessmen planters utilized the system of advances to furnish their tenants and sharecroppers and instituted the crop lien, and several of them operated commissaries and issued doodlum coupons.[64] Between 1900 and 1910 the land ownership rate in Poinsett County's delta shifted dramatically downward (table 2.3), and the number of chattel mortgages tied to a cotton crop rose sharply. Although the number of farmers in selected delta townships increased from 124 to 369, the total number reporting that they owned their farms decreased from 70 to 45.[65] This continued into the second decade of the twentieth century and marked the emergence of major differences between the delta and the rest of the county. For example, while an impressive decline in landownership occurred throughout the county, only in the delta did it drop to 6.9 percent. And while planters in the delta leaped into the cotton market, farmers along the ridge experimented with vegetable and orchard crops, and farmers in the prairie began to sow rice.

Table 2.3. Land tenure of farmers who were heads of households in selected Poinsett County townships, 1900–1920

	Owners		Renters		Total		Missing cases*
	No.	%	No.	%	No.	%	
1900							
Ridge	112	63.6	64	36.4	176	100.0	
Prairie	39	81.2	9	18.8	48	100.0	
Delta	70	56.5	54	43.5	124	100.0	3
1910							
Ridge	108	38.7	171	61.3	279	100.0	
Prairie	26	60.5	17	39.5	43	100.0	
Delta	45	12.2	324	87.8	369	100.0	
1920							
Ridge	115	35.8	206	64.2	321	100.0	6
Prarie	44	54.3	37	45.7	81	100.0	4
Delta	60	6.9	808	93.1	868	100.0	25

Source: Federal Manuscript Census, Arkansas, Poinsett County, Schedule of Population, 1900, 1910, 1920.

* Did not report whether owned or rented their farms.

Although ridge farmers continued to grow some cotton, they did not become as dependent as the delta planters on the cotton crop and did not establish links with cotton factors in Memphis. Still, they experienced a stunning decline in landownership rates between 1900 and 1910. Their problems emanated, at least in part, from the mixed blessing of the arrival of the railroad which opened up a larger market for them but made them far more vulnerable to the fluctuations of that market. The farmers most likely to retain ownership of their farms over the first decade were those who lived in the county's prairie, where the landownership rate remained well above 50 percent. Like farmers on the ridge, prairie settlers did not establish links with Memphis creditors. Instead, they borrowed locally or from implement dealers in the Midwest.[66] Indeed, a large percentage of the prairie settlers had been midwestern farmers who purchased land in the Poinsett County prairie, sometimes sight unseen, and retained links with creditors from the Midwest. Some of them initially attempted to grow cotton along with a variety of other crops but found the soil there unsuitable and turned finally to the cultivation of rice, relying on wage laborers rather than tenant farmers to harvest their crops. They drew their laborers from among the tenants and sharecroppers from surrounding counties, and perhaps from the ridge rather than from the more remote delta in their own county.[67] Only in the delta, where certain individuals amassed considerable acreage, did the cultivation of cotton and the way of life associated with it take hold. Only in the delta, where links with cotton factors in Memphis firmly rooted farmers to one-crop agriculture, did the proportion of landowning farmers drop to 12.2 percent in 1910 and down further to 6.9 percent in 1920.[68] Farmers in the delta operated in a system that differed sharply from that of the rest of the county. By 1920 over 90 percent of those who labored in delta fields were not themselves the owners of the land they worked.

Marked Tree merchants owned many of the farms and plantations of the Poinsett County delta, and access to the market was the key to their success. The pattern of road building in the delta, however, contributed to the isolation of the county's delta from the ridge and prairie (map 4). Central to the establishment of a plantation economy, in fact, was the construction of roads that led not only to local towns like Marked Tree but also to larger metropolitan areas. Marked Tree businessmen played an important role in establishing new roads in the delta. Indeed, a

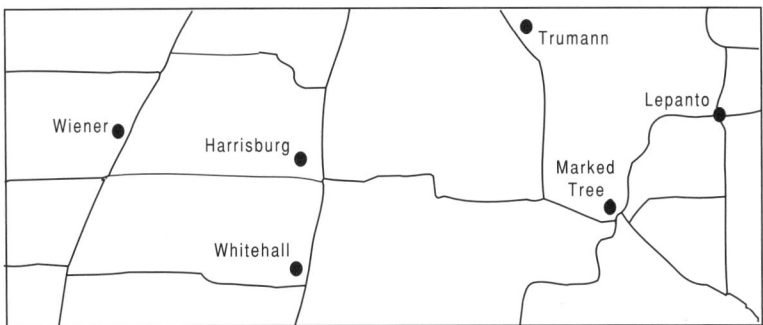

Map 4. Roads in Poinsett County, 1910–20

frenzy of road building occurred in the county in 1901 and 1902 shortly after the state passed a law enabling the county governments to create road districts.

Two important features of road building in the delta stand out. First, Marked Tree was the center from which all roads ran; second, the manner of road building in the county as a whole heightened the isolation of the delta from the ridge and prairie. This isolation reflected not only geographical peculiarities that made it difficult to maintain a road across the notorious swamp in the western delta but the preference of the delta planters for the Memphis market. Delta men petitioned the county court for the construction of seven roads in 1901 and 1902. Only one led to the ridge. Of the remaining six, three connected Tyronza Township in the southeastern section of the delta with Marked Tree and linked up with a fourth road leading to Crittenden County and eventually Memphis. A fifth road led from Lepanto in the northeastern delta to Marked Tree, and the sixth stretched from Marked Tree north to the Craighead County line.[69]

The fact that Marked Tree was the nexus for all delta roads was not coincidental. Marked Tree was the center of commerce and credit and therefore drew in the people from the surrounding countryside. In addition, those organizing the roads were often either Marked Tree residents or related to them. Ernest Ritter did not petition for the roads himself, but he won the contracts and built them. His brother, Louis Ritter, was one of the petitioners for the Tyronza roads leading to Marked Tree and to the Crittenden County line and Memphis. The roads on the ridge and prairie, meanwhile, showed a different pattern. Two ridge roads led to the prairie, and two additional roads linked the three prairie towns to one another. The six roads that were built exclusively along the ridge led directly to Harrisburg or connected with a ridge road that did. These roads also linked the ridge to the two counties bordering Poinsett to the north and south. The southern intercounty road led to Forrest City and the Little Rock and Memphis Railroad. The northern intercounty road reached Jonesboro, where the railroad from St. Louis to Memphis passed on its way through the delta and Marked Tree. In either case, the route to Memphis from Harrisburg and the Poinsett County prairie was roundabout, and weather conditions and the poor state of the one road leading from the delta to Harrisburg made the route to Marked Tree sometimes just as indirect.[70]

Perhaps nothing more aptly demonstrates the importance of roads to the businessmen-planters of the Poinsett County delta than a race between two of them in the fall of 1917. Anticipating the opening of the Harihan Bridge over the Mississippi River, Memphis merchants offered a prize to the first man to bring a bale of cotton across from Arkansas to Memphis. Sam Stuckey, a businessman-planter in the small northeastern Arkansas town of Lepanto, heard about the opportunity and loaded a bale grown by tenant Lonnie Ray onto his truck. The two of them then started down the rough dirt road that led out of town. They were just minutes ahead of businessman-planter D. F. Portis, Sr., who had secured his first bale of the season from tenant Joe Goodin and just had it ginned at Albert Seymour's cotton gin. Portis, Goodin, and Layton Seymour, Albert's son, struck out after Stuckey, and the two businessmen-planters were soon engaged in a forty-mile race toward the river. "The trucks passed and repassed each other until the Stuckey truck started to enter the Harihan Bridge and the Portis truck shot by, winning by 15 seconds."[71]

The roads symbolized the ties between Memphis merchants and the businessmen-planters in the small towns of northeastern Arkansas, and the network of roads had a significant correlation with another prerequisite for the formation of a plantation system in the delta: the establishment of credit. An analysis of the mortgages on file in Poinsett County reveals that delta creditors turned to the Memphis market for their credit needs. In fact, the sources of credit for the three regions of Poinsett County were distinct (table 2.4). An examination of a sample of the real property mortgages filed in the county in 1909 and 1910 reveals that each section of the county tended to provide the necessary credit for its own inhabitants.[72] Ridge borrowers showed the greatest loyalty to their local creditors (86.8 percent), while prairie borrowers were more likely than delta borrowers to go to the ridge for credit. Delta debtors were almost as unlikely to borrow from ridge creditors as ridge borrowers were to go to the delta. Delta farmers did, however, show a tendency to turn to Memphis, especially for chattel mortgages.[73]

By the time Ritter and his compatriots began developing the Poinsett County delta, the Arkansas legislature had clearly defined the lien laws applicable to the relationship between landlord and tenant and sharecropper. That body gave sanction to the crop lien on July 23, 1868; on

Table 2.4. Sources of credit for lenders and borrowers of real property mortgages in selected locations, 1909–10

Borrowers	Lenders				
	Ridge	Prairie	Delta	Other Ark.	Other states
Ridge	53	2	3	1	1
Prairie	5	24	1	1	2
Delta	3	0	36	1	11
Total	61	26	40	3	14

Sources: Poinsett County Mortgage Records, 1909–10, Poinsett County Courthouse, Harrisburg, Ark.; Federal Manuscript Census, Arkansas, Poinsett County, Schedule of Population, 1910. These two sources were compared and individuals matched.
Note: The townships examined were Bolivar, Prairie, Little River, and Tyronza.

January 8, 1875, it gave the landlord's lien precedence over that of other creditors. Finally, on April 6, 1885, the assembly encumbered the crop grown by a tenant or sharecropper if he had taken an advance from the landlord. In 1893, separate acts by the legislature worked to prohibit tenants and sharecroppers from disposing of personal property they had mortgaged to their landlords.[74] The existence of a separate body of law relating to personal property helps to explain why landlords entered into formal chattel mortgage arrangements, for the acts pertaining to personal property stipulated that they were governed by written encumbrances only. The crop lien, however, might be explicitly spelled out in a written agreement but also was recognized as implied by virtue of the landlord and tenant and sharecropper relationship.

Most crop liens were not formalized in writing, but much can be learned from an analysis of the chattel mortgages on file. A similar pattern to that identified for real property mortgages emerges. A sample of seventy chattel mortgages contracted by delta farmers in 1909 and 1910 reveals that nineteen involved Memphis lenders. Fourteen of these nineteen were cotton factors like A. F. Oliver and Company, which frequently placed advertisements in delta newspapers designed to attract plantation owners who would then finance their tenants. All nineteen of the Memphis-based loans included provisions stipulating the exact cotton acreage to be planted, and eleven of them required that the seed and lint be marketed through them at the "fair market value." Of the fifty-one local loans, thirty-four were tied to the cultivation of cotton. The chief creditors in the delta were either in Marked Tree or in the small town of Tyronza located northeast of Marked Tree. A number of individual landowners, many of whom had contracted loans with Memphis cotton factors or lenders in Marked Tree or Tyronza, contracted chattel mortgages with their tenants. The four principal creditors in Marked Tree were Ernest Ritter and his business partner C. A. Dawson, John Krier and Company, Chapman and Dewey, and the Marked Tree Bank. Ritter was one of the founding stockholders of the Marked Tree Bank, and the bank's president, M. W. Hazel, was on the board of directors of the newly incorporated E. Ritter and Company. The bank and Chapman and Dewey typically financed real property rather than personal property mortgages. Five of the Ritter and Dawson mortgages and four of the Krier mortgages required the borrower to grow cotton.[75]

The main lenders in Tyronza were J. C. Hooten, who had contracted with the town of Marked Tree for convict labor in 1908, and Louis Ritter and J. A. Emrich, who operated as merchants in Tyronza while serving on the board of directors of the Marked Tree Bank. Louis Ritter, who was in partnership with Emrich, was Ernest Ritter's brother, and the ties between the two included speculation in real estate. The enterprises of Hooten and of Louis Ritter and Emrich contracted eight chattel mortgages with tenants from the surrounding countryside. Four of the eight required that the borrower grow cotton. The only other single lender who operated on a comparable scale was J. T. Lee, a merchant (and soon to be banker) in Lepanto, located in what would become Greenwood Township in the northeastern section of the delta. Lee owned 101 acres of land in Poinsett County alone as well as property in neighboring Mississippi County and signed chattel mortgages with fourteen individuals. At least five of those fourteen borrowers were his own tenants and sharecroppers. Lee himself had signed a contract with A. F. Oliver and Company of Memphis which provided him with the necessary capital to fund his tenants and required that he grow cotton and market it through them. Thirteen of the fourteen contracts he signed as lender, consequently, also stipulated that cotton be grown. The remaining local chattels were held by individual landowners who were apparently engaging their tenants in contracts. None of the delta mortgages were contracted with merchants or landowners on the ridge or in the prairie. The delta had acquired a life and focus of its own, independent of the Harrisburg establishment. The credit market followed the lines of the transportation network and tightened the cotton economy's hold on the delta.[76]

Between 1900 and 1920 the town elders, with the help of the local women, more or less transformed what had been little more than a primitive settlement. The town council passed ordinances to control the activities of the single millworkers and to limit the number of vagrants in town. Those unattached vagrants, in fact, provided convict labor for both the town and the countryside. Even though the millworkers continued to defy the prohibition statute by frequenting illegal saloons, the town's prohibitioners had, with the help of the state, closed the legal saloons where white and black men socialized together. The statewide prohibition statute had been secured through the prodding of legislators such as Poinsett County's L. C. Going. The town council, meanwhile, had also forced the railroad to adhere to health standards imposed for the good of the community, and it had enacted taxes to help defray the costs incurred in engaging in various civic improvements. By 1910 the town economy was booming. Several new factories hired hundreds of workers whose wages helped to support the flourishing mercantile establishments. The local bank extended credit for the construction of homes and commercial buildings and also funded the purchase of farms in the countryside. By 1920 the mill no longer dominated the town; the needs of the emerging plantation sector had supplanted it in importance. Farmers flocked to town on Saturdays to do business and to socialize, and in the fall they often brought their cotton to one of the town's gins. The local elite felt they had "civilized" their town by making a number of improvements, and they were pleased at their success in attracting new businesses and new capital. Even as they focused on improving their town, however, they began to turn their attention to developing a plantation system in the surrounding delta. Busy with mercantile establishments in town, planters were only too willing to recruit tenants and sharecroppers from elsewhere to come to northeastern Arkansas. But they would discover that tenants and sharecroppers would constitute a major challenge to their hegemony.

3

LABORING IN THE "AMERICAN CONGO"

The failure of white authorities to apprehend whitecappers in Arkansas in 1898 prompted a Lonoke County African-American man to write a letter to the president demanding that he send U. S. troops to protect blacks from attacks by whites who "take us and kill us from our own homes" and who post "signs . . . saying don't let the sun go down on you." He challenged the president to "send your troops to relieve us from the point of their beyonets for [otherwise] we will spill blood."[1] In Phillips County "a gang of hoodlums" fired into the cabins of black renters and "actually succeeded in driving the majority of them out of the neighborhood." The whitecappers professed "to be actuated by a desire to rent land at reasonable prices, and say they cannot do so because the negroes will take it at the prevailing prices." Planters found it impossible to identify the perpetrators and thus rid the area of the plague of nightriders.[2]

A pitched battle between blacks and whites in Calhoun County led to the arrest of a band of whitecappers, some of whom were "prominent in the social and business affairs of the county." More often, however, blacks found it too dangerous to take the law into their own hands. When a group of black farmers in Arkansas County confronted whitecappers in 1904, nine of the black men were arrested and subsequently lynched.[3] Because state officials found it difficult to identify and then successfully prosecute suspected whitecappers, the *Helena World* was moved to suggest drastic measures when it became clear that local law officers were powerless to act in the face of nightriding activities in Phillips County. Whitecappers there burned barns, killed livestock, destroyed crops, and succeeded in driving many blacks from the area.

Planters who opposed the nightriders suffered retaliation, and the difficulty of securing reliable witnesses against masked men led the newspaper's editor to declare, "The *World* does not counsel unlawful methods in dealing with these people, but believes if a few of them were shot to death when they are out marauding, it would have a wholesome effect."[4] The situation in Phillips County was rapidly approaching social chaos, and unrest soon spread into northeastern Arkansas.[5]

Almost commonplace in the South in the late nineteenth and early twentieth centuries, attacks on blacks by whitecappers were rooted in competition between blacks and whites over rental contracts with plantation owners. Such encounters occurred from Georgia to Mississippi and surfaced in Arkansas in 1898 as the plantation system there underwent considerable expansion. Because blacks were more impoverished than whites, planters could secure a larger share of the crop by contracting with blacks. Thus, planters often preferred making arrangements with black farmers, and many whites took to wearing white caps and operating under the cover of darkness in attempts to drive blacks away and secure a place for themselves on Arkansas plantations.[6]

While white and black landless farmers competed fiercely with one another, white planters, anxious to bring their plantations into operation, struggled for control over labor. Planters, in fact, mounted a challenge to the whitecappers, and one of the most notorious whitecapping cases occurred in northeastern Arkansas just as the plantation system there was going through a period of rapid expansion. When a Campbellite (Christian Church) minister began a revival in 1902 at White Chapel, a community of Baptists and Methodists four miles south of Wynne in Cross County, he fanned into flames the smoldering resentments of a class of white men who were competing with the newly arrived African Americans for rental contracts on the expanding plantations in the area.[7] The minister may have been preaching racial toleration or speaking to an integrated assembly of worshipers. The Christian Church, the most liberal wing of the splintered Disciples of Christ, had long advocated a kind of limited integration of church services. Blacks were to enter by a separate door, sit in a separate section of the church, and drink the communion wine after white worshipers. In some cases, blacks were to hold separate services and use the church building on Sunday afternoon. By the beginning of the twentieth century, however, the Christian Church had come to terms with segregation after incidents that suggested their tolerant approach conflicted too much with community sentiments in the South. Nightriders in Texas, for example, drove an African American from a Texas church.[8] It is likely that the men who organized themselves into the White Chapel Club were similarly motivated. Tensions ran so high, in fact, that they burned to ashes a Christian Church attended by planter R. R. Fallis in the fall of 1902. No one was brought to trial for destroying the church, however, for "there was never sufficient evidence gathered to convict any one." According to the Memphis *Commercial Appeal*, the dispute spilled over into the plantation operations of Fallis and other planters: "When the time arrived for the renewal of contracts between the land owners and the renters for the crop of 1903 the negroes were the successful bidders, and the white renters became without jobs. Soon after the organized band sprung into active existence and rioting began in earnest."[9] They now directed their depredations against the rapidly expanding black population in the area. Tacking up notices warning blacks to leave, they "spread dismay and caused almost a reign of terror" by shooting into cabins, burning homes, and destroying crops. One source even suggests that some blacks were lynched. The nightriders succeeded in driving away approximately two hundred blacks who "fled for their lives," and only one black sharecropping family remained on the R. R. Fallis plantation.[10]

White planters responded by contributing to a fund to hire Captain J. H. Brown, a private detective from Memphis, to lead a squad of men and apprehend the nightriders. One evening about March 10, 1903, Captain Brown led his four men to the Fallis plantation and stationed them in hiding. Hoping to surprise the whitecappers, he instructed his men to remain quiet until the nightriders rode into view and then await his signal before making "a rush to capture the whitecaps." But when the men rode into view, Brown hesitated before giving the signal to attack. The nightriders opened fire and mortally wounded Brown.[11]

Tensions ran high in Cross County after the assassination of Brown, and the local sheriff deputized several extra men to secure the twelve alleged whitecappers arrested in the days following the incident. Officials feared for the safety of the remaining private detectives, whose identities and whereabouts were "being kept secret" while they waited to testify at the coroner's inquest. As for the nightriders, the community was bitterly divided. "Both friends of the movement against the band of whitecappers and sympathizers of the crusade against the negroes are in town."¹² Officials had to be prepared to protect the nightriders from lynching by one side and from being freed by the other.

Even as members of the White Chapel Club awaited trial in the federal district court in Helena, another group of whitecappers, protesting the hiring of only blacks at a new sawmill in the ridge community of White Hall (south of Harrisburg) in Poinsett County, "converged on the mill" bearing firearms, carrying torches, and demanding "that all blacks be fired and replaced by whites." The millowner, James Davis, who had only recently come to the county from Wynne (in Cross County), appealed for help from the local justice of the peace, John Harrington, but Harrington "not only refused to help keep the peace, but joined the mob." Although Davis capitulated and "discharged all his black workers," the situation caught the attention of federal officers and led to the arrest of fifteen men.¹³

Beginning on October 9, 1903, the trials of the twenty-seven men held in the two cases attracted widespread attention in Arkansas, but only three of the Poinsett County men were convicted. The charges against the members of the White Chapel Club were dropped because securing reliable testimony against them proved impossible. They had worn masks, thus obscuring their identities, and they produced witnesses who testified they had been elsewhere at the time Brown was assassinated. Although the judge, the prosecutor, and many observers were convinced of their guilt, the Cross County nightriders went free. Three of the fifteen men who had forced blacks to leave the Davis sawmill in Poinsett County, however, were not so fortunate. Even though they produced their own witnesses, the case against them was apparently stronger. William R. Clampet and Reuben Hodges "were sentenced to a fine of $100.00 each and to imprisonment at Atlanta a year and a day." Wash McKinney "was sentenced to pay the same fine, and to 60 days imprisonment in the jail of Pulaski County, Arkansas."¹⁴

Although the government secured convictions in only one of the two cases, the same reasoning was employed in fashioning the indictments against all twenty-seven defendants. The nightriders in both cases were charged with violating Section 1 of the 1866 Civil Rights Act by driving African Americans away from their occupations. When counsel for the defense "demurred on the ground that congress did not have the authority to enact the statute under which the indictments were returned," prosecuting attorney William Whipple responded "that such authority was given by the thirteenth amendment." Judge Jacob Trieber agreed with Whipple and overruled defense counsel's demurrer. His charge to the jury outlined a provocative interpretation of Thirteenth Amendment protections, in which he declared that the amendment provided blacks with a constitutional right to employment.[15]

The attorney representing the whitecappers, L. C. Going, challenged Trieber's decision. Going, an ambitious man who would go on to become a state senator and even serve as an acting governor, had been raised just a few miles from White Hall. He almost certainly knew the accused whitecappers personally, and the fact that his representation of them did not ruin his political career reflected the complex forces at work within northeastern Arkansas. While some white planters and farmers decried the depredations against the black community, perhaps with mixed motives, others supported the effort to drive blacks away. Going had ties to both planters in the delta and farmers on the ridge and in the prairie. His defense of the Poinsett County whitecappers may have helped him secure election as prosecuting attorney for the state's third judicial district in 1904, a position to which he was reelected in 1906. During his long career he also served as attorney for the Marked Tree Town Council and as a member of the politically powerful St. Francis Levee Board. It was while he was state senator from the Twenty-ninth District (1910–16) that he wrote the amendment to the local option law that excluded blacks from petitioning for open saloons.[16]

Going won his appeal to the U.S. Supreme Court in the whitecapping case in *Hodges v. U.S.* in 1906, for the court reasoned that the Thirteenth Amendment had provided blacks with citizenship and in so doing had established that they were not "wards of the nation." Since the Thirteenth Amendment restricted only the actions of states, not those of individuals, blacks, as citizens, would have to take "their chances with other citizens in the states where they should make their homes." Congress was thus "not empowered . . . to make it an offense against the United States" for private citizens "to compel negro citizens, by intimidation and force, to desist from performing their contracts of employment."[17]

Lawlessness returned to the Arkansas delta after the Supreme Court decision; indeed, the violence which first surfaced in 1898 continued unabated throughout the first two decades of the twentieth century. In 1908 another case of whitecapping occurred in Cross County a few miles away from White Chapel. Nightriders engaged in a sustained campaign against black sharecroppers farming in "the enclosure," an area covering thousands of acres in Cross and St. Francis counties. Mary Lee Rhodes, who owned "the enclosure"—part of a huge Spanish land grant—lived in Memphis and hired overseers to run her plantation. Although testimony was taken and warrants were issued, no one was arrested.[18] In 1909 the General Assembly passed Act 12, as "an act to suppress and punish nightriding and other riotous conspiracies," which resulted in some arrests in Mississippi County and at least one well-publicized conviction. But lynching and lawlessness continued, and whitecappers like those who threatened Ernest Ritter's black tenants and sharecroppers in 1912 continued to elude capture and prosecution. The nightriders who plagued the black farmers working the Ritter plantation reportedly began their depredations after it was rumored that fifteen black families Ritter brought from Mississippi had carried boll weevil–infested cottonseed with them. Ritter insisted they brought only a few sacks of shucked corn and little else from Mississippi. Five years later nightriders again visited black sharecroppers on the Ritter place and told them to leave the county. None of the whitecappers in either of the incidents were ever apprehended.[19]

In 1915 law enforcement officials used bloodhounds to track down five nightriders who had posted notices threatening blacks and promising to burn the homes of white landowners hiring blacks in Mississippi County. But the apprehension of nightriders was so rare that in 1916 Judge William Driver was moved to make a suggestion reminiscent of that made by the *Helena World* editor in 1898: "If anyone ever loses his life while engaged in night riding, the man who kills him in defense of his home shall never go to jail."[20] The 1909 state law clearly had been ineffective in halting nightriding activities, and although passed at the behest of planters who felt powerless to conserve their labor force, the law was used as often to quell vigilante violence perpetrated by whites against other whites. In at least one case it was used to suppress striking workers. It was called into service a few times against whitecappers, but "comprehensive protection against racial discrimination in employment was not provided until the enactment of the Civil Rights Act of 1964." In 1968 the Supreme Court formally overruled *Hodges*.[21]

The emasculation of the Thirteenth Amendment in *Hodges v. U.S.* was a final capitulation to the forces of redemption in the South. The

Supreme Court participated in this drama by a series of decisions in the late nineteenth century that weakened the Fourteenth Amendment, voided the KKK Acts, and essentially abrogated the 1866 and 1874 civil rights laws.[22] Judge Trieber, in fact, had fashioned his unusual argument regarding the Thirteenth Amendment in the nightriding cases precisely because it was the only possible legal argument remaining for those who wished to protect the rights of black citizens. It is not surprising that the case, and thus Trieber's argument, arose in an area where the plantation was relatively late in arriving and rapidly expanding. That some planters paled at Trieber's open invitation to the federal government to meddle in the South's labor and racial relationships is not to be doubted. That others found it acceptable is indicative of the lengths to which the social chaos prevalent in the Arkansas delta had driven them.

Nightriding persisted, and the "champions" of the black victims, the planters, were more concerned with an adequate supply of labor than they were with justice for African Americans. Even Judge Jacob Trieber, who devised the ingenious argument in the Cross and Poinsett county cases and is generally regarded as one of the most persistent defenders of black rights in Arkansas, may have had mixed motives. A German-Jewish immigrant and son of a dry goods merchant in Helena, young Trieber became active in Republican party politics and involved in the town's civic affairs. A well-respected man in this prosperous plantation center, he served on the town council and as county treasurer, and he founded and became the first president of the First National Bank of Helena in 1893. Although he was appointed U.S. district attorney for the Eastern District of Arkansas in 1897 and to the federal bench in 1900, he retained his business interests in Helena at the same time that he established ties with prominent planters, businessmen, and manufacturers in Little Rock. So prominent had he become by 1911, that he was invited to become a member of the exclusive XV Club. Made up of prominent men from both major political parties and of various religious persuasions, the club prided itself on providing a forum where men of different opinions could gather and amicably discuss usually divisive issues. The issues considered included the rights of African Americans. After one exchange of opinions over black voting rights, a member was moved to remark: "Seldom, if ever before, in Arkansas, have so many men, of so many parties, from so many sections, discussed so calmly and dispassionately, a subject which heretofore has occasioned so much bitterness and so much discord. It betokens a bright day for our town, state, and nation."[23]

It was a bright day for conciliation among elites but a dismal day for blacks aspiring to vote, hold property, or seek employment in certain places and occupations. Trieber's concern for the rights of African Americans may have been genuine, but his own economic interests were intertwined with those of a prospering plantation economy, and his inclusion within the fold of the XV Club confirms that his identity rested with the upper class. Some among that class were most certainly disturbed at Trieber's federal prescription for the nightriding malady, but they were nevertheless determined to conciliate with one another and present a united front. Even Trieber thought that there were limits to the federal government's role in southern racial affairs, and he may have had doubts about the wisdom of his ruling in the *Hodges* case. Writing to Theodore Roosevelt in 1905, after the *Hodges* convictions but before the Supreme Court overruled him, Trieber indicated that blacks should be left to the care of the better class of white men in the South. His reference to the nightriders as "vicious and ignorant" men further established his credentials as a man of the upper class and exemplified the attitudes of that class.[24]

After the *Hodges v. U.S.* decision, the federal government provided blacks with little legal defense against the wrath of the white community and essentially placed them under the protection of the southern planters who employed them. But the planters had their own agenda, and while they objected to the activities of nightriders who sought to drive African Americans from the plantations, they instituted a system of debt peonage which challenged the Thirteenth Amendment in yet another way. Although technically free, many sharecroppers and tenants found it difficult to remain free of debt. In the economics of the plantation system, debt bound laborers to planters, and thus planters were able to institute a system of exploitation so pervasive that, according to NAACP field secretary William Pickens, it rendered the Thirteenth Amendment virtually inoperative. Pickens found the inequities in the system so profound that he dubbed Arkansas, Louisiana, Mississippi, western Tennessee, and eastern Texas the "Congo of America."[25]

The key to the web of debt that tenants and sharecroppers faced in the southern plantation system was the commissary system. Planters advanced both tenants and sharecroppers the supplies necessary to carry them through the winter. Planters maintained a record of the charges, added an exorbitant interest, and when they marketed the cotton crop in the fall, they subtracted the amount owed by the tenants and sharecroppers from the amount due them. The Department of Agriculture estimated in 1909 that the typical sharecropper could hope to clear $175 before settling his account at the plantation store, but after settlement the sharecroppers and tenants would have to pay for the coming year's "meat, meal, tobacco, molasses and other things necessary." Thus, when they did "succeed in getting out of debt at the end of the crop year," tenants and sharecroppers became indebted "again almost immediately"; after settling their debt at the commissary, most had no money to purchase the necessary items to carry them through the winter. And so the cycle continued. According to freedman Henry Lee, brought to Arkansas with his family by his former master immediately after the Civil War, sharecropping was not itself the problem. "Things went on very well till the commissary come about." In many cases tenants and sharecroppers found themselves unable or barely able to pay out at the end of the year. As Henry Lee put it, they "got figured clean out" by their landlords at the commissary stores.[26]

Debt peonage became widespread enough in the South in the early twentieth century to prompt the federal government to send secret service agents into the field to investigate. Agents probed into peonage allegations in Arkansas and secured convictions in two cases in 1905. In one northeastern Arkansas case, the jury deliberated only thirty minutes after hearing one hundred witnesses and then acquitted the defendants of all charges. That was one of only a few cases to surface in northeastern Arkansas, however. The plantation was still too new to Poinsett County, and either the phenomenon of debt peonage had not yet arisen, or federal investigators concentrated their attention on counties in Arkansas where the plantation was more firmly established and peonage more clearly obvious.[27]

The plantation system arising in Poinsett County attracted African Americans precisely because of its relative youth and the fact that planters had yet to fully establish their control over labor. Elsewhere in the South planters had been able to erode the compromise that freedmen had forced upon them in the years immediately following the Civil War. In a hard-won battle over the structure of the postwar plantation system, freedmen had accepted the sharecropping arrangement because it gave them maximum control over their own lives and implied a partnership with planters in the plantation enterprise. But by means of their influence with state legislatures throughout the South, planters enacted a series of lien, vagrancy, and antienticement laws that circumscribed freedmen. After the populist threat of the late nineteenth century demonstrated the political power that could accrue to an alliance of black and white voters, southern legislatures disfranchised blacks (and some whites), played upon racist sentiments, and drove the final wedge between blacks and whites by enacting segregation statutes. Thus, by the beginning of the twentieth century, many blacks found themselves in dependency relationships with planters too reminiscent of slavery. Gone was the semblance of partnership and independence. Many of them decided to move to the newly reclaimed countryside in northeastern Arkansas. Poinsett County was no safe haven for landless black farmers, but the nascent plantation system there offered opportunities unavailable elsewhere. They found it easier to maneuver within the system just coming into operation in northeastern Arkansas, enabling them to resist the complete dominance of planters and maintain the semblance of partnership.[28]

The kind of resistance threatened by the Lonoke County African-American farmer in 1898 was rare, however, for open defiance against either whitecappers or planters was almost always suicidal. If a violent encounter occurred between blacks and whites, and the black man was left standing, his only recourse was flight, for he would get no justice from a court system poisoned with the racist ideology of the era. Blacks came to realize the futility of openly challenging whites and drew on time-tested strategies inherited from slavery and from the earliest freedmen. Adopting the artifice developed by blacks slaves in the antebellum period, they put on "a mask of deception," assuming a demeanor of compliance, ignorance, and docility. Freedmen elaborated on this stratagem and, like peasants elsewhere, employed "the ordinary weapons of relatively powerless groups: foot dragging, dissimulation, false compliance, pilfering, feigned ignorance, slander, arson, sabotage, and so forth." These ploys worked well for blacks who were often isolated from one another on their own farmsteads in remote countryside locales. As James Scott argues, "These Brechtian forms of class struggle have certain features in common. They require little or no coordination or planning; they often represent a form of individual self-help; and they typically avoid any direct symbolic confrontation with authority or with elite norms."[29]

Although these everyday forms of resistance were commonplace, violent encounters did occasionally occur, especially in places where the plantation was expanding. In northeastern Arkansas, for example, many planters were as new to the area as the sharecroppers and tenants who worked the land for them, and the nature of the plantation system that emerged there was negotiated in this context. Planters had not yet firmly established their ascendancy, and African Americans who came to northeastern Arkansas had high expectations and sought to stretch the plantation system beyond the limits imposed elsewhere. Thus, numerous violent encounters occurred between landlords and tenants in northeastern Arkansas, especially in the first decades of the twentieth century when the plantation made its first appearance. Such encounters almost always led to the imprisonment or lynching of the African Americans, but that they occurred at all is indicative of the propensity of black tenants and sharecroppers to test the limits and the determination of white planters to fashion a system which gave them maximum control over labor.[30]

The specific grievances that led to violence varied. Violence sometimes resulted when planters attempted to evict tenants from plantations. In March 1910, for example, planter William Sidle and his nephew, Claude Burtinett, approached the tenant shack of Steve Green for the purpose of evicting him from Sidle's Crittenden County plantation. Words were exchanged in front of the shack, and then Green "went into the house and returned with a Winchester rifle," shot Sidle once, and forced Burtinett to leave. Green then escaped, and white citizens contributed to a fund to encourage his capture.[31]

Tenants sometimes wanted to leave plantations before they secured the crop, and planters often objected. For example, in July 1903 tenant Will Hughes shot planter Arthur Sheddan after Sheddan discovered that Hughes intended to leave his Mississippi County plantation. Concerned that Hughes owed him forty dollars for supplies advanced, Sheddan went to Hughes's shack, discovered that he was not at home, and appropriated some of his household goods. Later, upon learning that Hughes had returned, Sheddan, accompanied by his brother Will went to the shack and confronted Hughes, but Hughes opened fire, killed Arthur Sheddan, and drove Will Sheddan off. Hughes then fled for his life.[32]

Hughes's desire to leave the Sheddan plantation precipitated the violent confrontation between the men, but key to the conflict was a dispute over property. One of northeastern Arkansas's most tragic confrontations between an African-American tenant and a planter occurred forty miles from Marked Tree in Mississippi County in 1920 and reflected a kind of grievance that was probably the most common: disagreement over settlement of the crop. On Christmas Day tenant Henry Lowry went to the home of planter O. T. Craig demanding, for the second time, a settlement. Craig instead struck him with a piece of firewood, and Craig's son then shot him. Lowry, though wounded, pulled his own gun and fired, killing Craig and his daughter and wounding Craig's two sons. Lowry then escaped to El Paso, Texas, with the help of black friends he had made through the Odd Fellows Lodge. He tried to get a letter back to his wife and six-year-old daughter, but it was intercepted, and Lowry was arrested and extradited to Arkansas. Before the train carrying him reached Arkansas, however, a posse of white vigilantes intercepted it and carried him back to Mississippi County, where two thousand people gathered at Nodena, a landing on the river, and watched as he was set afire and burned to death.[33]

The burning of Henry Lowry is a particularly gruesome example of the extent to which white supremacy ideology had infected the population. Although useful to planters in keeping black and white tenants and sharecroppers divided and thus at least partially conquered, white supremacy was a difficult mechanism of control. Planters found it almost impossible to prevent whites who felt blacks were competing unfairly for rental contracts from mounting campaigns against the blacks. So widespread became the notion that northeastern Arkansas was "a white man's country" that the editor of the *Osceola Times* was moved to write a long editorial decrying that attitude in 1904.[34]

Yet when conflict broke out between planters and black tenants, such as that between Lowry and Craig, white tenants could be counted upon to join with planters in dealing "justice" to blacks who challenged the system. The burning of Henry Lowry, in fact, was a carefully orchestrated event designed to serve as a lesson to the black community at the same time that it drew the white community together. The mob on the landing at Nodena cut across class lines, and if the sheriff of Mississippi

County is to be believed, "every man, woman and child wanted the burning to take place."[35]

The conflict between Craig and Lowry exemplified the lengths to which planters were willing to go to protect their interests and the extremes to which some tenants went to pursue theirs. It also demonstrated the complicity of white law enforcement officials and their predilection toward supporting the landlords and white supremacy. But perhaps even more revealing was the way the incident was misrepresented in the press. As William Pickens indicated in an account published in the *Nation* in 1921, the facts of the case were deliberately obscured so that Lowry was depicted as a "Negro fiend," a "maniac," or "drunk." The press quoted the planters as being mystified about the reason for the confrontation between Lowry and Craig.[36] Planters were clearly unwilling to engage in a public discourse which would reveal the inequities in the plantation system operating in the Arkansas delta.

The plantations of Poinsett County constituted a battleground where blacks and whites competed for rental contracts, where tenants of both races strove to secure their own interests, and where planters struggled for control over labor. In this arena blacks learned that open resistance was futile, and thus they donned the mask of compliance "for purposes of survival."[37] Another form of resistance, however, tread a thin line between defiance and compliance: mobility. Although the attempt to leave a plantation sometimes led to violence, as in the case of Will Hughes, the spectacle of families on the move after settlement time every year was common. They migrated "from plantation to plantation whenever they have an opportunity of bettering their condition."[38] Fanny Johnson described what life was like after she left Tennessee and moved to Arkansas with her mother: "We farmed. . . . We worked at lots of places. One time we worked for a man named Thomas H. Allen. He was at Rob Roy on the Arkansas near Pine Bluff. Then we worked for a man named Mimbroo. He had a big plantation in Jefferson county. For forty years we worked first one place, then another."[39]

Historians Jonathan Wiener, Jay Mandle, and Pete Daniel have suggested that tenants and sharecroppers were not moving around with any realistic expectations of bettering themselves but were futilely searching for a way out of the poverty that bound them to the South and to the plantation. These historians fail to comprehend fully the role that mobility played as a form of resistance, and thus Jonathan Wiener has engaged in a strident debate with Robert Higgs, who emphasizes the salutary effects of mobility. Higgs regards the mobility of freedmen as exemplifying the manner in which the free market worked to the advantage of freedmen. Yet Higgs's findings that freedmen significantly improved their circumstances are suspect, and he, too, fails to perceive the true meaning of mobility: defiance of planter control.[40] Wiener, Mandle, and Daniel are closer to the truth when they argue that blacks gained little materially in moving around, but that was not entirely the issue, and, in any case, the fact that they did not better their situations does not make their attempts to do so meaningless.[41]

By putting pressure on the plantation system, by keeping the planters constantly off balance, blacks both resisted complete domination and at times were able to secure better situations for themselves. Even though debt often traveled with them and interfered with their ability to progress beyond tenancy and sharecropping, mobility introduced a certain instability into the plantation system and provided a means by which particularly shrewd blacks could maneuver themselves into an advantageous situation. As the Henry Lowry case demonstrated, if a black man challenged the planter's figures and demanded a fair settlement, he put himself at great risk. George Stith, an African-American sharecropper who later became a union organizer, did on occasion challenge a planter concerning his recordkeeping, and his experience indicates that sharecroppers could maneuver within the system using a combination of the "mask" and the implied threat of departure. "I made a crop with a man and I kept all my figures so when I went up to him to settle, he give me a statement on how much I owed, and I said 'this is not right.' And he said 'what do you mean its not right'? And I said 'I don't owe that much.' He looked at me, and he said 'do you know who keeps my books?' I said 'no sir,' and he said 'my wife keeps those books. You don't mean she lied?'" Understanding that to challenge the planter's wife was dangerously close to insulting southern womanhood, an infraction punishable by lynching, Stith responded: "'No sir, I didn't get a thing from your wife. Everything I got was from you.'" He pulled receipts from his pockets to demonstrate that he had kept tabs on his account. The planter in the end relented and accepted Stith's accounting because he regarded Stith as a "good farmer" and needed him to keep his operation in good running order. "That was the way it operated. Once in a while you find [yourself] on a farm [where you] could get a little bit better break." The mobility of plantation laborers allowed them to find those situations where planters were willing to compromise.[42]

Like all forms of compliant resistance, however, mobility also worked to the advantage of planters. By adopting strategies of passive resistance, blacks helped maintain the plantation system and, as Gavin Wright has suggested, enabled the plantation to expand into new areas like Arkansas.[43] Yet mobility was the best offense for blacks, and it was not accepted with equanimity by planters. The refusal of blacks to contract with certain planters left those planters desperate for labor and led to failed experiments with foreign labor and, ultimately, to the enactment of laws that attempted to limit mobility.[44] African Americans were attracted to the newly developing plantation sector in Arkansas in the early twentieth century precisely because the system elsewhere had become so encumbered with mechanisms of control that flight to Arkansas represented their last best hope for independence from planter dominance.

African Americans who were dissatisfied with their circumstances elsewhere often responded eagerly to the overtures of labor agents. These agents canvassed the southern states east of the Mississippi River and bought train tickets for those willing to make the move west. Black emigrants saw themselves getting a free ticket and a new chance, and they spoke of coming to Arkansas because of the fertility of the land rather than the abundance of jobs. Thus, they were not motivated by the desire to work for the white man on the plantation but to farm for themselves on the remarkably fertile land for which the area was becoming famous. Yet landlords were waiting for them when they reached the Arkansas delta. Henry Anthony of North Carolina reported that he came "on immigration ticket to Mr. Aydelott. . . . Train full of us got together and come. One white man got us all up and brought us here to Biscoe." George Johnson's father responded to the promises of a labor agent when he decided to bring his family to the Arkansas Red River Bottom in 1869. But things did not turn out well for George, who was seven years old when they emigrated to Arkansas, or for his father. According to George, "The way my father happened to bring me out here was, Burton Tyrus came out here in Richmond stump speaking and telling the people that money grew like apples on a tree in

Arkansas. They got five or six boat loads of Negroes to come out here with them. Father went to share cropping on the Red River Bottom on the Chickaninny Farm. He put in his crop but by the time he got ready to gather it, he taken sick and died. He couldn't stand this climate." After his father's death, young George was "bound out to Henry Moore and his wife. I stayed with them about six years and then I ran off."[45]

Labor agents painted vivid pictures of the abundance of the Arkansas delta that proved compelling to African Americans who were finding soil exhaustion in the older South another handicap to advancement. The portraits the labor agents painted were designed to appeal not only to men eager for fertile land to cultivate but to women and children whose lives had been dominated by hunger and privation. Thus, when a labor agent told Talitha Lewis, who came to Arkansas in 1889 at their behest, that the place she was to report to "had fritter trees and a molasses pond," she knew they were speaking in metaphor, but she found the message irresistible.[46]

The blandishments of labor agents attracted many, but others came on their own because of the rumors they had heard about the fertility of the farms in the Arkansas delta. These were men who were out for the main chance and who were willing to go to any length to secure better lives for themselves and their families. Isaac Crawford, who left Mississippi to settle near Brinkley, Arkansas, put it most succinctly: "I been farming all my life. I come here to farm. Better land and no fence law." The land's reputation for high productivity attracted a number of immigrants who echoed Crawford's sentiments. Emmet Beal's mother came to Arkansas from Tennessee in 1880, and he followed her in 1881 when he heard "there was better land out here." Wade Dudley walked part of the way from Mississippi because the "land was better here." They regarded themselves as farmers rather than as farm laborers, and although rarely successful in moving into land ownership, some of them vacillated in status. Mose Evans, who worked plantations all over the Arkansas delta, sharecropped until 1908 when he managed to acquire tools and stock. He then "rented on thirds and fourths" until he became ill, had to sell his stock, and returned to sharecropping. According to Evans, "It's a lot easier to get behind than it is to catch up."[47]

While the freedmen who came to Arkansas almost without exception initially came to farm in the plantation sector, other opportunities existed. The availability of employment in laying track for the new railroads in the state and in denuding the forests that covered much of eastern Arkansas proved an added advantage to emigration to Arkansas. It was another way that African Americans could maneuver within an oppressive social and economic system. Many reported that they worked for the railroad or in the lumber industry during slack times on the farms; thus they could resist the dominance of the plantation owner by earning money elsewhere. Although many preferred farming to any other occupation, the majority of the blacks interviewed by WPA workers in Arkansas in 1937 and 1938 reported that they had worked off the farms for at least some portion of their work histories. Only 81 (31.6 percent) of the 256 men who reported occupations failed to mention employment outside farming. In contrast, 175 (68.4 percent) indicated that they had spent some time working in another industry or enterprise in addition to having spent some time as farmers.[48]

Despite the opportunities apparently available in Arkansas to augment plantation wages with work in other industries, African Americans confronted numerous obstacles. Not the least were the oppressive heat, disease, lack of artesan wells, and reluctance of planters to screen the shacks occupied by tenants and sharecroppers, leaving them to the not-so-tender mercies of malaria-laden mosquitoes.[49] Those who labored in the delta, which fell within the "malaria zone," confronted disease and risked heat exhaustion. Although blacks were not so susceptible as whites, George Johnson's father apparently succumbed to malaria or swamp fever in southeastern Arkansas; conditions were no better in Poinsett County.[50] Emigrants often found life in the delta brutal and even deadly. The first to experience the ferociousness of the swamp, however, were not tenants and sharecroppers who came in after the land had been reclaimed but convict laborers from the Marked Tree jail employed to clear the swamps and dig the first drainage ditches.[51] Given the preoccupation with black vagrants and the suggestion that they be arrested or made to work, it is likely that many of the convicts were black.[52]

Clearing land in the best of circumstances was arduous labor, but clearing land in mosquito-infested swamps involved additional hazards. The convicts worked from sunup to sundown, and in 1907 a black convict died of heat stroke in August, one of the hottest months in the delta.[53] Those who contracted to build the drainage ditches and purchased convict labor intended to make a profit on the enterprise and so the riding bosses who supervised the convicts were not known for their benevolence. In 1909 a J. C. Going faced a justice court for "cruelty to county convicts." J. C. may have been related to state legislator L. C. Going, or the paper may have misprinted the name and the perpetrator may have been L. C. Going himself. Regardless of his identity, the man barely avoided the consequences of his riding boss's actions.[54] Clearly, the life of a convict laborer in Poinsett County was a difficult one and sometimes a short one.

The tenants and sharecroppers also suffered from dietary deficiencies. Foods necessary to health were scarce because planters required tenants and sharecroppers to devote nearly all of their acreage to cotton, and few tenants and sharecroppers kept a large enough kitchen garden to carry them through the winter. By the end of winter each year many of them were on the verge of pellagra, a grave dietary deficiency. Red Cross representative C. M. Hubbard discovered during a disastrous flood in 1912 that many tenants and sharecroppers suffered from the disease.[55]

Despite certain unpleasant realities awaiting blacks in Arkansas, the fact that the businessmen-planters of Poinsett County were desperate to bring their land into cultivation and unfamiliar with the plantation system was auspicious for the tenants and sharecroppers who flocked into the area in the first two decades of the twentieth century. The county's plantation system grew out of the town of Marked Tree at the behest of businessmen who sought to be planters, and the most important of these businessmen-planters had no previous connection to or experience with the system that evolved out of the antebellum plantation. They simply adopted its economic infrastructure because it suited their purposes. Ernest Ritter, the son of German immigrants, came to Arkansas in the 1880s from an Iowa farm, and although white tenancy was at that time becoming a problem in Iowa, black sharecropping was not.[56] Yet Ritter, having determined that the plantation system provided an efficient means of developing his land, chose to import whole families of white tenants and black sharecroppers from Mississippi to staff his plantations.

The unwillingness of Poinsett County planters to adopt the paternalistic responsibilities associated with the traditional southern planter reflected both their unfamiliarity with the traditional plantation system and their status as businessmen first.[57] The failure of paternalism to emerge also contributed to uncertainty about the nature of the relationships that existed between planters and those who cultivated their fields for them. Even in areas where the plantation had long existed, great ambiguity attended the unwritten contracts between planters and both tenants and sharecroppers. The struggle between them was over which form of tenancy, sharecropping or share-tenancy, would prevail, and the stakes were high. While both are recognized as a kind of tenancy, share-tenants are usually referred to as tenants, and sharecroppers are generally known as either sharecroppers or croppers. The tenant farmer furnished his own livestock and implements and paid the planter one-third of the crop, ending up with two-thirds of the crop. The sharecropper, who furnished nothing but his labor, received in payment for his services only one-half of the crop. Aside from getting more of the crop, the tenant farmer enjoyed another distinct advantage. According to law, the tenant owned the crop and paid a share to the planter. The sharecropper, however, did not own the crop; ownership was vested in the planter who then paid the sharecropper. In the case of any dispute, the tenant farmer had much greater standing in court than the sharecropper, who was relatively powerless.[58] The sharecropper was essentially reduced to the status of a wage laborer. To be sure, the sharecropper did not regard himself as a wage laborer but as a farmer, and he hoped to improve his situation and become a landowner himself. Impoverishment explained a man's willingness to enter a sharecropping arrangement. Yet the legal differences between tenancy and sharecropping, although clear in the statutes, "tended to break down in practice" since planters marketed the crop for both tenants and sharecroppers and then divided the proceeds.[59]

Nevertheless, planters usually preferred the sharecropping arrangement because it gave legal sanction to their taking a larger share of the crop, but the businessmen-planters of Poinsett County were so eager to bring their land into cultivation that they accepted either arrangement and were not always clear in their own minds about the nature of the relationships in which they were engaged. And the size of the first plantations in the Poinsett County delta created greater autonomy for the tenants and sharecroppers. There were no huge business-plantations like the Delta and Pine Land Company in Mississippi, where to be a sharecropper in the twentieth century meant working in gangs under close supervision. Although there were such enterprises elsewhere in

Arkansas, nothing like them had developed in the early decades of the twentieth century in Poinsett County. While the typical plantation elsewhere in Arkansas was in excess of 500 acres, the average landholding in Poinsett County's delta in 1900 was 410 acres. Yet even a man with a mere 100 acres might choose to cultivate his farm by placing sharecroppers upon it rather than working it himself, especially if he was otherwise engaged in some mercantile enterprise in Marked Tree.

Between 1900 and 1920, while the land in crops in Poinsett County increased from 91,365 to 127,124 acres, cotton cultivation rose from 3,681 to 26,532 acres. As more land was brought into cultivation, plantations increased in size, but whatever the size of the plantations, the tenants and sharecroppers employed on them worked approximately the same amount of land as their counterparts elsewhere. Estimates of how much the individual tenant and sharecropping families farmed in the South range from ten to forty acres. The difference in the acreage cultivated doubtless depended on the size of the family.[60] Jim Kirkendall, a Poinsett County black sharecropper from Mississippi, had a large enough family to work at least a forty-acre farmstead. Sometime in 1916 or 1917 Kirkendall reached the Arkansas delta where the last of his four children was born. He also supported his mother-in-law, two nieces, and a nephew. Two of his children and the nieces and nephew ranged in age from eleven to fifteen and so were old enough to help work his Poinsett County farm.[61]

Although the specific arrangement, whether sharecropping or tenancy, was not always explicitly understood, and even though the legal distinctions "tended to break down in practice," distinct material differences existed between tenants and sharecroppers. Although the thirty-nine-year old Kirkendall might have been able to farm a good amount of acreage with the help of his family, his economic circumstances were precarious. He had no chickens or cows, which might have better sustained his family, and he held no livestock or tools, which might have allowed him to move into tenancy and gain a larger share of the crop. He listed only $50 worth of property on the personal property tax records for 1920, all household goods.[62]

Wiley Walker, on the other hand, was probably a tenant farmer. A rarity among Poinsett County blacks, only 15.1 percent of whom were tenants rather than sharecroppers, he was better able to provide for his growing family than Kirkendall, for in addition to three mules, a wagon, and some farm implements, Walker had two cows and some poultry. He also claimed $50 in household and kitchen goods. Like many of the African-American immigrants to the Arkansas delta, Walker came to Arkansas from Mississippi sometime after 1917. He brought with him a wife and four children, and a fifth child was born in Arkansas in 1919. Wiley's total personal property worth was $272 in 1920, well above Jim Kirkendall's $50 worth. The average black farmer in the Poinsett County delta had personal property holdings valued at $169.85. The typical black farmer was landless, forty-two years old, had a wife and two or three children, and had come to Arkansas from Mississippi. He was unlikely to have a horse, but he had one chance in three of having a cow and one chance in four of having a couple of mules and four hogs. In other words, fully three-quarters of the black landless farmers owned no mules; thus, they almost certainly worked as sharecroppers. Although the printed census of population for 1920 does not differentiate between sharecroppers and tenants by race, it does establish that only 42 (6.6 percent) of the 636 black farmers in the county in 1920 owned their farms. The 1930 census establishes that the number of black farmers had increased to 736, but the number of black farm owners had decreased to 21 (2.9 percent), and of the 698 tenants, 583 (84.0 percent) were sharecroppers.[63]

Reflecting the fact that they were far more likely to be tenants than sharecroppers, white farmers were somewhat better off. The typical white landless farmer in Poinsett County in 1920 was thirty-eight or thirty-nine years old, had a wife and three children, and had come to Arkansas from Tennessee or Mississippi. He was twice as likely as his black counterpart to have a cow or two. His chances of having hogs, horses, and mules were also greater. He claimed $56 in household and kitchen goods and $185 in livestock. His total personal property wealth was $353.25.[64]

While there were very real differences between black and white sharecroppers and tenants in Poinsett County, both were economically vulnerable, and thanks to legislation originating in the 1870s, the law armed planters with the means to enforce their control over those who farmed for them. Planters also prevailed over other creditors, thanks to other legislation of the 1870s. The new planters of Poinsett County, however, were not entirely cognizant of the extent to which the law worked in their favor, and an incident in 1908 indicates the sometimes near-tragic consequences of their ignorance. Tragedy was narrowly averted when certain businessmen-planters, law enforcement officers, and an African-American doctor from Marked Tree, S. L. Mitchum, learned that planter liens took precedence over those of other creditors. The lesson in plantation economics began late one evening in June 1908 when Mitchum was called to the home of a dying black woman in Tyronza Township. After ministering to her, Mitchum took two cows as payment for his services and returned to his own farm just outside Marked Tree. He penned the cows and went to bed only to be awakened sometime during the night of June 9 by Constable Gracey and his deputy from Tyronza, who by authority of an arrest warrant requested by the dead woman's landlord took Mitchum to the Tyronza jail. They charged him with theft and returned the cows to the landlord. The doctor, however, spent only one hour and fifty-five minutes in jail. As a major spokesman for the black community in Marked Tree, he had protectors among the white elite who sent Marked Tree's constables to Tyronza to arrest Gracey on a charge of kidnapping. In the end, all charges against both parties were dropped. But the landlord retained possession of the cattle.[65]

The return of the unfortunate woman's cattle to her landlord not only illustrates that the Arkansas legal system served to protect the interests of landlords over those of other creditors but provides clear evidence that the plantation system had arrived in Poinsett County's delta. By arresting Mitchum, however, Constable Gracey demonstrated an ignorance of the lien laws that had grown up around the plantation system in the late nineteenth century. When the local justice of the peace in Marked Tree heard the case, he viewed it simply as a dispute between two creditors—he was probably encouraged to do so by John G. Waskom, the prominent white attorney representing Mitchum—and sanctioned the return of the cattle to the landlord.[66] By the time the plantation system was established in Poinsett County's delta, the legal system in the South had subordinated the claims of merchants to the claims of planters.[67] Yet in a county such as Poinsett where the most prominent delta planters were merchants and businessmen, the law was rarely challenged and perhaps little understood. Mitchum's decision in 1912 to sell his farm and return to his birthplace in Illinois may have had a great deal to do with a growing awareness of the limitations he faced as a black man in a society increasingly dominated by white businessmen-planters.[68]

Although Dr. Mitchum suffered the humiliation of arrest, his situation was preferable to that of both sharecroppers and tenants. Neither the black nor the white farm laborers who flocked to Poinsett County's delta would own the land they worked (see table 2.3). Only 18.5 percent of the white heads of households reported to the census taker in 1910 that they owned the farms they operated. Only 5.2 percent of the black household heads so reported. Taken together, the white and black heads of household who claimed to own the farms they operated represented only 12.2 percent of the total household heads engaged in farming enterprises. The situation had worsened by 1920. Only 6.9 percent of those operating farms in the county's delta owned the farms. The percentage of farm owners, in fact, had dropped precipitously since 1900. Although the total number of black farm owners had increased—coincident with the increase in acres in production in this period—the majority of the blacks who settled the delta worked for planters.

Yet at the beginning of the century, tenancy and sharecropping were not firmly established in the county, and the implications of their coming to the area were not lost on the whitecappers in Poinsett and Cross counties. Although described by Judge Trieber as "ignorant and vicious men," they had no trouble perceiving the threat that the plantation presented to them.[69] In 1900 fully 60 percent of the white farmers in Little River and Tyronza townships (in the Poinsett County delta) owned their farms; by 1920 this had dropped to only 10.8 percent. These figures accompanied the rise of the plantation system and a considerable increase in the acres in production in the delta. In other words, the decline in landownership rates there must be understood in terms of the movement of already landless farmers into the area. On the ridge, where the acres in production likely remained fairly steady, the drop in landownership rates was less severe but probably reflects a more straightforward loss in status. In Bolivar Township, located next to Scott Township where the Poinsett County whitecappers resided, the landownership rate dropped from 63.9 percent to 38.1 percent between 1900 and 1920. A profile of the whitecappers in the two cases provides evidence that they were motivated by more than racism.

A superficial analysis of the *Hodges* case might obscure the motivations of those who attempted to drive the African Americans from the county. The exclusive use of blacks at a sawmill might not seem a direct threat to small farmers, but, in fact, the policy of hiring only black workers represented a direct challenge to white farmers. The previous decade had been marked by considerable economic instability on the ridge, and the farmers' ability to acquire cash in the sawmill industry often made the difference between solvency and bankruptcy. The nightriders might have feared that the African Americans who had come to work at the new sawmill in White Hall might soon compete with them in the farming sector. Certainly Dan Shelton, one of the black men driven from the sawmill, demonstrated the capacity of at least one of the victims to move into landownership. By 1920 he owned and operated a farm in Scott Township near where three of the whitecappers owned farms.

A profile of the Poinsett County nightriders establishes that they were almost entirely from the farming sector and almost certainly nervous about competing with African Americans in a depressed economy.[70] Eleven of the fifteen indicted men could be identified on the manuscript census of population,[71] and only one of those eleven, Reuben Hodges, was not a farmer but a sawmill worker. Six of the eleven were farm owners, and two were farm renters. One, William Stafford, was a farm laborer in 1900, a farm renter in 1910, and a farm owner by 1920; he clearly had ambitions for farm ownership and eventually was able to fulfill those dreams. But by 1920 only two of the six men who had owned farms in 1900 continued to do so. Except for those two and Stafford, the other nightriders had disappeared from the census. One of the surviving nightriders was Stafford's stepfather, William Clampet, whose family had settled in Poinsett County before the Civil War. Although he was able to maintain ownership of his farm despite the economic chaos in the area and despite his conviction and brief imprisonment for nightriding, Clampet's family's fortunes had declined steadily in the late nineteenth and early twentieth centuries. The Clampets had been small slaveholders who held wealth in personal and real property in excess of the median in 1860. The 1870 census demonstrates that they had been hit hard by the war, and the proliferation of landless Clampet households according to the 1880 and the 1900 censuses indicates that there was a distinct narrowing of opportunities in the county.[72]

Motivated by economic insecurity and probably inspired by the white supremacy rhetoric that dominated Arkansas politics at the time, these farmers sought to drive away the competition and keep Poinsett County for white men only.[73] And Judge Trieber's description of the nightriders as "ignorant and vicious" does not fully describe who they were or why they took to nightriding. They were probably more desperate than vicious. Ranging in age from twenty-seven to forty-eight, nine of them were married men with children, one was a widower, and only Reuben Hodges was single. Six were Arkansas natives, two were from Alabama, one from Kentucky, and one from Mississippi. In other words, like most of the white farmers in the delta and on Crowley's Ridge, they were native southerners (table 3.1). At least some of the men were related to each other: three were brothers, and two were William Clampet's stepsons.[74]

A great deal of similarity existed between this group of nightriders and those who formed the White Chapel Club in Cross County. Eight of the twelve Cross County nightriders could be identified on the man-

Table 3.1. Nativity of all delta farmers in selected townships including owners and farm laborers, 1900–1920

	1900		1910		1920	
	No.	%	No.	%	No.	%
Arkansas	65	28.0	183	23.3	186	16.6
Other southern	137	59.1	565	71.8	884	78.8
Nonsouthern	25	10.8	38	4.8	50	4.5
Foreign	5	2.2	1	.1	2	.2
Total	232	100.1	787	100.0	1,122	100.1
Missing cases*	3	8	1			

Source: Federal Manuscript Census, Arkansas, Poinsett County, Schedule of Population, 1900, 1910, 1920.

Note: The townships examined were Little River and Tyronza.

*See table 2.1 note for problems concerning the 1910 figures.

uscript census of population. They ranged in age from twenty-three to forty-nine, and their average age was thirty-one. The average age of the Poinsett County nightriders was thirty-two. Five of the eight were married, one was a widower, and two were single. Four of the married men had children. Seven were Arkansas natives and one was from Tennessee, making them all native southerners. Four of the eight men were of the Hall family (brothers and cousins), an old and respectable Cross County family which had seen better days.[75] Despite these similarities with the Poinsett County nightriders, there was one significant point of departure: economic status. Only three were farm owners. Three were farm renters, one was a farm laborer, one was a carpenter, and one cut crossties for the railroad. By 1910 one of the farm owners had joined his brother in the carpentry trade. They were clearly economically vulnerable, and it was likely that this economic vulnerability contributed to their nightriding activities. Although racism undoubtedly played a role, it was economic insecurity and the challenge of the plantation that stimulated nightriding against African Americans in both counties.

Not only did few among the general population of farmers in Poinsett County manage to maintain ownership of their farms after 1900, most of them disappeared from the area entirely. Only 9.5 percent of the white men and 14.8 percent of the black men who worked farms in Little River and Tyronza townships in 1900 remained there throughout the decade. The situation between 1910 and 1920 was worse: only 4.3 percent of the black farmers and 8.8 percent of the white farmers remained in the county by 1920. It is worth noting that those who persisted through the 1910–20 decade were more likely than the average 1910 farmers to have been farm owners in 1910 (19 percent of the persisters were farm owners compared to 12.2 percent for the total population of farmers in 1910), and several of them had become farm owners by 1920. While the percentage of farm ownership in 1920 was 6.9 percent, 30.8 percent of the "persisters" were farm owners in 1920. Clearly some of those who chose to stay in the county did so because they either already owned farms or were successful in moving into farm ownership. But not all of the persisters were so lucky. Most moved down the scale, and those who persisted between 1910 and 1920 were, in any case, a very small percentage of the farm operators there in 1920 (6.1 percent). The vast majority of those who came to farm the plantations migrated into the county in the intervening years, and 54.3 percent of them were blacks from the upper South. Indeed, despite the whitecapping incidents, the black population of farmers in the Poinsett County delta, unlike the black population of millworkers in Marked Tree, increased by a striking percentage. They helped to transform the delta countryside from a thickly forested wilderness to one of the most productive cotton areas in the South (tables 3.2 and 3.3).

While most of those who would work the county's plantations arrived in the county after the turn of the century, others had been living in the vicinity for many years. The area around Marked Tree had long been the abode of a transient river and backwoods culture dominated by white men who did not wish to farm any more than necessary to feed their families. They hunted and fished rather than farmed, and they sometimes supplemented their meager incomes through bootlegging. The spread of the plantations threatened the habitat of the fish and animals upon which they subsisted and brought law enforcement officials to bear upon their illegal moonshining activities. Such people were few in number and transience was their greatest problem, for they failed to regularize their holdings and thus had little or no legal standing. As civilization encroached they receded further into the backwoods, but the rapid development of the region seriously circumscribed their ability to find isolated, undeveloped pockets. Some of them ended up working plantations and lived to regret it. James Weeks, for example, determined in the spring of 1922 to stop rafting and working timber, sell his

Table 3.2. Race of all delta farmers including owners and farm laborers, 1900–1920

	1900		1910		1920	
	No.	%	No.	%	No.	%
White	208	88.5	399	50.2	513	45.7
Black	27	11.5	396	49.8	610	54.3
Total	235	100.0	795	100.0	1,123	100.0

Source: Federal Manuscript Census, Arkansas, Poinsett County, Schedule of Population, 1900, 1910, 1920.

Note: See table 2.1 note for problems concerning the 1910 figures. The townships examined were Little River and Tyronza.

Table 3.3. Race and land tenure of delta residents who classified themselves as farmers and who were heads of household, 1900–1920

	1900		1910		1920	
	No.	%	No.	%	No.	%
White owners	69	60.0	36	18.4	42	10.8
White nonowners	46	40.0	160	81.6	346	89.2
All white farmers	115	100.0	196	100.0	388	100.0
Missing cases	3		0		11	
Black owners	1	11.1	9	5.2	18	3.8
Black nonowners	8	88.9	164	94.8	462	96.2
All black farmers	9	100.0	173	100.0	480	100.0
Missing cases	0		0		14	
All owners	70	56.5	45	12.2	60	6.9
All nonowners	54	43.5	324	87.8	808	93.1
All farmers	124	100.0	369	100.0	868	100.0
Missing cases*	3		0		25	

Source: Federal Manuscript Census, Arkansas, Poinsett County, Schedule of Population, 1900, 1910, 1920.

Note: The townships examined were Little River and Tyronza.

*Three individuals in 1900 did not report race; twenty-five individuals failed to report in 1920 whether they owned or rented their farms.

houseboat, and try his hand at farming in the Poinsett County delta. Since he could not buy mules and farm implements, he arranged to sharecrop for the Haverstick family, despite a "warning from an old colored gentleman that had farmed with the Haversticks the year before and said he came out with nothing due to the fact he didn't know how to figure for himself." At settlement time Weeks "collected about $100.00 from the Haversticks." He later discovered that, among other things, the scales used by the Haversticks "had been tampered with." Weeks subsequently alternated between working plantations and plying his old trade on the rivers. Many other white backwoodsmen probably did the same.[76]

But backwoodsmen and "river rats" were too uncertain a supply of labor to satisfy the needs of the developing plantation sector. Planters often turned to labor agents to recruit for them. The boll weevil, ironically, may have contributed significantly to solving the planters' labor needs, for by 1910 the appearance of the weevil had prompted a migration from infested areas to the western cotton-growing regions. Although southeastern Arkansas itself faced the weevil by 1908, the northeastern delta was relatively free of the pest. The legal sanctions imposed in the late nineteenth century against those who hired labor away from plantations, always difficult to enforce, wilted in the face of the boll weevil crisis. Ernest Ritter probably used labor agents to locate the forty white tenant families that he moved onto one of his plantations in 1908 and the fifteen black sharecropping families he brought to the county in 1913. Yet neither Ritter nor his compatriots in Marked Tree seem to have faced any difficulty from the antienticement legislation. Generated by individual states, such laws were hard to enforce across state lines.[77]

Although white tenants and sharecroppers were often in desperate economic circumstances, they did have the advantage of being white, an advantage that carried considerable weight in a racist society. Economic competition complicated by racism had driven a wedge between the white and black sharecroppers and tenants and prevented them from recognizing or acting upon their common grievances. And when black sharecroppers banded together on their own to attempt to address their problems, they faced insurmountable odds. The Elaine race riot of 1919 is a case in point. Occurring about one hundred miles south of Marked Tree, the riot in Elaine sent shock waves through both the white and black communities. It served notice to the planters that they should be ever more vigilant in controlling black labor, and it demonstrated to African Americans the danger of open defiance.

In some ways the Elaine riot was simply the Henry Lowry incident writ large. It epitomized a landlord-tenant conflict turned violent; it exemplified the network that had grown up among African Americans in the Arkansas delta, it illustrated the complicity of law enforcement officials, and it demonstrated the uses to which white supremacy could be put. And just as in the Lowry case, the press sought to obscure the nature of the conflict. This was somewhat more difficult to do in the Elaine situation, for the African Americans had formed the Progressive

Farmers and Household Union of America and hired an attorney, O. S. Bratton, the son of a prominent white Republican attorney located in Little Rock. The older Bratton had been an assistant U.S. attorney at the time of the peonage investigations in 1905, and the father and son were working together in the case.[78]

The union had hired the Brattons to represent them in suits they planned to file against landlords who failed to settle fairly with tenants and sharecroppers. As the union was holding a meeting at a church near Elaine to discuss the proposed suits, however, a shooting occurred outside. It remains unclear who fired the first shot, but African Americans standing guard outside the church killed white Special Agent W. A. Adkins of the Missouri Pacific Railroad. Adkins's companion, a black trusty named Charles Pratt, alerted white citizens in Helena, and "a posse of eight hundred men, including an armed detachment from the local American Legion post," rushed to the church. As black families took refuge in the canebrakes to avoid massacre, whites burned the church where the meeting had been held. Thus began the Elaine race riot.[79]

Governor Charles Brough requested and received federal troops from Camp Pike to "restore order," and the troops arrived to find scattered battles between blacks and whites occurring. Brough himself traveled to the county and "was even fired upon as he walked down a country lane." By nightfall of the day after the shooting of Adkins, "the fighting had been suppressed by the troops, and authorities began to count the toll." Over twenty-one blacks were killed, and over one hundred were arrested. In trials that resembled kangaroo courts, fifty-four were sentenced to life in prison, and twelve were sentenced to death. Even though all of the African Americans were eventually freed, including those sentenced to death, white supremacy had been restored to the Arkansas delta.[80]

Despite clear evidence that the union had formed to pursue its goals through legal action against planters over settlement, the press reported the incident as a race riot and indicated that the African Americans were planning insurrection. No one mentioned that the union was initiating suits through the courts. The planters claimed that the black organization was planning to murder white planters and their families and seize their lands. As in the Lowry case, the facts were obscured to conceal the actuality that the dispute reflected the injustice in the economic system prevailing in the delta.[81]

While planters struggled to establish their ascendancy over the tenants and sharecroppers, landless African Americans tested the resolve of the planters at every turn. So determined were they to pursue their interests that they sometimes confronted planters openly to demand settlement or restitution when they felt they were being cheated; they sometimes responded violently when they met with intransigence or violence on the part of the planters. Incidents like the Henry Lowry lynching, however, demonstrated that individual defiance was suicide; the Elaine race riot illustrated that collective action was no better. Both exemplified the way that white supremacy could be employed as a powerful weapon against African Americans in the Arkansas delta. Mobility rather than open defiance proved to be the most effective maneuver for blacks, especially since they could establish no common ground with white tenant farmers. The strategies employed by both blacks and whites had limited success and sometimes resulted in disaster. Fighting each other at the same time that they struggled against the planters, they divided their energies. Each group had its own ways of resisting the planters: blacks donned the mask of docility, employed mobility, resorted to direct confrontation, and turned to unionizing; whites sought to drive blacks from the area and thus force planters to engage in share-tenant rather than sharecropping arrangements. Even as black sharecroppers tested the limits of the plantation system in the first decades of the twentieth century, white homesteaders in the Arkansas delta engaged in a bitter struggle with planters over thousands of acres of newly reclaimed land in northeastern Arkansas. This struggle pitted some of the most powerful forces within the delta against one another and involved both state and federal governments in determining the course of economic development in the region.

1. Railroad work crew. (Courtesy of the Special Collections Division, Picture Collection no. 4413, University of Arkansas Libraries, Fayetteville)

2. High-water marks on cypress trees near the White River in Phillips County, ca. 1913. (Courtesy of the Howe Collection/UALR Library Archives)

3. Ritter store, ca. 1900. (Courtesy of Mary Ann Ritter Arnold)

4. Ritter family with Ernest Ritter seated in the center, ca. 1905. (Courtesy of Mary Ann Ritter Arnold)

5. Mules dragging a car out of the water, in Mississippi County, ca. 1910. (From Elliott B. Sartain, *It Didn't Just Happen* [Osceola, Ark.: Drainage District No. 9, 1975], 24)

6. Charlie B. Greenwood. (Courtesy of the Lepanto Museum)

7. Eliga ("Kidd") Abbott and family, homesteaders. (Courtesy of Dorothy Abbott)

8. Surveyors encounter snake in a Mississippi County swamp. (From Elliott B. Sartain, *It Didn't Just Happen* [Osceola, Ark.: Drainage District No. 9, 1975], 8)

9. Cabin in the swamps, Phillips County, ca. 1935. (Courtesy of the Charlton Collection, Special Collections Division, University of Arkansas Libraries, Fayetteville)

10. Moving time, Phillips County, ca. 1935. (Courtesy of the Charlton Collection, Special Collections Division, University of Arkansas Libraries, Fayetteville)

4

SUNK LANDS AND LOST HOPES

When homesteaders dynamited a passage through the right hand chute of the Little River in August 1913, they prepared to face the obstructionist tactics of one of the most powerful men in the Poinsett County delta. Using motorboats, they intended to move through the passage "as soon as we can handle timber without molestation from E. Ritter and Company in Marked Tree." That same summer, A. W. Cheatham of Lepanto, claiming to represent 141 settlers, wrote to the secretary of the interior to complain about violence against homesteaders by "lumber companies and others" and even went to Washington to try to meet with the secretary. Two years later Fred Greenwood of Lepanto published an open letter in the *Marked Tree Gazette* reporting that "men have been overheard saying that they received $20 per day for the burning of squatters houses." While many settlers denounced the companies for the violence, at least one homesteader, Della Abbott, blamed African Americans for displacing white settlers: "Rome Hays got beat out of his homestead. The negro beat him."[1]

Violence and intimidation continued uninterrupted throughout the decade. In 1917 Hannah Embrey, president of the Homesteaders Union of Lepanto, wrote to the Interior Department to complain that settlers were being burned out of their homes. In November of the same year, Hilliard C. Smith sent a letter to the editor of the *Osceola Times* in nearby Mississippi County opposing demands made by some plantation owners that homesteaders pay rents on lands claimed by both. On his way to a U.S. Army training camp and then to Europe to "best serve my country," Smith expressed his faith that the U.S. government and its agency, the Department of the Interior General Land

Office, would never endorse those demands, and he praised the homestead laws that sought to protect settlers. Smith regarded his letter to the editor "as a last chance perhaps" to air his claims fully before he left for military service. In one of his final sentences he remarked, "The party not toting fair with the Government will be appropriately dealt with."[2]

The controversy involving settlers and planters in the northeastern Arkansas delta raged for at least another decade. In the end, Smith's faith in the government's willingness to represent the interests of homesteaders was not warranted. Through stubborn persistence some settlers in the disputed land were able to certify their claims, but by 1921 the government awarded planters preferential rights through legislation introduced by a delta congressman and his senatorial counterpart. That legislation was supported by the Department of the Interior, which oversaw the General Land Office and homestead legislation.

From 1908 to 1921 the Department of the Interior followed a policy designed to enhance the public domain and promote homesteading. In so doing, the federal government worked against the power of plantation-style farming. Yet there were serious obstacles to homesteading in the emerging plantation belt of northeastern Arkansas at that time. Entrepreneurs were aggressively implementing the plantation model. And Warren G. Harding's "Teapot Dome" administration repudiated the government policy started by Theodore Roosevelt's progressive Department of the Interior. Yet the vagaries of national politics alone were not enough to thwart the Arkansas delta homesteaders. The entrepreneurs were forced to launch court battles over title to thousands of acres of public domain. Although they lost in the courts, they succeeded in the political sphere and gained federal legislation to protect their interests, securing the plantation model of development in northeastern Arkansas.

The battle between homesteaders and planters in northeastern Arkansas revolved around acreage made available through comprehensive drainage projects that uncovered some of the richest uncultivated land remaining in the South. The New Madrid Earthquake of 1811–12 had created the sunk lands, formed many lakes, and left much of the rest of northeastern Arkansas swampy and difficult to penetrate. Most of the disputed sunk lands were located in Poinsett County,[3] and most of the lakes involved in the controversy were in adjacent Mississippi

County. Draining the sunk lands and lakes became a major focus of efforts by those who wished to develop the area. Ernest Ritter and other prominent Marked Tree men were instrumental in forming drainage districts in the Poinsett County delta after the state passed an enabling act in 1902. But the first drainage enterprises were orchestrated by an agency created by the state in 1893, the St. Francis Levee District. Planters, not homesteaders, dominated both the St. Francis Levee District and the county drainage districts. Those planters envisioned the development of cotton plantations and stood ready to employ tenancy and the crop lien, two tools of plantation farming, to further their designs. The land at the center of the controversy was located in four northeastern Arkansas counties: Poinsett, Mississippi, Craighead, and Green. Most of the disputed acreage was concentrated in Poinsett and Mississippi counties (map 5). Although estimates of the total land area involved in the dispute ranged from sixty thousand to two hundred thousand acres, the government eventually claimed approximately ninety-six thousand acres.[4]

Hopeful settlers began to flock to Poinsett and surrounding counties after the Interior Department announced in 1908 that it intended to open the sunk lands to homesteading. That announcement encouraged land-hungry farmers to leave their native states and join hundreds of others who had already homesteaded in northeastern Arkansas, some of whom had lived quietly in the sunk lands for many years. They located parcels, sometimes with the help of land agents who then defrauded them,[5] and they hoped to register their claims once the General Land Office regularized the process. Despite harassment by "the companies" and warnings by the General Land Office that ownership of the sunk lands was in dispute, settlers established homesteads there and prepared to do battle with some very powerful entities.

The battle lines, initially drawn in the local, state, and federal courts, involved three basic sets of litigants. First was the state of Arkansas and its agency, the St. Francis Levee District; second were prominent planters and "the companies," who were themselves sometimes at odds with one another; and third was the federal government through the Department of the Interior. The Department of the Interior's General Land Office also served as de facto agent for the homesteaders until the early 1920s, attempting to claim the land for the federal government so that it could be opened for homesteading.

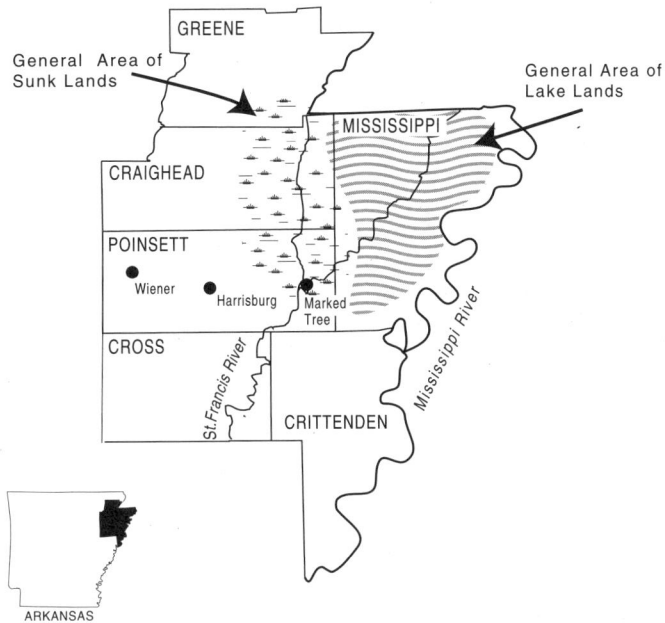

Map 5. Counties involved in sunk lands and lakes dispute

The state of Arkansas through the St. Francis Levee District was primarily interested in promoting drainage in the northeastern Arkansas delta. One way the district hoped to fund these projects was through the sale of state lands within its boundaries. The district encompassed eight counties, including Craighead, Poinsett, and Mississippi counties where most of the sunk lands and lakes were located.[6] The legislation creating the district officially transferred ownership of the sunk lands and lakes, as well as other "swamp and state lands," to the St. Francis Levee Board. The total land area granted to the board was approximately 200,000 acres and was part of what remained of 8.6 million acres donated to the state by an 1850 act of Congress.

From the outset, the legitimacy of the state's claims to the sunk lands and certain lake lands was open to question. Thus, the chief engineer of the district, H. N. Pharr, suspecting that the sunk lands and lakes donated to the district by the state might, in fact, belong to the federal government, applied to the commissioner of the General Land Office in 1894 to have the sunk lands and lakes surveyed. After the survey he intended to petition Congress to donate to the district or to the state the land in question. He assumed that the sunk lands and lakes had never passed from the federal government to the state government for he knew that they had never been surveyed and only surveyed land was subject to the 1850 act. But Secretary of the Interior Hoke Smith's land commissioner ruled that the title to the sunk lands and lakes resided with the individuals or entities that owned the surveyed land surrounding them, citing the long-recognized tradition of riparian rights. In 1894 Hoke Smith through his commissioner declared that the federal government had no claim to the sunk lands or the lakes in question and that the St. Francis Levee District could assert riparian rights.[7] But in 1895 and 1896 Governor James P. Clarke negotiated a compromise agreement between the state of Arkansas and the federal government that inadvertently reopened the question of the status of the sunk lands. Intended to redeem the state's credit and facilitate economic development, the compromise called for Arkansas to relinquish to the federal government the unsold swamplands.[8] Over the next twenty-five years, litigation ensued, over the precise definition of these "swamplands," finally leading to a U.S. Supreme Court decision that the sunk lands and lakes were unsurveyed swamplands that had never left the public domain. In the meantime, though, the St. Francis Levee District, relying on the 1894 ruling by the Interior Department and unaware that Clarke's compromise had placed its claims in jeopardy, began selling options on some of the sunk lands to a speculator at fifty cents per acre. The district later sold thousands of acres of land adjacent to the unsurveyed lands to other speculators and used the proceeds for more drainage enterprises.[9]

While the district was disposing of land it believed it owned, a report was making its way through Interior Department channels challenging the St. Francis Levee District's claim to the land. In 1902 a special agent of the General Land Office, apparently assuming that the Clarke compromise included the sunk lands, charged that "purchasers from the St. Francis Levee District are trespassing upon Sunk Lands" and asked the commissioner of the General Land Office to recommend to the secretary of the interior that the district "be enjoined from disposing of the lands and its purchasers be enjoined from cutting timber thereon." The commissioner, believing his special agent's report had merit, passed the recommendation along to the Department of the Interior, but Secretary E. A. Hitchcock disappointed the commissioner and repudiated the special agent's report. He reaffirmed Hoke Smith's 1894 opinion and reasserted the principle of riparian rights.[10]

The matter might have ended there had it not been for the emergence of a dispute over who could claim riparian rights. Chapman and Dewey of Marked Tree and the R. E. L. Wilson Company of Mississippi County claimed such rights to sunk lands that were adjacent to land they had purchased from the St. Francis Levee District; they may well have been the "trespassers" the General Land Office special agent complained about in 1902. They then became involved in litigation with the St. Francis Levee District over the question of whether they or the district had the riparian rights to these sunk lands or lakes.

Meanwhile, a sharp reversal in policy by the Interior Department in 1908 threw the claims of the state, the St. Francis Levee District, purchasers of land from the district, and any other claimants of riparian rights into jeopardy. When James R. Garfield, who was appointed secretary of the interior in 1907, ruled in 1908 that the sunk lands and lakes belonged to the United States, he nullified all the transactions carried out between the St. Francis Levee District and land purchasers, and he denied that the principle of riparian rights applied to the land in question. Ernest Ritter's 1906 purchase of several thousand acres from the district was thus compromised. Ritter, along with two associates from Memphis, had negotiated that purchase through his old friend W. B. Miller of Marked Tree, at the time the president of the St. Francis Levee Board. Miller had the authority to make such transactions, and when the Interior Department announced its stunning change in policy, he vigorously defended his actions and appealed to the Arkansas Supreme Court for a ruling. That court ruled in the district's favor, but the Interior Department persisted in its claim. In 1914 the U.S. attorney's office filed suits against Ernest Ritter, Chapman and Dewey, R. E. L. Wilson, and one other litigant in an effort to determine once and for all the ownership of the land in question. The suits were explicitly filed to quiet title to acreage claimed by Ritter and the others, and the federal district court ruled in favor of the federal government. Two of the defendants, Ritter and Wilson, appealed to the U.S. Supreme Court, which decided in 1917 (in the *Wilson* case) and 1918 (in the *Ritter* case) that the U.S. government had a legitimate claim to the disputed land.

While his appeal to the U.S. Supreme Court was in progress, Ritter sued the St. Francis Levee District in order to retrieve the sum of about six thousand dollars he had expended in 1906 to buy the land from the district if the U.S. Supreme Court's decision went against him. He filed this suit in the state court, and as might be expected, he lost. A great deal was at stake for Ritter. He stood to lose not only his initial investment in the land and the legal fees and court costs he had incurred but also some very valuable farmland. Even though Ritter operated a lumbering enterprise within the disputed acreage until court injunctions prohibited further cutting, it is unlikely that he ever recovered all costs associated with the disastrous purchase.

The lumbering enterprises operated by "the companies" were one of the pivotal focuses of the litigation.[11] The U.S. attorney in 1914 secured injunctions against Ritter, Chapman and Dewey, and Wilson forbidding them from cutting timber in the land identified in the suits making their way through the federal courts. The St. Francis Levee District sought to enjoin Chapman and Dewey and R. E. L. Wilson from cutting timber on land they claimed by riparian right. The "companies," in turn, secured injunctions through the local chancery courts preventing settlers from cutting timber on their homesteads, and the Interior Department eventually issued an edict prohibiting settlers from disposing of the timber on their claims while the suits were in progress.

Aside from the violence they confronted, settlers faced two basic problems. First, the Interior Department prohibited them from cutting timber on their homesteads. Second, it decided to withhold final certification of all homestead claims in the sunk lands. The riparian claimants had petitioned the department to withhold the land involved in litigation from entry by prospective homesteaders, but the department had declined to go that far. Instead, the first assistant secretary of the interior directed the commissioner of the General Land Office on September 24, 1913, to "suspend all final proofs which may be offered on any entries or filings of any of the so-called sunk lands and to withhold the issuance of final certificates of entry and of patent thereon until further advised by this Department."[12] Department of Interior officials cited the cases then pending in the Supreme Court and in the federal district courts as justification for the suspension.

Even though the Department of the Interior gradually revoked suspension of certification on various claims in the sunk lands, the homesteaders continued to feel the negative effects of its second policy on cutting timber. When attorney R. E. L. Johnson wrote to the secretary of the interior on April 22, 1914, he welcomed as "good news" the department's decision to revoke the suspension on final certification in certain cases, but he pointed out that the homesteaders were more interested in a change in the department's policy regarding cutting timber. The inability to cut timber was no small matter, for it hindered the homesteaders efforts to improve their claims, as they were required to do if they hoped to perfect their titles. One homesteader, H. C. Hall, writing to the secretary of the interior on February 17, 1911, complained that the riparian claimants "secure an injunction against the 'squatter' cutting, removing, or otherwise interfering with the timber. These cases are set for some stated term of the court, but then the time arrives, the case is continued for cause. This works various hardships on the enjoined—stops all work and, of course, causes the injured party to move off the land in order to sustain himself and his family, which, we are told, creates an abandonment of the land."[13]

Attorney Johnson, who would ride the issue to election as judge of the second circuit (1918–23), pointed out that the 1914 Department of the Interior order "greatly embarrassed the homesteaders as they have been utterly unable to do anything except to exist as best they could and remain on their claims." Johnson asked the secretary of the interior to reconsider his order so that the homesteaders could "clear and improve preparatory to perfecting their homestead claims." An assistant secretary in the Department of the Interior responded to Johnson on May 22, 1914, with a polite but firm refusal to alter the policy.[14]

By the time the U.S. Supreme Court ruled in favor of the government in its 1917 and 1918 decisions and thus vindicated the claims of the settlers, some of the settlers had given up their claims. Outright violence drove many of them away, and the prohibition against cutting timber made it impossible for others to improve their homesteads and thus receive final certification. Those who remained faced a final obstacle: legislation passed in 1921 giving preferential rights to the speculators and "the companies." Whether intentionally or not, the wording of the 1917 decision provided Wilson and other riparian owners with an argument for legislative action that would allow Wilson, having begun to make improvements upon the disputed acreage when the Interior Department was ruling that he and others like him might claim riparian rights, to have preferential right to purchase land he had improved.[15] Harding's acting secretary of the interior, E. C. Finney, first officially recognized this implication of the court's decision. Signaling yet another major change in Interior Department policy, Finney wrote a letter on June 17, 1921, in support of the legislation giving riparian owner R. E. L. Wilson the "preferential" right to purchase the acreage that had been the subject of his recent litigation in the federal courts.[16]

Even before the Supreme Court's 1917 ruling in the *Wilson* case, riparian claimants had attempted legislative remedies. The earliest legislative effort on record surfaced on January 28, 1910, when Arkansas congressman William A. Oldfield introduced House Resolution 19637, described as a bill to quiet and confirm title to certain "overflowed" lands in Arkansas. The Oldfield bill was referred to committee and never seen again. His opposition to it later won attorney William J. Driver a U.S. congressional seat; a decade later in the House he would promote legislation that would do for a different set of claimants what the Oldfield bill attempted to accomplish. And Joseph T. Robinson, who served on the House Committee on Public Lands, to which the Oldfield bill was referred and who apparently made sure the bill was never reported out, later became a U.S. senator from Arkansas and the sponsor of legislation in the Senate similar to that introduced by Driver in the House. Obviously, neither Driver nor Robinson was opposed to legislation assisting riparian owners generally; they were opposed to legislation designed to serve the interests of the individuals involved in the Oldfield bill.[17]

The Oldfield bill sought remedies for specific individuals who were in conflict with powerful figures in northeastern Arkansas. The dispute involved a long-standing struggle for control over the St. Francis Levee District, a battle which had been waged vigorously during Jeff Davis's governorship of the state earlier in the decade. Davis had won that battle and secured the position of his man on the board of directors of the district, W. B. Miller of Marked Tree, who became president of the board on May 9, 1905.[18] In 1906 Miller sold to several parties nearly eighty thousand acres of sunk and lake lands. Ernest Ritter and two associates purchased several thousand acres in Greenwood Township in northeastern Poinsett County, and the St. Francis Land Company of Forrest City in St. Francis County purchased nearly forty thousand acres in Mississippi County.[19] The two transactions involved land that was the subject of the Oldfield bill.

Miller lobbied vigorously for the passage of the Oldfield bill, for the best interests of the St. Francis Levee District lay in having its right to sell the lands it claimed to Ritter and others confirmed. Miller even traveled to Washington to meet with Assistant Attorney General Oscar Lawler. Lawler told Miller that he was opposed to the bill and would "report unfavorably" on it. Miller then discussed the bill with James P. Clarke, then Arkansas's senior senator and shortly to become president pro tem of the Senate. Clarke recommended that the bill be withdrawn, given Lawler's unfavorable report, but he expressed sympathy for the purchasers of the land in question and indicated to Miller that he would support the bill if the Public Lands Committee reported it out of committee favorably. But Clarke was far more ambivalent about the measure than Miller suspected, and during hearings over the bill, he denied that he favored it.[20]

The hearings held by the Public Lands Committee in February 1910 over H.R. 19637 reveal the deep divisions that existed over the issue and provide insight into a dispute which pitted settlers in the new Greenwood Township in northeastern Poinsett County against Ernest Ritter in adjacent Little River Township. The land described in the Ritter suit was all within Greenwood Township, so that in addition to the larger battle over control of the St. Francis Levee District, the sale of the Greenwood Township land to E. Ritter and his associates created controversy within Poinsett County. The first man to testify before the committee was an attorney representing Ritter and others who had purchased land from the St. Francis Levee District.[21] Levi Cooke began his testimony with a statement which implied that his clients played a key role in drafting the legislation. The settlers were represented at the hearings by J. A. Tellier, who had been a field agent operating out of the federal land office in Little Rock before resigning to begin a law practice there. Although he testified that he had no connection to the homesteaders in the sunk lands while he was employed by the government, he brought to the situation an intimate knowledge of homestead law and revealed a thorough understanding of the situation confronting settlers in northeastern Arkansas.[22] While Tellier directly represented the homesteaders, Senator Joseph T. Robinson revealed his sympathies for them and presented numerous petitions, affidavits, and telegrams he had received that were signed by approximately one thousand persons who opposed the legislation. Some of them identified themselves as merely residents of the area, and others claimed to be settlers on the disputed land in Poinsett and Mississippi counties.[23]

A close examination of the signatures from Poinsett County reveals that the supporters of the settlers included some of the most prominent men in Greenwood Township. The mayor of Lepanto, Charles B. Greenwood, who had homesteaded and laid out the town site of Lepanto earlier in the decade, and for whom the township was named, was the first signatory on a petition which included ninety-one other names. This petition requested that the legislation be amended "so as to exempt from its provisions any and all lands upon which citizens have entered with a view to homesteading same." J. T. Lee, a Lepanto businessman-planter who would later organize the town's first bank, was the second signatory on a petition which included sixty-two other names. Lee, who identified himself as a land claimant, went further than Greenwood. While Greenwood had asked only that the legislation be amended, Lee's petition requested that "no legislation affecting this land be passed by Congress." He argued that "it is in our interests that this land be doled out in small portions, and that the government supervision of this land be maintained until we shall acquire patent under existing laws."[24]

While attorneys for Ritter and the other purchasers from the levee district argued that few or no "legitimate" homesteaders resided within the disputed area, the petitions, telegrams, and affidavits provided ample evidence that hundreds of settlers had built homes and begun to raise crops there. Ruben Conaster sent a notarized affidavit attesting that he had settled on a tract of land a few miles from Lepanto with his family and had lived there for two years.[25] Conaster had moved into Poinsett County with his mother and siblings in 1890. In 1900 he lived in what was then Little River Township (in what may have been a portion of what became Greenwood Township a few years later) with his father and two brothers, James and John. Ruben and his brother James indicated on the manuscript census of population for that year that they were day laborers.[26] By 1910 Ruben had married, started a family, and moved across the county line into Mississippi County to homestead his acreage. His brothers helped him clear the land and also sent affidavits swearing to the accuracy of Ruben's affidavit. All of their affidavits were sworn to and attested in Lepanto, indicating that the brothers still did business there. They very likely traded with J. T. Lee, who operated a mercantile establishment, and they were undoubtedly acquainted with Greenwood, who was the mayor of Lepanto in 1910.[27]

Lee was one of only four merchants operating in Lepanto in 1910. The sparsely settled, isolated town had been very appropriately named after a Greek city that was surrounded by water. The Arkansas Lepanto had mud streets and was known locally as "lamp city" because there was no electrical service. But this very isolation worked to the advantage of the small town; it was often too difficult to travel to Marked Tree because the roads were too uncertain and frequently washed out. David Seely Buck, who operated the first mercantile establishment in Lepanto out of his two-deck houseboat, grew "weary of going to Marked Tree fifty miles by boat or through dense woods by horse" and requested that a post office be established. He served as postmaster from 1894 to 1910, with his houseboat functioning as a post office and a dry goods and grocery store. The area seemed so unpromising when Charles Greenwood homesteaded there in the early 1890s that when he "traded part of the land for a wagon and team of mules to Marion Jackson of Etowah, the latter returned to take a second look and figured he had been out-traded and asked to be released from the so-called bargain.... Greenwood returned the wagon and mules." Although refusing to be a homesteader himself, Jackson did turn up as one of their supporters.[28]

Greenwood had engineers survey the town site in 1903,[29] and the lumber industry began to thrive in the area. The rise in the number of those employed in that industry between 1910 and 1920 paralleled a similar increase in Marked Tree a decade earlier (table 4.1). W. C. Dawson opened the first sawmill in Lepanto in 1905, and the Grismore-Hyman Mill was established in 1911. Chapman and Dewey opened a logging camp between Lepanto and Marked Tree and in 1905 built a railroad linking Lepanto to its base of operations in Marked Tree.[30] Incorporated in 1909, Lepanto grew from a population of 154 to 987 between 1910 and 1920. Although the lumber industry continued to dominate the town, farms and plantations slowly began to replace the swamps in the surrounding countryside, and the contest over who was to own that land became particularly heated after 1910.

Lepanto became a center of homesteader activism. Letters published in the Harrisburg *Modern News* in 1913 and 1914 and in the

Table 4.1. Occupations of adult men in Lepanto, Arkansas, 1910–20

	1910		1920	
	No.	%	No.	%
Mill and timber workers	11	19.6	127	35.2
Merchants	4	7.1	24	6.6
Professionals	3	5.4	14	3.9
Manufacturer			1	.3
White collar	4	7.1	50	13.9
Skilled and artisan	8	14.3	38	10.5
Railroad workers			10	2.8
Unskilled	15	26.8	30	8.3
Service			18	5.0
Farm-related	11	19.7	41	11.3
Other			8	2.2
Total	56	100.0	361	100.0
Missing cases*	6		15	

Source: Federal Manuscript Census, Arkansas, Poinsett County, Schedule of Population, 1900, 1910
* Individuals who failed to report occupations.

Marked Tree Gazette in 1915 identified many of the settlers, announced the creation of a Homesteaders Union, and demonstrated that they had the support of prominent citizens in Lepanto.[31] Greenwood and Lee were no doubt particularly uncomfortable with the intrusion of the Ritter and the Chapman and Dewey enterprises into Greenwood Township. The township had been carved out of the northeastern section of Little River Township in 1903, and while Chapman and Dewey owned considerable acreage there already, they expanded their acreage as a result of their claims to the sunk lands. Ritter, meanwhile, claimed ownership of land in the township through his 1906 purchase from the levee district. By 1910 the two enterprises claimed 69.1 percent of the land within the northeastern one-half of Greenwood Township.

It is no wonder that men like Lee and Greenwood supported the settlers. Greenwood had laid out the town site of Lepanto hoping to establish a trading center in the area and expand his own enterprises. He had sons and nephews in his household whose own expectations would be limited if "the companies" were allowed to dominate. After Greenwood died in 1912, his nephew Fred carried on with the crusade and wrote a letter to the *Marked Tree Gazette* in 1915 in support of the homesteaders. J. T. Lee, himself a land claimant according to his petition to the Public Land Committee in 1910, may even have been a member of the Homesteaders Union. Indeed, the union apparently included "doctors, lawyers, teachers, carpenters, fishermen, farmers, railroaders, and furnace men—men from almost every vocation." Lepanto merchant H. L. Finn, who appeared on a list of union members and who—like R. E. L. Johnson—later would win political office (Arkansas General Assembly, 1919–23), held 617 acres in real property in 1920.[32] Even the county quorum court responded to a petition from the Homesteaders Union by voting it the funds necessary to construct a small drainage ditch in 1913.[33]

Although African Americans were attempting to homestead in the area, no evidence has surfaced, indicating that the union had black members. Assuming that Della Abbott's sentiments were shared by other white homesteaders—that blacks "beat out" whites—the union probably did not welcome black members.[34] Some of the whitecapping that occurred in Poinsett and Mississippi counties in the period may have been designed to discourage blacks from settling homesteads, but efforts to prevent black migration to the area failed. In 1910 only 2 (1.3 percent) of the 152 farmers operating farms in Greenwood Township were black. By 1920 the number of black farm operators had increased to 75 (43.1 percent). All but 2 of them, however, were landless farmers. Consistent with the emergence of the plantation system, the rate of landownership dropped sharply. In 1910, 45 (29.6 percent) of the Greenwood Township farmers owned their farms; in 1920 only 29 (18.2 percent) of them owned farms. This was still considerably higher than in Tyronza and Little River townships but lower than might be expected given homesteading opportunities. Although the perception that blacks were "beating out" white homesteaders was clearly mistaken, tenancy rather than landownership awaited most of those who migrated into the area between 1910 and 1920.

While hopeful settlers flocked into the region from the states of Illinois, Kentucky, Missouri, Indiana, Texas, Tennessee, and North

Carolina, the Conaster family history indicates that some of the homesteaders resided in the area even before the controversy over the sunk lands arose. They worked as day laborers, sawmill workers, timbermen, or fishermen prior to settling their claims.[35] J. A. Huddleston managed a farm for a plantation in 1900, and Grant Music was a sawmill foreman in 1910. Both appear in a list of homesteaders in 1913.[36] They carved out homesteads and built rude shacks like that occupied by Eliga ("Kidd") Abbott, husband to Della. Abbott left his home in southern Illinois in 1892 and over the next decade and a half wandered from sawmill to lumber camp looking for work. He often sent money back to his mother and stepfather, and his letters left a trail of his wanderings. Before landing in Lepanto, he worked in Poplar Bluff, Missouri, Arkansas City, Arkansas, Ernest Pan, Mississippi, and Marion, Arkansas. Sometime after the turn of the century, he came to Lepanto and worked at Sam Warren's sawmill before moving to Memphis to take a job as a cowpoke on the railroad. When he was grievously injured in an accident, losing one leg above the knee and the other below the knee, the railroad gave him his first wooden leg and paid his hospital bills. His old employer, Sam Warren, brought him back to Lepanto and loaned him the money to buy out a homesteader claim.[37]

By 1920 several of the homesteaders laying claim to acreage within Poinsett County had managed to perfect their claims, and a few of them seem to have been prospering. They included Jim Nichols, the secretary-treasurer of the Lepanto Homesteaders Union in 1914. Although Nichols may have owned land across the county line in Mississippi County, just a few miles east of Lepanto, he owned only 39 acres in Poinsett County in 1920. He held livestock, including horses, cows and mules, considerably in excess of the norm for the Poinsett County delta at that time. While the average personal property value claimed by individuals in Greenwood Township in 1920 was $220, Nichols held $500 in personal property. H. L. Finn's personal property holdings amounted to $300, with $200 being in merchandise, but he also held 617 acres in Poinsett County. Kidd Abbott, meanwhile, claimed ownership to only 249 acres but had personal property holdings amounting to $742. Abbott's success was due largely to a sawmilling enterprise he established on his homestead earlier in the decade, and his personal property holdings included "materials and manufactured articles," which were undoubtedly machinery located in his sawmill.[38]

Even as homesteaders like Abbott and Nichols were thriving, legislation came before Congress in 1921 that worked to the benefit of certain riparian claimants. It was at this point that Acting Secretary of the Interior Finney wrote his letter in support of two of the four bills presented to Congress. Finney argued that the riparian claimants "should be granted preferential rights to purchase the lands at a reasonable price where they can show that, prior to the initiation of valid claims by qualified homestead settlers, they have in good faith taken possession of the lands and have placed valuable improvements upon or reduced them to cultivation."[39]

Four bills were presented for consideration by Congress. The first three bills very narrowly defined the owners and the acreage to be effected by the legislation,[40] and two of them passed. One allowed R. E. L. Wilson preferential rights to purchase land involved in the case he lost in the 1918 U.S. Supreme Court decision. The fourth and final bill, H.B. 6863, gave the land commissioner in the Interior Department broad discretionary powers; introduced by William J. Driver on June 6, 1921, it did not designate particular acreage and owners.[41] The bill represented the final legislative remedy. Thereafter, it would be the secretary of the interior's responsibility to evaluate the merits of the claims by different individuals, for the bill authorized the secretary, at his discretion, to sell land to individuals claiming preferential rights to specific acres. In direct contrast to his position on the Oldfield bill in 1910, Representative Driver spoke at great length in support of this bill, lamenting the problems confronting riparian owners who had improved and cultivated the land under dispute only to have a "cloud on their title." He described the process that the secretary of the interior would follow: "On petition of the riparian owners, a survey is to be made, the value of the land to be ascertained by the appraisal and the riparian owner pay the appraised value, that appraised value to include the value of any timber removed, but excluding the value of the improvement placed on the land by the riparian owner." Only a few questions touched on the homesteaders in the area under dispute, and Driver easily disposed of those. He assured his fellow congressmen that the rights of homesteaders who claimed some of the land in question would be guarded and indicated that he had been in touch with attorneys representing those homesteaders so that they would have an opportunity to air their concerns. Aside from establishing that the homesteaders were being afforded an opportunity to be heard, the discussion concerning the rights of the homesteaders focused on whether or not there were any such men on the land "improved" by the riparian claimants. This was an important point, for a combination of government policy and harassment by the riparian claimants made it difficult for the homesteaders to remain on their claims.

In the end, even a division among the elite like that which occurred in 1910 did not permanently preclude legislative remedies working to the advantage of the riparian claimants. The fact that some planters were more successful than others in lining up congressional support for their claims bears some scrutiny. Planters like R. E. L. Wilson, who had the requisite connections to the Arkansas Democratic party machine, were the first successfully to avail themselves of the legislative remedy. Planters like Ernest Ritter, on the other hand, got caught in a war among Democrats over the control of the powerful St. Francis Levee Board. Significantly, Ritter was not himself a Democrat but a Republican. Consequently, he was particularly dependent upon his Democratic friend, W. B. Miller, the president of the St. Francis Levee Board, to influence legislation. But in 1911 Miller fell victim to a power struggle over control of the board and was replaced as president of the board by a Cross County attorney, businessman, and planter, O. N. Killough, a very well connected man who had been president of the board before Miller. Originally from White Hall in Poinsett County, where the nightriders in the *Hodges* case resided, Killough was L. C. Going's co-counsel in that case. He served as prosecuting attorney in the Second Judicial District (1896–1900), and while in the state Senator (1903–1905), he became president pro tem. He also served a term in the Arkansas General Assembly (1907). He was a formidable opponent and his defeat of Miller left Ritter with a hostile board and little recourse.[42] Nevertheless, the September 1922 act of Congress, House Resolution 6863, might well have worked to Ritter's advantage. With the decision making now in the hands of a Republican Department of Interior, Ritter might have hoped for more than he got from Arkansas's Democratic congressional delegation. But Ritter had disposed of his interest in the land purchased from the levee board in 1906 and involved in the 1918 Supreme Court decision, selling it to Chapman and Dewey sometime between 1918 and 1920. When Ernest Ritter died in 1922, his son, Louis V. Ritter, took over the company and during 1920–

30, increased its real property holdings within Poinsett County alone from 1,697 to 5,001 acres.⁴³ All but 46 acres were within the sunk lands of Poinsett County, but none of the acreage was the area involved in the 1918 Supreme Court decision. Whether by reason of the Supreme Court decision, the opposition of the Lepanto elite, or the stubborn defiance of the homesteaders, E. Ritter and Company had ceased its involvement in that part of Greenwood Township by 1930.

Meanwhile, the prosperity Kidd Abbott and Jim Nichols enjoyed in 1920 was short-lived. They managed to survive the 1921 legislation favoring the riparian owners like Wilson, but the disastrous decade of the 1920s hit both of the homesteaders severely. Not only did they have to contend with the agricultural depression of the 1920s and the great flood of 1927, they were burdened with drainage taxes. By 1922 Abbott had lost part of his original homestead "for failure to pay" these taxes. The plans for a new drainage district involved closing off the lefthand schute of the Little River, which Abbott's homestead abutted, thus leaving him unable to float his logs downriver. He was prevailed upon to surrender the rest of his homestead in exchange for fifty-nine acres located three miles south of Lepanto. While Abbott still owned some land at the end of the twenties, Nichols lost all his land due to failure to pay drainage taxes, a phenomenon common among landowners throughout the delta in the 1920s.⁴⁴ Of the 20,242 acres listed on the real property records for the northeastern half of Greenwood Township, fully 43.3 percent was seized between 1920 and 1930 by the state for failure to pay taxes, and 62.2 percent of the seized land was owned by individuals; companies surrendered only 29.8 percent of the seized lands. Before the tax default, individuals claimed approximately 44.8 percent of this land and companies 50.7 percent; afterwards, individuals owned 31.8 percent and companies 67.7 percent.⁴⁵ Clearly, companies like Chapman and Dewey could better withstand the disastrous twenties and could better afford to pay the drainage taxes.

Even as the suits made their way slowly through the state and federal courts and the settlers continued to endure harassment, the businessmen-planters of the Poinsett County delta proceeded with their plans for drainage of the swamps and lakes. Ernest Ritter and other prominent men in Marked Tree played key roles in the establishment of county drainage districts and the reclamation of the delta. The prospering landowners around Lepanto eagerly participated in the formation of a massive drainage district there. Those districts imposed a heavy burden of taxation upon the settlers and small farmers, as Abbott and Nichols discovered, and accrued to the benefit of those whose enterprises were large enough to warrant the costs.

The St. Francis Levee District's canals deserve the credit for early successes in draining the land of excess water, but the formation of county drainage districts took up where the levee district left off. When the Arkansas legislature passed an act in 1902 enabling individual counties to organize drainage districts and fund them by selling bonds, it hoped to facilitate the development of the state's eastern delta. The process of establishing a drainage district was a simple one. Landowners interested in creating such a district banded together and petitioned the county court. They had to designate the exact boundaries of the district and were required to own land within those boundaries. Only landowners within the proposed district were to be apportioned drainage taxes to pay for the improvement. Once receiving a petition from qualified landowners, the county court then appointed viewers and commissioners who were not themselves landowners in the proposed district. The viewers would then appoint a drainage engineer and begin a preliminary investigation of the proposed project to determine its feasibility. Landowners who opposed the district could present their opinions when the viewers reported to the county court a few months after the process was initiated. If no opposition arose, the commissioners could begin advertising bonds for sale. They were also permitted at that point to take bids from contractors and then to award the contract to the construction company selected to dig the ditches. Often the county would advance the drainage district the money to begin operations and then reimburse itself once the bonds were sold.[46]

There were three likely sources of opposition to the formation of drainage districts. First, settlers on government land, or what they considered government land, might object because they had no voice in the process. They feared that the changes made by a proposed project might actually work to their disadvantage. Many of them had settled on high ground decades before and were accustomed to the periodic flooding, having adjusted their lives to the rhythm of the swamplands. They played a minimal role in the establishment of drainage enterprises, and, indeed, they typically objected to paying the drainage taxes for projects they had not initiated and which did not directly benefit them.⁴⁷ The settlers, however, were a minor inconvenience to those wishing to establish the drainage districts. Since many of them had not certified their claims, their status as landowners was not clearly established, even though they were required to pay the taxes on the land they claimed. Thus, they experienced the disadvantage of having to pay drainage taxes but had dubious standing in court on the issue.

Small landowning farmers unwilling to bear the burden of taxation were the second most likely source of opposition. Those few who lived in the swamps resembled the settlers in that they had more or less made their peace with the region, but they did have standing in court and theoretically could hold up the process of establishing drainage districts.⁴⁸ No record of opposition from either the homesteaders or the small farmers in the Poinsett County delta exists, at least not during the founding of the county's first districts in 1902 and 1906. The December 1917 letter from Homestead Union president Hannah Embrey, however, included a complaint about the drainage taxes settlers were forced to pay and indicated that the taxes were especially unfair because of the difficulty settlers had in maintaining their claims.

Lumber companies and land speculators represented the third most likely opponents of the establishment of drainage districts, for obvious reasons.⁴⁹ Although some lumber companies like Chapman and Dewey expanded their enterprises to include plantation farming, most were only interested in the land for one purpose—lumbering—and were unwilling to pay for the development costs designed to bring the land into cultivation. Some land speculators were equally unwilling to bear a heavier tax burden on land they were holding on speculation. The lumber companies and the land speculators were the ones, moreover, who could mount the most effective opposition to the creation of a drainage district. Unlike the homesteaders and small farmers, they often had substantial financial and legal resources, and they sometimes succeeded in either delaying or killing a proposed project. Speculator Douglas Alexander and the Southwestern Land and Timber Company, for example, prevented the formation of Drainage District Two in Poinsett County. The proposed district would have encompassed an area covering part of the ridge and part of the western delta. The businessmen-planters of the eastern delta had made no effort to develop the western delta, which was immediately adjacent to the ridge, because of an impassable swamp that bisected the delta.⁵⁰

The attempts to establish Drainage District Two marked the efforts of certain ridge elites to participate in the development of the delta. There were nine petitioners involved in the effort, and one of the objections raised by the nonresident opposition, the Southwestern Land and Timber Company and land speculator Alexander, was that four of the petitioners did not own land in the area defined as Drainage District Two, an important and revealing challenge. The five who did own land there held 1,960 acres, with the smallest landowner being one of the only two farmers on the petition, C. M. Thompson, who owned 80 acres. The other petitioners included a merchant, a lawyer, and a bank president who owned 1,500 acres among them. Among those who held no acreage was the county sheriff. The Southwestern Land and Timber Company, meanwhile, owned 14,243 acres and claimed that it would be liable for $24,428 in drainage taxes if the project was approved. Alexander owned 11,919.88 acres and claimed he would have been required to pay $19,601.72 in drainage taxes. They described the land as "wild lands valuable almost solely for the timber which grows upon them" and stipulated specifically that these were not agricultural lands.[51] Yet, once stripped of trees and drained, these were far more than "wild lands" and could very well become rich agricultural lands. These two speculative enterprises merely did not wish to engage in the improvements necessary to make that a reality.

In their efforts to establish Drainage District Two, the ridge politicians representing development-oriented business interests rather than farmers faced opposition from outside the county that proved impossible to outmaneuver, thwarting their attempts to participate in the development of the delta immediately east of them. Douglas Alexander and the Southwestern Land and Timber Company evidently had expert legal advice and the resources to fund concerted opposition to the formation of Drainage District Two. Between 1906 and 1908 their attorneys joined to defeat the creation of the district. The fact that the ridge elite never managed to acquire sufficient land in the delta, even that part of the delta closest to them and most remote from Marked Tree, continued to hamper their efforts to participate in the farming bonanza the rich delta seemed to promise.[52]

The businessmen-planters of the Marked Tree area succeeded where the ridge petitioners failed. The formation of Drainage District One, which covered an area east and southeast of Marked Tree, demonstrated how elites working together initiated and completed the construction of the county's first drainage enterprise. On October 4, 1904, several prominent delta landowners, Ernest Ritter among them, presented the first petition to establish a drainage district in Poinsett County. Other petitioners included W. B. Miller, who was later to become chairman of the St. Francis Levee Board; Charles Greenwood, who would later defend homesteaders in his township against the likes of Ritter; and S. L. Mitcham, the landowning black doctor and farmer who would face jail within a couple of years for his failure to comply fully with the legal ramifications of the tenancy system. M. W. Hazel, who would be elected that year to the state House of Representatives, and his brother N. J. Hazel, an unsuccessful candidate for county judge in 1904, were two of the three viewers appointed by the county court to oversee the formation of the district. Those who orchestrated the county's first district, therefore, were some of the most influential men in Marked Tree, and Ritter's association with at least some of the other principals in the original petition was close. W. B. Miller had served on the first Marked Tree Town Council with Ritter. M. W. Hazel, vice president of the Marked Tree Bank, would be on the board of directors of E. Ritter and Company within a few years. Ritter, moreover, was an initial subscriber of Hazel's bank and one of the chief depositors.[53]

On January 9, 1905, the engineer made his report regarding Drainage District One, as it had been named, and by March the county repaid the Bank of Marked Tree for funds it had advanced for the survey.[54] All seemed to be proceeding well, but on October 19, when the viewers made their final report and asked the county court to declare the drainage district official, opposition arose from an unlikely source. Ritter now opposed the district's formation on the basis of the apportionment set out by the viewers. He owned 454.68 acres outright and another 160 acres with R. F. McFessel in the district and was granted an appeal when the court ruled against him.[55] Whatever Ritter's motivations for opposing the district, disassociating himself from it rendered him eligible to bid for the construction contract.[56] In July 1906 he placed the best bid, winning a contract worth $41,640.25. When he bonded himself with the county court, guaranteeing his intent to proceed with construction, his cosigners included W. B. Miller, S. P. Thompson, and M. W. Hazel. Hazel was also the signatory for the bank a few weeks later when that institution agreed to buy the $45,000 in bonds that were to be issued the following year.

The bank, moreover, held the funds at Ritter's disposal as work on the drainage district proceeded.[57]

Despite the creation of this drainage district and one other in the delta, Drainage District Six, drainage remained a problem which became acute in 1916 and 1917 when some Missouri projects threatened (map 6) to inundate the entire St. Francis basin with overflow. This would have affected not merely the disputed sunk lands but thousands of acres of land to which Ritter and the other landed interests in northeastern Arkansas held clear title. Two major drainage projects in Missouri created the "perilous situation" facing eastern Arkansas. The Mingo District proposed to drain a huge reservoir located in southeastern Missouri and the Inter-river Improvement District, also in southeastern Missouri, planned to drain an area from the Mississippi River to Crowley's Ridge. Both of these projects involved diverting water into the already troubled St. Francis River.[58]

Two years before, in 1915, Marked Tree residents had been disappointed by a decision of the U.S. Engineer's office in Little Rock not to enlarge the channel of the St. Francis River. That decision held new significance in the face of the Missouri projects. When Congressman T. H. Caraway sounded the alarm in January 1917 "in letters to prominent men in eastern Arkansas," the initial response was to "hold meeting[s] to define plans for self-protection." By February a general mass meeting was held in Paragould, below Poinsett County, to discuss filing an injunction to delay the Missouri projects. But the planters and farmers of eastern Arkansas were not all of one mind on the issue. The Forrest City Commercial Club met in late February and announced that its "attitude toward the Missouri schemes was liberal rather than antagonistic and there was no disposition on the part of the meeting to take any harsh or drastic steps to hinder or force the citizens and taxpayers of Missouri to abandon important work looking to the improvement and development of their fertile farming area." The members argued that the solution was to develop more comprehensive plans in eastern Arkansas and to secure the assistance of the War Department, which had responsibility for drainage enterprises through the Corps of Engineers. What they wanted from the War Department was "the necessary data . . . from which to determine the exact extent of the damage, if any, which would result to the farming lands of the St. Francis basin." The idea was to formulate projects of their own to cope with the effects of those in Mis-

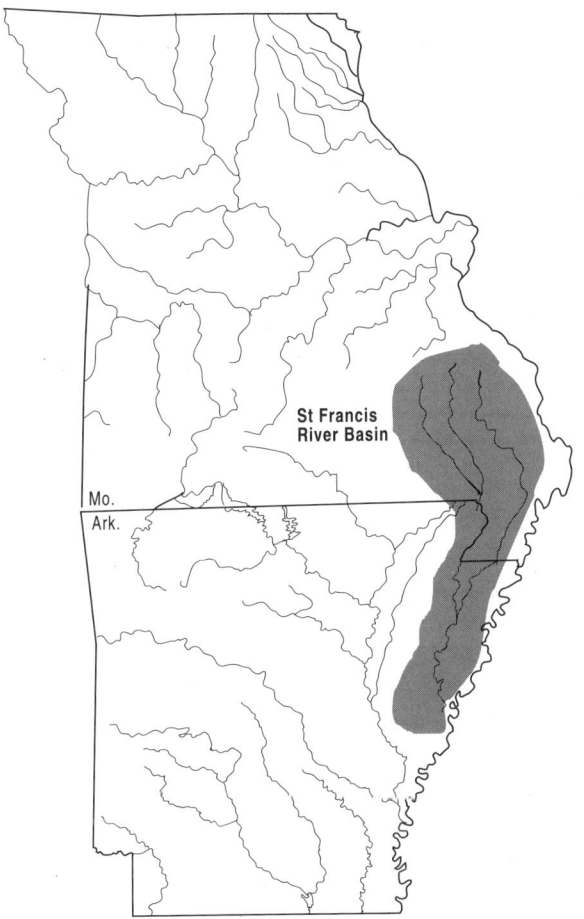

Map 6. St. Francis River basin and major rivers of Missouri and Arkansas

souri and to involve the federal government, at least on an advisory level, in their coordination.[59]

Landowners in Poinsett County took the view of the Forrest City Commercial Club. On the one hand, they again pressed for aid from the Corps of Engineers in widening and deepening the St. Francis River. Congressman Caraway convinced the General Land Office to withhold certification of homestead entries in the area around the St. Francis River in order to facilitate the expected work on the river. On the other hand, in early 1917 the Poinsett County delta and ridge elites, acting in concert, laid plans for an ambitious and very expensive drainage project to cover an area that reached from the Mississippi County line to Crowley's Ridge. Drainage District Seven, as it turned out, helped mediate the political turmoil within Poinsett County that was raging in 1917 after the courthouse burned down. It illustrates the importance of drainage in eastern Arkansas not only as an economic but also as a political issue.[60]

The new drainage enterprise involved citizens from both the ridge and the delta even as the controversy over the location of the new courthouse swelled.[61] The project went far beyond the aid rendered to homesteaders in the "Scatters" above Lepanto just a few years before when the Poinsett County Quorum Court voted funds to help the settlers build a modest canal to drain the area around their claims.[62] Both the delta and the ridge had something to gain from cooperating on the drainage question. Delta landowners were protecting their interests while ridge elites welcomed an opportunity to participate in the development of the delta. The one attempt made by men on the ridge to form a drainage district in the western delta, their effort to form Drainage District Two almost thirteen years earlier, had failed because they had not gained ownership of land in the western delta. Two non-resident landowners had effectively stopped the formation of Drainage District Two by demonstrating that some of those who petitioned for it did not own land in the western delta. By 1917, however, ridge men had more than doubled their ownership of land. For example, J. G. Gant, president of the Harrisburg State Bank, and Thomas Flournoy, an officer of the bank, purchased a 464-acre farm on the edge of the delta close to the ridge in November 1916. Meanwhile, the two opponents to Drainage District Two, Douglas Alexander and the Southwestern Land and Timber Company, no longer held land anywhere in the county.[63] In cooperating with the delta landowners to organize Drainage District Seven, the ridge developers achieved a goal they had long awaited.

Act 193 of the Arkansas General Assembly, passed on March 9, 1917, allowed the Poinsett County Court to abolish the defunct Drainage District Two and create Drainage District Seven. The act defined the boundaries of the district and appointed the directors. The four directors were prominent ridge men: C. M. Thompson, N. T.

Whitaker, J. G. Gant, and C. R. French. French was chairman of the county's minuscule Republican party. Their appointment to the board may have been a key factor in the mediation of the political dispute raging over the burning of the courthouse. Delta landowners were so anxious to launch the new drainage district, they were willing to give control over it to the ridge. Although some objections were raised over the tax assessments and the damages for surrendering right-of-way for the proposed drainage canals, no opposition to the creation of the district arose. Indeed, certain landowners in the delta objected to being left out of the district. Originally, Drainage District Seven excluded Greenwood Township and parts of Little River Township, but landowners in the excluded area filed a petition with the county court to have their lands included.[64] Hannah Embrey, who complained about drainage taxes in her December 26, 1917, letter to the secretary of the interior, failed to comment on the new district, although the engineers hired by the district undoubtedly were surveying the lines of Drainage District Seven even as she wrote. State assemblyman H. L. Finn, however, who appeared on a list of Homesteaders Union members earlier in the decade, was one among several petitioners at a meeting of the commissioners of the district on January 4, 1918, "to present a petition for annexation" of the Greenwood Township lands, but he did not state whether he was representing his own interests or those of the homesteaders.[65]

Whether the homesteaders supported the creation of Drainage District Seven, merely acquiesced in its creation, or protested in a manner that has not been preserved in the historical record, in the long run the district meant disaster for many homesteaders and other small farmers in the delta. Perhaps the prosperity that accompanied World War I caused them to be overoptimistic about their prospects. Drainage Districts One and Six had affected the settlers only minimally, but Drainage District Seven encompassed most of the Poinsett County delta and all of the sunk lands and so directly interfered with the settlers' struggle to survive.[66] The drainage taxes, insignificant to the development-oriented, worked a severe hardship upon the settlers whose claims faced even greater challenges than before.

Although Harrisburg men dominated the board of directors of Drainage District Seven at its inception, Marked Tree men were to assert themselves within a few years. In 1921 the Arkansas state legislature added three members to the district's board of directors. All three were prominent delta men, and two of them were from Marked Tree: H. C. Dewey, son of one of the founders of Chapman and Dewey Lumber Company, and M. W. Hazel, president of the Marked Tree Bank and long-time business associate and friend to Ernest Ritter. W. R. Payne, a planter in the western Poinsett County delta, was the third new member of the board. At the second meeting after the addition of the new board members, the board voted to move the district's office from Harrisburg to Marked Tree. By 1931 Dewey had become president, and L. V. Ritter, oldest son of Ernest Ritter and president of E. Ritter and Company, had been appointed to the board.[67]

The ability of the delta elite to capture control of the drainage district within a few years of its creation is a testament to their growing power and influence. Indeed, 1921 was a good year for the rich and powerful delta elite. Not only did the state legislature provide them with the opportunity to seize control of the most important drainage district in the county, but Congress also bestowed upon them "preferential rights" to purchase acres they claimed in the sunk lands. While it had taken the riparian owners almost twenty years of litigation to get their claims recognized, the Interior Department had come full circle. It is perhaps revealing that the controversy in the sunk lands began in the early years of the Progressive era and ended in that era's twilight. Under Theodore Roosevelt's administration, the General Land Office began to challenge the riparian owners. Under Harding's "Teapot Dome" administration, the General Land Office led the retreat.

5

DIFFERENT ROADS
TO DEVELOPMENT

Within days after the "firing of guns and the whistle from Bott Brothers stave factory in Harrisburg" announced the burning of the Poinsett County Courthouse in the early morning hours of May 4, 1917, a struggle between the ridge elite and the delta businessmen-planters over the location of the new courthouse emerged with dramatic force.[1] Ernest Ritter led the protracted and energetic effort to remove the county seat to Marked Tree, and the arguments made by delta men regarding the wisdom of relocating the courthouse replicated those made by some of the same men nearly ten years before during an earlier dispute over the courthouse. Marked Tree was the focal point for the delta, and after a decade of rapid growth, the delta population now far surpassed that of the ridge and the prairie combined. Aside from the growing numerical strength of the delta, there was a new urgency to the rivalry. The earlier assault on the Harrisburg courthouse focused on locating an additional courthouse in Marked Tree rather than moving the county seat to the principal town in the delta. But the 1917 struggle followed a contested election the previous year which saw the disappearance of a delta ballot box as it was being transported to the ridge for counting and the defeat of delta candidates for county political office. Apart from charges of political chicanery, delta planters hinted at economic fraud, questioned the integrity of the ridge politicians, and pointed to the huge county debt as demonstrating a lack of fiscal responsibility. Meanwhile, the state's fire warden and insurance commissioner, F. G. Lindsey, found the burning of the courthouse two days after the arrival of state auditors in Harrisburg rather suspicious.[2]

One principal issue behind the struggle over the location of the new courthouse in 1917 involved the control of tax assessments and county finances. Delta planters, who were rapidly expanding their enterprises, had succeeded in securing generous reductions in the tax rates imposed on their "swamp and wild lands," but some candidates for office cried foul and appeared to threaten this comfortable arrangement. At the same time, delta planters differed fundamentally with the ridge politicians in their approach to county finances. Ridge politicians continued to issue scrip, jeopardizing the credit standing of the county and thus making it more difficult to sell bonds to finance further improvements in the delta. In fact, delta planters tended to blame the financial mismanagement by the ridge politicians for the county's failure to pay bondholders who had invested in Drainage District Seven bonds and who then filed a suit in federal district court.[3]

Development was key to the future for delta planters, and they were quick to recognize the harmful effects of a bad credit rating. By the second decade of the twentieth century, plantation-style agriculture had become a prominent fixture of the delta's agricultural economy, and the high expectations of both businessmen-planters and small farmers hinged on cotton production. In 1911 the *Marked Tree Gazette* reported that "the fields are beautiful, everybody smiles, the money is plentiful and mortgages will be burned this season which have been carried for years. Marked Tree is busy, wagons loaded with the staple every day fill our streets, while every merchant reports a good business." E. Ritter and Company and S. P. Thompson were prospering, and in the summer of 1912, two new farming companies formed: Chapman and Dewey incorporated the Kansas City and Memphis Farm Company, and a consortium from Memphis and Illinois organized the Illinois Farming Company. The companies owned 10,042 acres between them and planned to cultivate cotton.[4] They did not put all of their acreage into production, however. They sold thousands of acres in both small and large parcels to interested parties. By the end of the decade, Poinsett County was still not dominated by the large agribusiness plantations in existence elsewhere. Only two farms ranged in size from 1,000 to 4,999 acres in the county's delta, which was precisely the situation in 1910.[5]

Even though the Poinsett County plantations did not rival the large agribusiness plantations arising elsewhere, they used tenants and sharecroppers to farm them and were dedicated to the cultivation of cotton. The expansion of the cotton economy in Poinsett County occurred despite a brief drop in cotton prices which accompanied the outbreak of war in Europe. By 1916 cotton was again bringing a good price. Ritter had by that time opened a cotton compress and was often the man through whom the local farmers marketed their crops. The first bale of 1915 was ginned on August 22 and had been "raised by John Martin, a colored farmer on the L. D. Williams plantation" and purchased by E. Ritter and Company for fifteen cents per pound. The second bale sold just a half an hour later to the firm of Louis Ritter and John Emrich of Tyronza for sixteen cents a pound. By the middle of September the gins in Marked Tree, Tyronza, and Lepanto were "running full blast."[6]

As cotton established a firm hold on the delta, the business of buying and selling land and farms increased, and the price of both cutover land and producing farms rose substantially. Smaller farms in the area attracted buyers from other states, who paid relatively high prices. In 1912 W. G. Yates of Poplar Bluff, Missouri, bought Newt Fisher's 267-acre working farm just a short distance from Marked Tree for $16,000. The Mitchum farm of approximately 160 acres sold for $10,000 just a month later to E. M. Duren of Adamsville, Tennessee. The price of cutover land—still bearing the stumps of the felled trees—rose from fifty cents to twenty dollars an acre. In January 1915 a Memphis man, J. K. Mitchell, purchased from S. P. Thompson 1,072 cutover acres at the higher price. The transaction involved land located west of Lepanto in one of the richest delta areas of the county. Newt Fisher of Marked Tree sold 288 acres of cutover land in 1915 for $11,500 to two men from Iowa.[7]

In 1917, after a year of very high returns for cotton, the price of Poinsett County unimproved acreage near Lepanto had risen to nearly $150 per acre, and improved farms were selling at even higher prices. In 1918 L. D. Williams sold his 100-acre farm near Marked Tree to another local farmer, T. B. Hughes, for $225 per acre, or $22,500. Just a week later the biggest real estate deal in the history of the county was consummated when Homer Sloan sold to a Little Rock man his "fine Willbeth plantation of 1600 acres, together with his handsome home, cotton gin, commissary and tenant houses, for $175,000." Just three years earlier the Willbeth plantation had been little more than a "vast wilderness, where nothing but swamp fever and catamounts could thrive, but under the able management of Mr. Sloan, half of this tract has been cleared and yields a bale of cotton to the acre. The Tyronza Central Railroad runs through this land, which is situated midway between Marked Tree and Lepanto, and is one of the very best plantations in Eastern Arkansas."[8] The editor may have been seized with hyperbole, but Sloan's plantation was unique in Poinsett County; he apparently possessed far more capital than was typical for Poinsett County businessmen-planters.

Development was the key to Sloan's success with the Willbeth plantation. Draining the swamps made land valuable both for speculators who merely wished to sell it at a profit and for agriculturalists who wanted to take advantage of the high price cotton was fetching. But development was expensive, and the rise in land prices worked more to the advantage of speculators and businessmen-planters. Small farmers were not the chief beneficiaries of the improvements, and the rising price of delta land made it difficult for them to compete. In a group of 160 landowners in two townships in the county's delta in 1920, companies not only declared more in personal property but also owned a greater share of delta land. The twenty-five companies, representing 15.6 percent of the personal property holders, held 84 percent of the personal property wealth. Of the individuals, 135, representing 84.4 percent of the personal property holders, held only $75,641, or only 16 percent of the personal property wealth. At the same time, the delta companies in the sample held 57,907 (69.2 percent) acres in land while the individuals held only 25,759 (30.8 percent).[9]

In contrast, small farm owners dominated both personal property wealth and landowning on the ridge. Out of a group of 236 ridge landowners, 214 representing 90.8 percent of the sample held 28,805 acres, 77.5 percent of the land.[10] Twenty-two companies held only 8,352 acres, or 22.5 percent of the land. Unlike the delta, where individuals held only 16 percent of the personal property wealth, ridge individuals claimed 58.4 percent of the personal property wealth. The small landowners on the ridge, however, struggled to survive in a market which fluctuated wildly. Their ability to capitalize on the forest products industry beginning in the late nineteenth century had discouraged agricultural diversification and had hidden the weaknesses in their economy. Although Harrisburg itself experienced economic growth and successfully attracted some industries and settlers, the surrounding countryside suffered serious setbacks in the years immediately following World War I. Harrisburg could not hope to match Marked Tree in forest industries because its population did not expand enough to compete.

Although earlier in the decade Harrisburg was involved in an economic revival, those bringing new industry to the town were not native to the ridge. In 1913 and 1914 five new industries began operation there. Catlett and Foley of Illinois founded three mills, opening its band mill in the fall and over the following winter putting in a stave barrel factory and a hickory handle mill. F. G. Foley, who ran the operations, incorporated the company with $40,000 in capital. A circular sawmill was opened by S. Walz of Union, Missouri, and a cooperage plant was erected by Bott Brothers, who had "an extensive cooperage business at Alexandria, Missouri, and Warsaw, Illinois." These enterprises were not on the scale of the Chapman and Dewey mills of Marked Tree, where the population supported larger endeavors, but they did contribute to Harrisburg's limited economic expansion.[11]

The ridge countryside went through a transformation of its own. Real estate agents like William Ainsworth sold thousands of acres to both large and small holders. Indeed, between 1910 and 1920 more than fifty thousand acres came onto the market, and most of that found its way into the hands of small farmers. Individuals rather than companies, however, held most of the real property on the ridge in contrast to the delta where more than two-thirds of the land was in the hands of companies.[12] The ridge was not suitable to plantation agriculture. Not only was the soil overworn, but the ridge was also hilly and broken up by streams and valleys. Small farmers were attracted to the area because the price of land was much lower than elsewhere, and some may also have preferred to avoid the controversies that confronted homesteaders in the delta.[13]

Even as the delta closest to them began to develop, ridge men played only a minimal role in fostering the enterprises that arose there. The development taking place in the western delta between Harrisburg and Marked Tree was largely the result of the efforts of outsiders. The Singer Sewing Machine Company purchased the Poinsett Lumber and Manufacturing Company in Truman on the northern edge of Poinsett County at the Craighead County line and operated lumbering and milling operations in a section of east Bolivar Township. The Weona Land Company, founded by a consortium of individuals from Jonesboro in Craighead

County, operated a land and lumber company there and were, in conjunction with Poinsett Lumber and Manufacturing Company, responsible for the construction of a short-line railroad from Truman into the region. Referred to by the census taker as "Bolivar Township East of Big Bay," the households were recorded as though in a world apart from the old Bolivar Township on the ridge. And indeed, an analysis of Bolivar Township East of Big Bay establishes that it was nothing like old Bolivar Township. It bore a greater resemblance to Little River Township, which contained Marked Tree, at the beginning of its development in 1900. Fully 297 (60.5 percent) of the adult men were employed in lumber-related enterprises, and another 75 (15.3 percent) worked on the railroad. Only 20 farm owners operated in the vicinity, and only 33 men listed themselves as farm renters.[14]

Although Bolivar Township East of Big Bay appeared more like Little River in 1900 than Bolivar, its one village, Weona, was not to enjoy the prosperity that Marked Tree realized. At least two factors combined to limit Weona's horizons. First, its location off the main line of the railroad and, later, off the main highway system limited its growth. Second, the timing was wrong. Weona and its countryside were coming into development just as the farming sector faced a serious depression beginning after World War I. The area remained depressed and underdeveloped until after World War II, when businessmen from the eastern delta, principally Marked Tree and Lepanto, began to purchase land and bring it into cultivation.

While some ridge businessmen and farmers were venturing into delta investments, as their support of Drainage District Seven suggested, many others focused their attention on the small farming enterprises possible on the more traditional ridge lands. During the second decade of the twentieth century, they reasserted their commitment to diversification, stimulated in part by the arrival of the county's first U.S. Agricultural Extension Service farm agent. S. P. Dent, an agricultural college graduate and farmer from Mississippi, came to Poinsett County in 1916. He opened an office in the courthouse in Harrisburg on January 11 and began to travel on horseback to visit the local farmers. He had his greatest influence upon the farmers along the ridge. Although he held demonstration days in the principal towns in the prairie and the delta, he spent little time in those regions otherwise, apparently having little to offer the rice farmers of the prairie and encountering as much difficulty reaching the delta as almost everyone else did. His most notable achievement in the delta, ironically, was in convincing Ernest Ritter to open a cannery so that farmers near Marked Tree could can their own vegetables.[15]

Although Dent brought new life to the movement, the drive for diversification in Harrisburg began before his arrival. In the 1890s the forest products industries had helped the small farmers on the ridge to survive the agricultural depression of that era, but two decades later they could no longer ignore the weaknesses in their economy. They had not successfully diversified and still depended too much upon cotton to bring a little cash into their economy. In March 1913, the year the new mills were being established in Harrisburg, farmers along the ridge purchased "at bargain prices" fruit trees from a boxcar-load of "select nursery stock in Harrisburg" sent by the Eagle Nursery of Piggott, Arkansas. In 1916 the Harrisburg *Modern News* urged farmers engaged in "extensive farming on small tracts" to rotate their crops and experiment with strawberry farming as one alternative. Farmers on the ridge organized a Fruit Growers Association and pledged to grow at least 71½ acres in strawberries and 80 acres in fruit trees.[16] But cotton remained popular, especially among farmers who owned or rented land in the more fertile western delta where the cultivation of cotton was more productive than on the less fertile ridge soil.

Agricultural agent Dent's work to encourage diversification faced nearly impossible odds once the war in Europe created a greater demand for cotton. Cotton prices were up in 1916 and wheat prices were down.[17] C. W. Mink, who operated a farm in the western delta, brought "the finest stalk of cotton we have seen" to the office of the Harrisburg *Modern News*; he had 200 acres in cotton. The cotton gin in town baled "700 bales more cotton this year than it had ginned at this time last year."[18] Yet at the end of the year, Dent's diversification drive was praised and a "cattle and hog campaign" was fully under way. Livestock shipments increased steadily over the next three years. By 1920 cattle and hog raising on the ridge were far more widely dispersed than they were in either the prairie or the delta. Nearly three-quarters (74 percent) of the ridge taxpayers in 1920 owned cattle, and 66 percent owned hogs. Only 53 percent of the prairie taxpayers and only 34 percent of the delta taxpayers raised cattle, and hogs were even more scarce in those regions.[19]

The farm agent coordinated the cooperative marketing of livestock, and his efforts largely explain the 50 percent increase in livestock shipments between 1916 and 1919. Not only livestock but chickens, eggs, cream, and vegetables were finding their way to market by this means. But the drive for diversification along the ridge had not garnered enough converts by the end of the decade to avoid problems in the farming sector when the price of cotton plummeted in 1919 and 1920. The cotton market in 1919 fell so low that ridge farmers considered holding their cotton for better prices. By this time the county was again without a farm agent, for Dent had resigned. But he stayed in the area farming and selling real estate, and it was Dent who went to Little Rock in an unofficial capacity in 1920 to arrange for the incorporation of a cotton warehouse. Local farmers organized the cotton warehouse in Harrisburg in October 1920 as "a way for the cotton farmer to borrow some money on his cotton so he could hold it for a better price."[20]

Local farmers were, however, largely on their own in dealing with the agricultural crisis that followed World War I. The federal government was unwilling to assume that authority or responsibility and few farmers wanted it to do so. The editor of the Harrisburg *Modern News* urged farmers to be "calm and conservative" in the face of the violence that threatened as cotton prices dipped further downward. Cotton farmers all over northeastern Arkansas directed their dissatisfaction at the local gins. "In various places about the country notices are being posted on gins to close until prices go up. Some have been burned when they don't comply and some farmers have been told not to pick any more." The newspaper warned farmers that the agitation "just makes matters worse"; it predicted that "in a few weeks it will all blow over. People will have to have the cotton."[21]

In the delta county sheriff W. D. Shelton and his deputy, P. D. Smith, went to Tyronza in the fall of 1920 to investigate "the posting of notices on the gin at that place. They said notices had been posted alright but there had been no demonstration toward carrying out any threats." The sheriff reported that there was little to be concerned about, and he returned to Harrisburg. But within weeks violence erupted in other parts of the delta. "The destruction by fire of the Earle compress on the 22nd [of October] with the loss of 7400 bales of cotton and the loss of the Farmers Union cotton warehouse at Warren with 900 bales, coupled with the whisperings of incendiarism and the overt acts of posting warnings on cotton gins in various sections of the state have created . . . [a] serious situation."[22]

For the small farm owner the situation was as bad in the delta as it was along the ridge. The delta planters could pass the burden along to their tenants and sharecroppers, but the smaller farm owners were unable to absorb the losses. While Ernest Ritter stored cotton in his warehouse and exhorted others in the Marked Tree area to have patience, J. W. Strickland, who operated a farm on Crowley's Ridge, brought in what he declared to be "his last bale of cotton ever."[23]

The prairie's farmers did not find the drop in cotton prices disastrous because they grew very little of it. They planted rice rather than cotton. Rice production in the prairie increased from 52,196 bushels on 986 acres in 1910 to 239,504 bushels on 6,484 acres in 1920.[24] And rice farmers were more similar to ridge farmers than they were to delta farmers in terms of their personal property and real property wealth. According to the subset generated by the match between the real property and personal property records, only 3 companies (41 percent) held personal property wealth in the prairie. They claimed $5,055 (11.3 percent) in personal property wealth, and they held only 1,445 (9.7 percent) of the acreage there. In contrast, 70 individuals held 13,514 acres (90.3 percent) and claimed $39,876 (88.7 percent) in personal property wealth. Significantly, the rice farmers in the prairie held their wealth differently. They invested in the equipment necessary to pump the water to their rice fields, and they bought farm machinery to aid them in harvesting the crop.[25]

Very real differences existed between farm operators in the prairie and those elsewhere in the county. In contrast to land in the delta, prairie land was solidly in the hands of small farmers rather than companies. Most prairie farmers operated 160-acre tracts, grew their own garden vegetables, and rarely cultivated cotton. After the initial experiments between 1907 and 1909 established that rice could be profitably grown in the prairie, farmers there cleared many acres, drilled wells, and used steam engines to pump the water necessary to flood the rice fields.[26] Indeed, the prairie farmers were strikingly different from the men establishing plantations in the delta. They were northerners, they mechanized, they avoided the crop lien system, and they used seasonal wage laborers rather than tenants and sharecroppers.[27]

Prairie farmers typically hired men with teams to harvest the rice in the fall, often drawing upon tenants on plantations in Green, Cross, and Craighead counties. Little is known of the racial composition of the workers. They were rarely from Poinsett County's own plantations because of the difficulty of traversing the swamp. In 1914 some prairie rice farmers began to use tractors, which allowed them to cover fifteen to twenty acres a day rather than the five acres a day a man with a team could cover. By 1920 prairie farmers owned far more in farm equipment than did farmers in the delta or on the ridge. Eighty-eight prairie farmers owned an average of $302 in farm equipment in 1920. Only twenty delta farmers owned an average of $244 in farm equipment, and only fourteen ridge farmers owned an average of $164 in farm equipment.[28] Although prairie farmers initially purchased their machinery from implement dealers from the Midwest, they soon turned to their local banker in Weiner more often. They weathered the beginning years of the agricultural depression that followed World War I somewhat better than farmers on the ridge. Yet while their farming sector grew steadily, if slowly, their principal towns remained secondary to Harrisburg. Weiner, for example, never had more than eight business establishments before 1920. Only one small sawmill operated in its vicinity, and the town's one attempt to establish a major enterprise there, the Weiner Rice Mill, was built in 1913 only to burn to the ground the next year. The refusal of insurance carriers to reimburse local farmers for rice they had stored at the mill forestalled the construction of a new one.[29] By the time the local farmers' claims were settled, a marketing network linking the Poinsett County rice industry to that around Stuttgart had been solidly established.[30] Stuttgart's mills had captured the local business, and it would be decades before Weiner would build a new mill.

Meanwhile, the relationship between the ridge and the prairie deepened between 1910 and 1920. Farmers and businessmen from the prairie regularly visited Harrisburg, and a number of ridge farmers engaged in rice farming on the edge of the prairie, close to the ridge. A greater number of ridge farmers were involved in farming in the prairie than ever before. In 1920 twenty-six of them held 22,397 acres there.[31] Access to the town of Harrisburg played the greatest role in encouraging ties between that town and the farmers in the prairie. It was much easier and much less expensive to build and maintain roads through the prairie than through the swamps, and by early 1916 the roads leading into the prairie from the ridge were in better condition than they had ever been. Scott Johnson, an absentee farmer from Rankin, Illinois, reported that the roads were "much improved" from when he last visited three years before. On his previous visit Johnson "drove from Harrisburg to Weiner and back and it took all day"; in January 1916 it took less than three hours to drive a few miles farther to Waldenburg.[32] In addition, the number of roads was increasing. Now four roads led from the prairie to the ridge while only one difficult-to-maintain road led from Marked Tree to Harrisburg.

While the ties between the ridge and prairie intensified, the delta drew closer to Memphis. When the citizens of Poinsett County were asked to contribute to the construction of a viaduct leading to the newly constructed Mississippi River bridge at Memphis in late 1915, ridge and prairie farmers and businessmen were not interested, but Marked Tree businessmen-planters were enthusiastic.[33] Marked Tree's orientation had long been toward Memphis, and the approaching construction in 1919 of a new federal highway, the Ozark Trail, which was to run through the heart of Marked Tree, helped local businessmen realize their goal. It contributed to considerable economic growth in Marked Tree, while Harrisburg enjoyed only moderate economic expansion.[34] In March businessmen in Marked Tree gathered to discuss the town improvements that would be necessary to accommodate the Ozark Trail, including not only paving the streets but also constructing a new hotel at "a cost of between $60,000 and $75,000." The local Ozark Trail commissioners, named at a meeting held the next week in the office of Ritter and Company, included E. Ritter, D. D. White, and J. A. Emrich. A few months later Ritter announced that he would erect "an entire block of new brick buildings" on land that he owned along the new highway.[35] That block of buildings became the main business street in Marked Tree.

Between 1900 and 1920 the new businessmen-planters of the Poinsett County delta emerged as a powerful force. Confronting a range of problems within the delta, they were poised to challenge the political ascendance of the Harrisburg elite. Their principal town, Marked Tree, was thriving and had become the center of commerce and credit in the delta, spawning the planters who would transform the countryside.

The differences emerging between the delta and the ridge flamed the controversy over the relocation of the courthouse in 1917. During an earlier debate in 1907-8, delta representatives had asked only for a district courthouse in the town of Marked Tree rather than the removal of the county government to the delta. The *Marked Tree Gazette* had focused on the savings to delta taxpayers that such a courthouse would mean and stressed the travel and lodging costs incurred by those wishing to serve on juries or as witnesses in the county court.[36] Indeed, the remoteness of Harrisburg had long discouraged delta men from serving on the petit and grand juries.[37] Editor T. D. Harris argued that having to pass through the swamp to reach Harrisburg worked too great a burden on "9/10ths of those who pay taxes from the east side." Indeed, it would prove impossible to maintain a road between Marked Tree and Harrisburg for another twenty-five years, and delta taxpayers continued to incur added expenses. Harris outlined the difficulties encountered by citizens from the delta town of Lepanto, for example: "They have to come to Marked Tree and if they are not 'jonny on the spot' the train leaves and an all night stand has to be made here at the cost of say $2.00. Then to Nettleton and a lay over of several hours at the cost of $1.50. Then to Harrisburg and a lay over at a cost of $2.00 making a grand total of $5.50. The same expenses incurred on the return trip making a grand total of $11.00. If you are a juror or a witness you receive the munificent sum of $1.50."[38]

Despite the difficulties faced by those in the delta having business at the courthouse, a spirit of conciliation reigned in the years before 1907–8. Although the delta planters put forward their own county ticket in 1904, letters to the editor of the Harrisburg *Modern News* that year addressing the issue of representation on the county level were highly conciliatory. J. A. Emrich, a prominent businessman-planter from Tyronza, wrote of the support the "east end of the county" had always given to the Harrisburg politicians and stressed the fact that the delta was solidly Democratic. There was no hint of reproach in Emrich's letter, but merely a plea for recognition. The one letter that did reveal more impatience with the dominance of "the Crowley's Ridge boys" enumerated no particular grievances. In fact, planter C. J. Watkins, its author, admitted that delta men "have been too busy to go office seeking on account of building fences and repairing homes destroyed by the disastrous overflows." Finally, a delta man writing under the pseudonym of W. Tookone, cautioned against "factional quicksands" and suggested there were no significant differences between the ridge and the delta candidates. Apparently, the voters in the delta agreed, for the delta ticket went down to defeat even though there were more registered voters in the delta than on Crowley's Ridge and the county's prairie combined.[39]

The defeat of the delta ticket in 1904 was the first indication of the inability of politicians there to capture all of the delta voters at the polls. Yet the desire of delta politicians for recognition was not entirely repudiated by the ridge elites, for in the same election the Harrisburg establishment gave unqualified support to the campaign of M. W. Hazel of Marked Tree who ran for state representative from the district that included Poinsett County.[40] The courthouse clique must have felt secure on the county level even in the face of the patronage power that delta politician Hazel would acquire as a state representative.[41] As a quid pro quo, the delta men gave their wholehearted support to a ridge candidate for prosecuting attorney, L. C. Going. Going's candidacy was an easy one for them to support, for he had been attorney for the first Marked Tree Town Council. Although a ridge-born man, his association with delta businessmen-planters would deepen in the years to come as his political ambitions took him to a powerful position in the state Senate and a stint as attorney representing the St. Francis Levee District (1915–17).[42]

It was in the first decade of the twentieth century, in fact, that delta businessmen planters came to dominate the county's delegation to the St. Francis Levee Board, probably the most influential political body operating in Arkansas. Founded in 1893 after years of political pressure by delta planters, the St. Francis Levee Board was vested with considerable power on the local level, which translated into even greater influence on the state.[43] The 1893 act creating the district also empowered the governor of the state to appoint three directors from each of the eight counties to be served by the district. From the beginning, "appointment as a member of the board was sought by politicians, sheriffs, judges and clerks of the respective counties, friendly to the governor who readily appointed them and created a strong political machine for his support in office." The potential for creating such a political machine was recognized before Jeff Davis became governor in 1900, but it was Davis who turned the St. Francis Levee Board into a powerful political engine.[44] Following his notion of the importance of a coalition between planters and yeomen, Davis appointed both to the board.[45]

By 1910, however, the planters and small farmers would engage in a bitter struggle for control of the board that reflected their different attitudes toward development. Planters typically favored ambitious and costly drainage enterprises that would uncover tens of thousands of acres of rich delta swamps. Small farmers simply wanted to protect their farms from the disastrous overflows that had plagued them for decades. Since they had no desire to engage in plantation agriculture, they did not want to incur the burden of debt that only such enterprises would warrant.[46]

The businessman-planter who directed the board and battled representatives of small farmers over policy was W. B. Miller of Poinsett County's delta. Only one of the three board members from Poinsett County was from Crowley's Ridge, and there is no evidence of his position in the dispute. What is clear is that Poinsett County planters not only dominated the county's delegation to the board but also exercised significant influence over its policies. This had not initially been true. The first three Poinsett County directors, appointed in 1893, were ridge men: Benjamin Harris, L. J. Collins, and J. B. Gant. Collins and Gant had settled in the county in the 1870s and become prominent citizens of Harrisburg, while Harris was the grandson of Poinsett's first county judge. It is unlikely that anyone from the county's delta was even considered for a position on the first board since the thinly populated delta was hardly developed economically in 1893 and had no political organization of any significance. Four years later, however, delta men were represented on the board, and six years later they dominated it and were clearly on their way to developing political influence on the state level.[47]

Coming to control the county's delegation to the St. Francis Levee District was not enough to satisfy delta men, however, and delta politicians jockeyed into position for an assault on the courthouse clique in 1908. In March of that year delta men began to dominate the affairs of the County Democratic Central Committee (CDCC). M. W. Hazel, who had been unable to defeat county judge J. R. Willis in 1906 when the judge ran for reelection, was named chairman of the CDCC in 1908; clearly by 1908, the population of the delta had reached the point where delta representation on the important CDCC gave delta politicians an edge. Yet politicians in one of the delta townships, Willis, named for Judge Willis, a longtime resident of Crowley's Ridge, were

Table 5.1. Population of Poinsett County by section, 1900–1920

	1900		1910		1920	
	No.	%	No.	%	No.	%
Ridge	3,454	49.2	4,828	37.7	6,706	32.2
Prairie	1,383	19.7	1,641	12.8	2,206	10.6
Delta	2,188	31.1	6,322	49.5	11,936	57.2
Total	7,025	100.0	12,791	100.0	20,848	100.0

Source: Fourteenth Census of the United States: 1920, Population 1 (Washington, D.C., 1922): 349.

Note: Since these data were taken from the published census rather than from my sample of specific townships, the figures are much higher than those reflected in my discussion of the three sections elsewhere.

Table 5.2. Population of delta townships in Poinsett County, 1900–1920

	1900	1910	1920
Little River	1,032	2,842	3,223
Tyronza	800	1,678	3,059
Greenwood*		935	1,783
Willis	356	867	3,871

Source: Fourteenth Census of the United States: 1920, Population 1 (Washington, D.C., 1922): 349.

* Greenwood Township was carved out of part of Little River Township in 1904.

not solidly in league with delta politicians east of the St. Francis River floodplain, inhibiting the ability of eastern Poinsett County delta politicians to secure election of their candidates (tables 5.1 and 5.2).[48] Nevertheless, the Marked Tree–Tyronza–Lepanto politicians did on occasion unite and came to wield considerable influence on the CDCC. The CDCC typically endorsed certain Democratic candidates for county offices, and in some counties that endorsement was tantamount to election.[49] Since 1904 the delta had presented its own slate of candidates but found it impossible to penetrate the courthouse clique. The CDCC was the key, for it appointed the judges who supervised the elections and ruled on grievances presented by Democrats within the county when elections were disputed. Committee members could, moreover, influence political appointments made by the county judge.

While control of the CDCC had certain advantages on the local level, it was of even greater importance on the state level. Whenever a dispute could not be resolved on the county level, for example, the chairman of the CDCC could filter information to the state body and influence its understanding of any given situation. The State Democratic Central Committee listened to and respected those who controlled the county committee. Significantly, the Poinsett County delegation to the State Democratic Central Committee meeting in 1908 was led by Marked Tree businessman-planter and chairman of the St. Francis Levee Board W. B. Miller. It was, perhaps, no coincidence that Miller was chosen to represent the county, for he had already established himself as a state power. As the leader of the county delegation he had to be accorded special notice, and as the chairman of the St. Francis Levee Board, he was a force to be reckoned with.[50]

In the very month that the delta politicians came to control the County Democratic Central Committee, however, they failed in yet another attempt to get one of their own elected to county office. Even in the face of factional disputes that divided the ridge vote, delta politician Crockett J. Hazel—M. W. Hazel's younger brother—failed to defeat County Judge Willis's reelection bid. Willis polled 786 to Hazel's 770 votes. J. W. Rooks, a descendant of one of the county's pioneer settlers on Crowley's Ridge, got 373 votes. The exact nature of the factional cleavage among ridge politicians is unknown, but Rooks and Willis reportedly had to be "stopped by friends from coming to blows" at a preprimary debate in the prairie town of Fisher. The only office won by a delta man was one held since 1904, that of state representative.[51]

Clearly, as early as 1908 politicians in the Poinsett County delta were flexing their political muscles. And even though they had proved unable to capitalize on factional disputes among ridge politicians, they were better positioned than ever to challenge the courthouse clique. This was accomplished even though the delta's most powerful businessman-planter was unable to avail himself of Democratic party favors. Although one of the letter writers to the editor of the *Modern News* in 1904 had stressed the preponderance of Democrats in the delta and suggested that "there are not enough Republicans in this part of the county to act as pall bearers at a funeral," he neglected to mention that one of the few Republicans in the delta was Ernest Ritter, Poinsett County's single most powerful delta businessman-planter.[52] Although a Republican in a Democratic stronghold, Ritter was not without influence. He served on a town council made up almost entirely of Democrats, and many of them were his friends and business associates. He had nominated Democrat L. C. Going as attorney for the first town council. One of his own business partners, M. W. Hazel, at one time or other chaired the County Democratic Central Committee, won election to the Arkansas General Assembly, and served several terms on the St. Francis Levee Board. His brother Louis's business associate, J. A. Emrich, a businessman-planter from Tyronza, would also occupy a position on the St. Francis Levee Board for several years.[53]

Nevertheless, Ritter probably played almost no direct role in county Democratic politics and had minimal influence in the state Republican party. Even in the delta his political power was circumscribed. His service on the town council ended in 1904; in 1912 he failed to win the mayor's office in Marked Tree after a particularly heated election. The power he exerted was economic rather than political. On the state level he essentially abandoned the regular Republican party and aligned himself with the Bull Moose Progressives, which placed him outside the Arkansas Republican clique and rendered him even less politically influential than he might have been otherwise.[54]

Ritter's power would have been immensely useful to the delta Democrats. Nevertheless, they had achieved a measure of political power by 1908 and seemed poised for a serious challenge to the ridge elite. But the delta elite would not launch that challenge for almost a decade. Two basic problems hindered the political ambitions of delta Democrats. First, it appears from the fragmentary sources available that the prairie farmers represented the crucial element in the struggle between the delta and ridge.[55] Politicians from both of the contending factions pitched their appeals to the prairie farmers, but the ridge elite had the advantage. They had formed economic ties with the prairie, and ties of kinship linked the two regions.[56] The prairie, moreover, was too remote from the delta and far too different a society to promote a commonality of interests. The second problem facing the delta elite struck much closer to home. They could not muster the votes within the delta to mount a serious challenge, for it appears that their ambitions were thwarted not simply by the ridge-prairie alliance but also by the disloyalty of certain delta citizens. Whether these were small farmers disenchanted with the massive drainage enterprises orchestrated by the planters; sharecroppers and tenants exercising an opportunity to strike out at those who attempted to dominate their lives; or competing elites within the delta such as those from Greenwood Township, they weakened the delta challenge to the ridge. And citizens within Willis Township, located between the ridge and the St. Francis River floodplain, demonstrated greater loyalty to the ridge and failed to unite with elites from Greenwood, Little River, and Tyronza townships. The population of Willis increased from 356 in 1900 to 3,871 in 1920, and their votes were immensely useful to ridge politicians seeking to retain control of county government.[57]

Given these political disabilities, delta politicians continued a policy of conciliation with the ridge by promoting and electing ridge men who were sympathetic to the needs of the expanding delta economy. J. C. Mitchell, for example, a member of the CDCC from the ridge who had landholdings in the delta, was elected county judge in 1910 upon Judge Willis's retirement. Mitchell immediately embarked on an inspection tour of county roads, an issue of great importance to the delta planters and to his own economic interests in view of his investment in delta land. Judge Mitchell's county court ruled the next year that jurors and witnesses could charge the court for travel expenses. His successor, Benjamin Cole, was another ridge politician with economic ties to the delta. Cole, who took office after Mitchell's accidental death in 1912, immediately honored Marked Tree by holding a session of the county court there. One of Cole's first actions as county judge was to conduct an inspection of roads in the delta; he also joined Ernest Ritter, John Krier, and M. W. Hazel on a visit to Mississippi to inspect roads there.[58]

The spirit of conciliation was so strong that in the year after the 1908 battle over the courthouse, editor Harris of the Marked Tree newspaper, who became a member of the CDCC in 1908, defended the ridge men who held office in Poinsett County in 1909 and even declared that "if we have a ring it must be a pretty good one if we were to judge it by the men who now hold office."[59] Nevertheless, the courthouse fight that erupted in 1917 had long been simmering. Some of the old arguments made in 1908 resurfaced: delta representatives complained about the difficulty of reaching Harrisburg from the various delta towns, and the complaints revealed what little progress had been made on the Harrisburg–Marked Tree road.[60] County judge S. T. Mayo responded by visiting Marked Tree during the heat of the 1917 courthouse controversy to attend to details concerning improvements to the Harrisburg–Marked Tree road, despite his energetic defense of Harrisburg's right to maintain its position as county seat.[61]

But the 1917 conflict over the courthouse went far beyond complaints of inconvenience. The dispute, in fact, took place against a background of political intrigue. A political feud had been threatening since the disputed election of 1916. Delta representatives charged that the "Harrisburg oligarchy" was guilty of both political and economic fraud. According to the delta men, the Harrisburg clique had deliberately misplaced the Marked Tree ballot box in that election and had also misappropriated county funds.[62] The two charges were linked only in the imagination of the delta elite, however, for the contested election involved county treasurer C. A. ("Dad") Blanton, who was not only a delta merchant but above suspicion, and county school superintendent H. B. Thorn, a well-respected ridge man who would have had little opportunity to misappropriate county funds. Blanton had served as county treasurer since 1912, and when he announced his intention to seek reelection in 1916, editor Harris of the *Marked Tree Gazette* reported that he had "a clean record." Harris's insinuation that the courthouse fire had been deliberately set in order to cover up a shortage of funds a year later was not aimed at Dad Blanton or H. B. Thorn.[63]

Although Blanton and Thorn were apparently not under suspicion for having stolen county funds in 1917, they had been accused of stealing the election of 1916. Blanton's reelection was bitterly contested by L. C. Griffin of the delta, who was making his first run for public office. Amid allegations of fraud, which included the charge that the Marked Tree township ballot box had been "lost" in the swamps on its way to Harrisburg, the CDCC ruled that Blanton had won the election. When Griffin appealed to the state Democratic Central Committee, that body ruled in Blanton's favor but agreed that there had been fraud in Little River, Scott, and Dobson townships, from the delta, ridge, and prairie respectively, and declared the offices of committeemen in those townships vacant.[64]

The CDCC's decision to validate Blanton's election probably rested on an informal agreement between delta and ridge politicians to share power in the county government. Dad Blanton was the candidate of the committee, and Griffin was an interloper. Although Griffin's candidacy illustrated a certain lack of consensus on the issue within the delta, the willingness of ridge and delta politicians to accommodate each other was maintained despite the election of 1916. Delta planter J. C. Hooten of Tyronza Township had taken over chairmanship of the committee in 1914 from M. W. Hazel and endorsed Blanton as county treasurer. The fact that Hooten had been county treasurer between 1910 and 1914 gave added weight to his endorsement of Blanton for that position. Significantly, the committee played no active role in the courthouse controversy of 1917. None of those listed as petitioners in favor of moving the courthouse to Marked Tree were members of the committee. While the interests of the delta committee members would have been served best by such a move, they remained silent on the issue, possibly from an unwillingness to endanger the spirit of accommodation that existed within the committee.[65] It may have been the committee that worked out the compromise giving the ridge elite dominance (temporarily, as it turned out) over Drainage District Seven when it was created in 1917.

Having chosen to remain mute during the courthouse controversy, the committee voiced no opinions as to whether the charges of the delta elite regarding the corruption of ridge officials were warranted and left it to editor Harris of the *Marked Tree Gazette* to make insinuations. Harris did not, however, direct his insinuations of misappropriation at the only party likely to have been guilty. While Blanton was above suspicion and Thorn lacked opportunity, county sheriff and tax collector Joe Hall was both suspect of character and well positioned to act. Harris's failure to name Hall probably rested on the fact that the evidence against him was circumstantial. Certainly the public particulars served merely to raise suspicions. In November 1915, for example, when Hall announced his intention to seek reelection, the *Marked Tree Gazette* reported that the county sheriff had "made clear" to his friends an "alleged shortage . . . and has a receipt to show that every cent owed to the county" was accounted for.[66] Yet when the courthouse burned in May 1917 a few days after the county auditors arrived, the personal property tax records were the only records lost. And a year after the courthouse fire and just weeks before Hall was to turn his office over to the new county sheriff and tax collector, the personal property tax records for 1917–18 disappeared while in his care.[67] Whether Joe Hall was dishonest, incompetent, or merely unlucky, he was repeatedly implicated in the loss of county records. While the county assessor was responsible only for placing names on the tax books, Sheriff Hall collected the taxes and gave the funds to the treasurer who deposited the money and handed the books over to the clerk for safekeeping. Hall usually took the books with him to different communities so that taxpayers could more easily have access to him. It was during one of these sessions that the 1917–18 personal property tax book disappeared.

It is unlikely, however, that the funds Hall might have misappropriated would have reduced the staggering county debt substantially. And the charges by the delta planters that the "Harrisburg oligarchy" had run the county's finances poorly seem misplaced, for much of the debt incurred by the county involved incidentals connected to the roads and drainage districts of the delta. Delta men, moreover, had exercised some influence on the county level during the decade in which the county debt rose to $100,000.

Whether or not the charges against the ridge politicians were warranted, the destruction of the Harrisburg courthouse in 1917 brought the debate over the county's finances into the open. Despite owing $100,000, the county proposed to incur further debt and impose a greater tax burden upon its citizens in order to build a new courthouse. The cost of building the new courthouse in Harrisburg was placed at $100,000, and the county proposed to issue scrip in order to fund construction. Landowners throughout the county would then be required to pay off the debt by means of additional taxes. Marked Tree men did not wish to fund construction of a courthouse in Harrisburg, and they questioned the ability of the Harrisburg clique to build the courthouse economically and handle the funds involved honestly. They offered, instead, to build the courthouse at no cost to the taxpayers if it was relocated in Marked Tree. Ernest Ritter pledged to donate the property and help provide the men and materials for construction.[68]

While Ritter's offer fell on deaf ears in Harrisburg, prairie taxpayers found it compelling. Within ten days of the burning of the courthouse, Marked Tree representatives met in Weiner, the principal town in the prairie, and secured a pledge of support "to help in putting the court house where it properly belongs." This was an unprecedented departure from the political allegiances that had helped to maintain the power of the Harrisburg clique. At a tumultuous meeting on May 20, just over two weeks after the courthouse burned, the prairie magistrates lined up behind those from the delta at a special session of the quorum court, and together they blocked the appropriation of funds for the construction of the new courthouse in Harrisburg. The delta delegation then was "driven to Weiner, where the people treated us in a most royal manner. A splendid dinner was served at the Trotter Hotel, and later a mass meeting of several hundred people, headed by the Weiner band, gathered in the auditorium of their beautiful new high school building, where several stirring speeches were made pledging support in the court house move in which new and lasting friendships were formed." The optimism displayed by the delta men was not warranted, however. On July 1 the prairie magistrates realigned with the Harrisburg contingent and voted the funds for the Harrisburg courthouse.[69]

The key to the change of heart was the promise of a district courthouse to be located in Weiner. By the end of the month the Marked Tree men secured a restraining order from the chancery court in Jonesboro to delay the construction of a new courthouse in Harrisburg. They presented a petition with over two thousand signatures in support of their plea. But many of the signatories were not voters, and once the Harrisburg contingent established that fact, the injunction was dissolved. Marked Tree lost an appeal to the circuit court, and an attempt to convince the Arkansas Supreme Court to hear the case and have the matter put to a popular vote met with defeat when that body determined that it had no jurisdiction. The Marked Tree contingent was now without recourse. Yet the *Marked Tree Gazette* printed a last, almost plaintive, letter from Dr. J. R. Black of Marked Tree in its December 7, 1917, issue. Black railed against the crooked deals cooked up by the "Harrisburg oligarchy" and suggested that it was "just such deals as this which have been pulled off in Poinsett county for years that have enriched a few men."[70]

The irony in Black's statement hardly needs pointing out. Although Dr. Black was not among them, certain delta men had been enriching themselves for more than a decade at the expense of settlers in the sunk lands, sometimes in ways that were no more ethical or legal than those employed by the so called Harrisburg oligarchy during the election of 1916. The settlers, in fact, probably had a very different view of which elite was more corrupt. They could hardly have expected the cooperation that would take place between the ridge and delta elites later that same year. The establishment of Drainage District Seven in late 1917 demonstrated clearly that economic gain was a strong inducement to cooperation.[71]

The voters of the county failed to elect a delta politician to the important county judgeship until 1920. Judge W. W. Warren of Tyronza won with the support of the Poinsett County Taxpayers Association, headquartered in Marked Tree, and once elected, he attempted to address the economic problems confronting the county. They were enormous. County scrip was worth half of its face value, and bondholders who held Drainage District Seven bonds had gone into federal court and secured a judgment. Although Warren negotiated a compromise which mitigated the worst effects of the judgment, he engaged in a massive tax reassessment which, while allowing the county to meet its obligations and attempt to rescue its credit, angered many residents throughout the county. Some citizens believed, apparently falsely, that certain large companies, like Singer Sewing Machine in Truman, were paying far less than their fair share.

In an election made controversial by charges of Ku Klux Klan (KKK) activity on the part of one candidate for county sheriff, Warren lost his bid for reelection in 1922 to ridge attorney, W. P. Maddox.[72] Another delta man, W. D. Shelton, held on to his position as county sheriff, which he had secured in 1918 after defeating the discredited Joe Hall. The puzzle is how one delta man, Shelton, could win so convincingly (1,530 to 649) while Warren went down to defeat. Shelton's opponent, W. D. Bradford of Harrisburg, in defending himself against an accusation that he was a member of the KKK, provides a clue. While complaining about the "dirty mud-slingers" attacking him, Bradford asserted that Shelton had "publicly shown" his own KKK membership card, thereby stirring "up a lot of prejudice" and exemplifying how "the present system" maintained itself. Bradford's overwhelming defeat suggests that his condemnation of Shelton's alleged KKK association worked as an endorsement rather than as an indictment. Just six week after Bradford publicly directed his charges against Shelton, a "big meeting of the Ku Klux Klan" took place outside Harrisburg. After a "supper, composed of a bountiful supply of the best eats," the attendees listened to speakers. Passersby reported seeing nearly five hundred cars, eighteen hundred men, and a burning cross. Under the light of that cross, "a large number of new members" took the "oath of the organization." The next year, two more large meetings of the Klan occurred near Harrisburg. One featured a "national Klan lecturer" and the other reportedly had somewhere between thirty-five hundred and five thousand people in attendance. Clearly, Bradford was out of step with the sentiments of much of the ridge population.[73]

The appearance of the Ku Klux Klan near Harrisburg coincided with the emergence of the Klan nationwide after World War I, and like the local branches elsewhere, the ridge Klan addressed issues that were especially relevant to those who attended. Its August 17, 1923, meeting, for example, included a speech by a deputy sheriff who had participated in the Elaine riot in 1919. White supremacy, something celebrated by nightriders from the ridge twenty years earlier, remained a potent force.[74]

No mention of Klan activity in the Poinsett County delta has surfaced in the historical record, although the fear stirred by the riot at Elaine makes it seem likely that the Klan appealed to the delta's white citizens. The strident racism that surfaced in Marked Tree in the first decade of the century suggests that at least certain citizens would have found the KKK message appealing. Citizens in Truman, who had a reputation for antiblack sentiments so harsh that they refused to allow blacks to settle in the town, probably found the KKK equally attractive. By the mid-twenties the organization lost force on the ridge, just as it did elsewhere. Whether or not he was a klansman, as Bradford claimed, Shelton's defeat in 1925 by a ridge man, A. F. Landers, coincided with the decline in interest in the Klan.

Whether joined in the brotherhood of the Klan or not, elites within Poinsett County attempted to accommodate each other in the first decades of the twentieth century even as they struggled for control over the county government. There was an impression among politicians in the delta that ridge politicians were trying to make up for their declining economic and political power by political chicanery and even outright criminality, and for them issues of competence and honesty became inextricably bound. Clearly, the ridge politicians were fighting against economic and demographic factors running almost uniformly to their disadvantage. The ridge elite seemed unwilling to accept the inevitability of their junior role, so a series of political and quasi-economic disputes broke out. In a sense, the disputes were crystallized geographically in the construction of the courthouse. The ridge politicians survived the delta threat because they could draw support from the prairie and from Willis Township. The weakness of delta politicians also contributed to the ridge success; they were disunited, and none of them were able to secure and maintain the loyalty of a sufficient number of delta voters. While the delta "companies" exerted considerably economic clout and their representatives were able to secure gubernatorial appointment to the St. Francis Levee Board, delta voters would not consistently vote them into control of the county government. Aside from the inability of competing elites within the delta to unite, the small farmers, homesteaders, sharecroppers, tenants, millworkers, and others disenchanted with the companies, demonstrated—by their behavior at the ballot box—the limits of economic power.[75]

6

DELTA BLUES

When W. D. Ezell arrived in Poinsett County in 1926 to fill the position of agricultural agent that had been vacant for three years, he found that it was "hard to do much with livestock for most of it is on [the] open range." Within two years of his arrival, voters in a countywide election approved a stock law. Ezell's successor, H. S. Hinson, remarked that "by the passage of a stock law over the County . . . we are hoping to improve the stock standard a considerable degree." For the county agents the issue was simple. Enclosing livestock within fences resulted in a higher quality of stock and a greater profit margin. In 1929 the third county agent in four years, A. Raybon Sullivant, put it succinctly: "Since the passage of the county wide stock law last spring farmers are asking about 'better livestock and pastures.' Farmers are paying more attention to the 'profit end' of their livestock than ever before."[1]

In encouraging farmers to support a stock law, county agents entered a debate which had raged in Poinsett and other counties in Arkansas for decades. Small farmers were accustomed to grazing their animals on the open range and resisted the efforts of larger farmers and planters to impose fencing laws. In addition to violating what they considered one of their traditional rights, stock laws required farmers to make a higher capital investment. Aside from the initial expenditure of erecting a fence, a farmer would have to maintain a pasture and provide feed for his animals. Twelve years before the enactment of the countywide stock law, a delta newspaper editor opposed a similar law for the area east of Crowley's Ridge, arguing that poor people could not buy the feed necessary to maintain their stock and needed "to let them graze freely."[2] Ironically, the decisive support for the 1928 law came from the delta. The vote was 2,113 to 1,091 overall, with ridge voters polling 458 against and 342 in favor while the prairie and the delta voted overwhelmingly in favor. Ridge farmers got a restraining order from the county judge which the Arkansas Supreme Court initially sustained but invalidated by the end of 1929. By early 1930 the stock law was in effect and enforced.[3]

In participating in the long-standing debate over the passage of a stock law, the county agents also injected themselves into the middle of a rivalry between the ridge and the delta. The principal opposition to the law arose on Crowley's Ridge where the greatest number of smallholders farmed. Ridge citizens voted against the measure by a margin of 116 votes. As might be expected, the voters of Harrisburg provided the largest margin of support; town dwellers, who had lawns and shrubs to protect, were the first to embrace stock laws. Some disgruntled ridge farmers delayed the enforcement of the new law by filing an injunction and appealing to the chancery court and finally the Arkansas Supreme Court. They, like farmers filing similar appeals in counties all over Arkansas, were disappointed by the court's decision in favor of the stock law. Voters in the delta had supported the measure in the election by a margin of 1,571 to 466, but perhaps most galling for ridge farmers was the fact that prairie citizens, on whom they had counted for decades to offset the growing numerical strength of the delta, voted in favor of the law 200 to 167. While the prairie's defection, which reflected the commercial orientation of the rice growers, disappointed those on the ridge who opposed the stock law, in fact, the prairie population was so small that even had they voted unanimously against the law, it would have made no difference. The election illustrated the power of the delta at the ballot box once its citizens united behind a particular issue.[4]

The county agents were slow to see the implications of the vote. Not until 1929 did a new county agent recognize that hospitality awaited him in the delta. Planters eventually welcomed the advice of the county agents because the agricultural depression of the period demanded expert attention and because two natural disasters struck in 1927 and in 1930–31. While the drop in agricultural prices made planters receptive to some innovative solutions to their economic woes, the natural disasters encouraged them to accept relief efforts mounted by the Red

Cross. The agricultural agent played a pivotal role, for not only did he introduce solutions to their economic problems, he also coordinated relief efforts during the natural disasters. Planters adapted to the economic and natural calamities confronting them and learned how to incorporate federal programs and national relief efforts into their way of doing business.

The economic problems originated with the drastic decline in the price of farm commodities in the 1920s and the "relative increase in the cost" of nonfarm goods.[5] Since local merchants in agricultural communities were largely dependent upon the health of the farm economy, they, too, faced hardship as prices continued to go down throughout the decade. Greatly complicating the crisis was a constriction of credit resulting from a drop in the number of cotton factors supplying eastern Arkansas with credit (from sixty-two to four). The problems that forced cotton factors out of the market did not abate. Prices rose slightly in the years between 1922 and 1926 but at their best remained 14 percent below the war years.[6] Signs of improvement were misleading. The drop in the mortgage debt in 1924 was actually the result of a wave of foreclosures, and a short crop in 1925 caused a mild price recovery that proved to be temporary. A record crop the next year drove agricultural prices down even further. The overproduction of cotton was one of the fundamental problems creating the economic crisis in southern agriculture, but southern farmers were not willing to accept federally mandated crop reduction programs and were generally "indifferent to federal aid" in the 1920s. They turned instead to more conservative solutions.[7]

Farmers' organizations began to court Washington politicians, and, indeed, the "Farm Bloc," which would grow to considerable influence over the next decade, was incubated in the early 1920s. Acting in concert, the American Farm Bureau Federation, the National Grange, and the National Farmers Union persuaded Congress to pass legislation providing new avenues of credit and protecting their marketing cooperatives. They were unable to unite, however, behind the protective tariff. Southern cotton growers, who exported a great percentage of the crop, opposed the tariff, but midwestern farmers and the Republican administrations in Washington believed that building a tariff wall around American agricultural products was an appropriate solution. Farmers were also unable to unite behind legislation that rejected these conservative approaches and instead would control surpluses by requiring the formation of a government corporation to underwrite farmers through an equalization tax placed on processors. It called for fixing prices domestically and dumping surpluses on the world market. The McNary-Haugen bill, as the legislation came to be known, was not fully supported by farmers and faced vehement opposition from business groups and the Republican White House of the 1920s. The more traditional remedies, therefore, were the only ones to be tried.[8] Higher tariffs failed to resolve the price deflation, however, and marketing cooperatives proved to be largely ineffective. The expansion of agricultural credit solved only the short-term credit needs of farmers and did little to relieve the ever-spiraling farm mortgage debt.

The efforts of the Arkansas Cotton Growers Association (ACGA) serve as an example of the weaknesses inherent in marketing cooperatives. The planters and businessmen who formed the association in 1922 believed that by pooling cotton they could drive a better bargain. They planned to eliminate the middleman and, if necessary, hold cotton off the market in the hope of better prices. Many members did not deliver their cotton, however. Some had signed contracts hastily after being persuaded by the appeals of an association promoter and often had second thoughts once the promoter left. Many other farmers needed cash at harvest in order to meet their obligations and could not afford to hold the cotton for a better price.[9]

While some farmers and planters worked with the marketing cooperatives, others sought loans through the agricultural credit associations. These associations, however, were not well funded and did not address the long-term mortgage debt facing planters and farmers. They did enable some farmers who might otherwise have been unable to plant a crop to stay in operation, but they merely answered the farmers' short-term credit needs, essentially providing annual crop loans.

Poinsett County farmers and planters experimented with marketing cooperatives, utilized the services of agricultural credit associations, and generally ascribed to the conservative approach outlined by the Farm Bloc. Delta landowners were better able to withstand the depression than their counterparts on the ridge or in the prairie until the flood of 1927 pushed them over the edge. After 1927 few delta landowners paid their taxes on time, and many failed to pay them at all. The governor signed into law a tax reduction in 1928, but by 1930, 747 landowners throughout the county forfeited 303,320.77 acres (62

Table 6.1. Land forfeiture in Poinsett County, 1920–30

	Acreage held in 1920		Acreage forfeited 1920–29		Acreage forfeited in 1930		Total acreage forfeited	
	No.		No.	%	No.	%	No.	%
Ridge	146,377.27		72,343.60	49.4	25,171.59	17.2	97,515.91	66.6
Prairie	148,897.80		84,367.53	56.6	43,940.61	29.5	128,308.14	86.2
Delta	193,592.78		13,445.60	7.0	64,051.12	33.0	77,496.72	40.0
Total	488,867.85		170,156.73	34.8	133,163.32	27.2	303,320.77	62.0

Source: Poinsett County Real Property Records, 1920, 1930, Poinsett County Courthouse, Harrisburg, Ark.

percent of the land held in private hands) to the state for failure to pay taxes. Only 185,547.8 acres remained in private hands. Businessmen-planters like Louis V. Ritter, Curtis Dewey, and J. A. Emrich, who operated on a large scale, withstood the crisis better than did the rice farmers of the prairie and the small farmers on the ridge (table 6.1).[10]

While larger enterprises like E. Ritter and Company not only survived the crisis but also increased their landholdings between 1920 and 1930, many smaller farmers in the delta faced bankruptcy and ruin. The number of tax defaulters in 1930 revealed a significant shift in land tenure relations taking place within the delta, not only in Poinsett County but in neighboring Cross, Mississippi, and Crittenden counties as well (table 6.2). The number of landowners in Poinsett County dropped by 6.7 percent between 1920 and 1930, and the number of sharecroppers increased by 14.3 percent. Although the percentage of tenants remained almost steady, cash tenants dropped by nearly 8 percent. These changes reflected the growing impoverishment of landless farmers, for fewer of them than ever owned horses, mules, or implements, and thus they found themselves in the least advantageous tenure arrangement with planters: sharecropping, a status more nearly resembling wage labor. The loss of the stock necessary to farming was accompanied by a reduction in the number of cattle and hogs kept by delta farmers and explains their acquiescence in the stock law of 1928 (table 6.3).

Planters shared in the staggering decline in livestock, and many of them defaulted on their drainage, road, and school taxes, as a result

Table 6.2. Land tenure in Poinsett, Cross, Mississippi, and Crittenden counties, 1920–30

	Poinsett Co.				Cross Co.				Mississippi Co.				Crittenden Co.			
	1920		1930		1920		1930		1920		1930		1920		1930	
	No.	%	No.	%	No.	%	No.	%	No.	%	No.	%	No.	%	No.	%
Owners	602	26.7	734	20.0	648	25.8	711	16.2	1,088	16.7	986	9.1	558	10.2	382	4.7
Managers	4	.1	7	.2	9	.4	7	.1	23	.4	36	.3	53	1.0	34	.4
Cash tenants	236	10.5	92	2.5	351	14.0	188	4.3	1,016	15.6	524	4.9	1,022	18.6	995	12.4
Croppers	752	33.3	1,750	42.6	844	33.7	2,385	54.2	3,385	52.0	6,473	60.0	2,378	43.3	5,402	67.1
Other tenants	663	29.4	1,093	29.7	655	26.1	1,107	25.2	999	15.3	2,764	25.6	1,485	27.0	1,231	15.3
Total	2,257		3,676		2,507		4,398		6,511		10,783		5,496		8,044	

Sources: Fourteenth Census of the United States: 1920, Agriculture 4 (Washington, D.C., 1922): 562, 565, 566, and *Fifteenth Census of the United States: 1930, Agriculture 2* (Washington, D.C., 1932): 1137, 1139, 1140, 1142, 1145, 1156.

Table 6.3. Livestock owned by residents of Tyronza Township, 1920–33

	1920	1933
Horses	123	5
Cows	249	25
Mules	194	82
Hogs	198	12

Source: Personal Property Tax Records, 1920 and 1933, Poinsett County, Tyronza Township, Poinsett County Courthouse, Harrisburg, Ark.

losing at least a portion of their real property. Although some sustained heavy losses during the 1920s, delta planters apparently did not organize marketing cooperatives and received little attention from the county farm agent until after the flood of 1927. By 1929 they welcomed the county agent's program into the landscape of possibilities. Inspired by Seaman A. Knapp and created by the Smith-Lever Act in 1914, the U.S. Agricultural Extension Service was oriented toward the more successful farmers and typically appealed to delta planters. That Poinsett County's delta farmers were late in embracing the extension service program bears scrutiny. Although its membership dues "screened out the poorer members of rural society," the placement of the local farm agent in the Poinsett County Courthouse oriented his program to the county's least prosperous farm owners, the dairy and orchard farmers of the ridge. The difficulties of traveling to the delta make it no surprise that Poinsett County's first county agents spent little time there. Thus, the county agents failed to affiliate with those who typically would have been most interested in their programs. Instead, they focused on the ridge, which had the most conservative farmers in the county. Many of them apparently resented the intrusion of an outside agency into their farming enterprises. With little local support for the program, the county had no agent present for several months in 1919–20, for the entire years of 1923, 1924, 1925, for four months in 1927, and for half of 1928. A county agent who recognized the importance of support from delta planters arrived in 1929, and never again was the county without the services of an extension agent.

After S. P. Dent ended his three years' as agent in November 1919, he was replaced by W. A. Owens, who took over on March 1, 1920, and remained until the end of 1921. During his tenure Owens lauded cooperative marketing, encouraged farmers to turn to poultry farming, called for diversification, and helped ridge farmers who planted cotton on the edge of the delta to fight an infestation of boll weevils. In keeping with the extension service's affiliation with certain farm organizations, he drew support from the local Farmers Union and helped organize an affiliate of the Farm Bureau which was described as serving the "central and western portions of the county."[11]

In 1922 county agent E. F. B. Sargent replaced Owens and, like his predecessors, focused his efforts principally on the ridge. His county committee was made up of members of the Farm Bureau and the Farmers Union, and he claimed these two organizations depended upon the county agent for their existence. He continued the efforts of Dent and Owen to encourage cooperative marketing, poultry farming, and diversification, and he helped ridge farmers fight another infestation of the dreaded boll weevil. His one innovation was the reintroduction of strawberry farming to the ridge. Turning his attention to the prairie, he battled anthrax in the rice section. His successes were apparently limited, however, and he lamented the failure of a "demonstration in sheep raising" which was "started in the spring but abandoned when the demonstrator 'laid-by' his crop, sold it, and went west." Even more serious, he appears to have incurred the outright disapproval of some farmers. In July 1922 County Judge W. W. Warren, a Tyronza businessman-planter and the first county judge from the delta, received "several large petitions" against his quorum court's decision to retain the farm agent. Warren was defeated later that year in a reelection bid, and his successor's quorum court refused to fund the extension program. Despite a telegram from the governor urging the county to retain the county agent, the program was discontinued for three years.[12]

A clue to the opposition to the county agent emerges from Sargent's annual report to his superiors. He included a narrative section on "successes" in the boy's club, and in two of the three success stories, Sargent referred to the resistance of the boys' fathers. Carl Reddit, the sixteen-year-old son of ridge farmer J. W. Reddit, was not allowed to attend a "short course" in Jonesboro because his "father needed him in the fields." The boy borrowed the money to pay for a replacement, but none could be found, and he missed the opportunity to learn how to maintain and repair tractors. Mabern Powell, son of ridge farmer Robert L. Powell, participated in club work over his father's protest. Only his mother's support enabled him to take part.[13]

The two fathers involved both reported personal property above the county average. J. W. Reddit owned a 176-acre farm on the ridge and claimed $309 in personal property in 1920. Robert Powell owned no acreage but claimed $334 in personal property in 1920, with $267 of it in livestock; he apparently lived in Harrisburg and rented acreage in the countryside that he farmed. He may have been one of the violators of Harrisburg's fence law in 1922. After the limited stock law was passed that year, many town residents tore "their fences away," believing that they no longer needed to protect their yards and home gardens, but Mayor Benjamin Harris had to impose "fines on people for letting their stock run out in town."[14]

These men were precisely the farmers that the extension service hoped to reach, yet they resented the intrusion of the program, and it is not surprising that they were among those who protested its continuation in 1922. The two boys were facing not only generational conflicts between fathers and sons but also the older generation's unwillingness to entertain new ideas. Yet even had the county farmers embraced the extension service program, it would not have been the panacea some believed. Perhaps the protestors intuitively understood the limitations inherent in the program.

By the time the extension service returned in 1926, the new county farm agent, W. D. Ezell, and the new home demonstration agent had to "sell our work to the people before we could do anything else." It was county agent Ezell who complained about the open range in Poinsett County, and he was the first agent to devote much time to efforts off the ridge. He continued the programs begun by his predecessors, but he focused considerable attention on helping rice farmers fight an infestation of insects caused by improper drainage. He did not keep track of the farmers' self-help groups, noting "The rice association and cotton association have several members, but I do not know how much business they did."[15]

Nothing in Ezell's 1926 annual report indicated that he had done any work in the delta. That would change dramatically the next year. The flood of 1927 required him to devote fifty days "to miscelaneous work with [the] Red Cross in helping to get food, feed, and seed, in addition to time helping farmers with draining and planting problems." The flood, which extended for almost one thousand miles from Cairo, Illinois, to the Gulf of Mexico, hit hard in Poinsett County. After two smaller floods in February and March, a break in the St. Francis River levee in early May caused more flooding in the county's delta. In fact, Poinsett County was struck more severely than any other county in the state. The flood rendered homeless over "2000 sharecroppers with families of two to five in a family" and put more than fifteen square miles of the delta under water. Over 200,000 acres were flooded during the worst phase, and 27,845 people registered in the county's Red Cross camps. Months later 127,381 acres of cropland remained flooded, and 4,905 people remained in three Red Cross camps.[16]

The distribution of food and feed in the county appears to have been accomplished with little controversy. The Red Cross cared for 3,470 whites and 1,435 blacks in its three refugee camps. Outside the camps it provided food to over 12,000 people, almost a third of whom were black sharecroppers. In all, Poinsett County's flood victims received $62,932.89 worth of food. Little loss of livestock occurred in the county, mostly because residents had enough warning to remove their livestock to Crowley's Ridge before the flood reached major proportions, but the Red Cross provided feed for the remaining stock. The food for the flood victims, the feed for the livestock, and the seed for the planting of crops were all funneled through the county agent and the local Red Cross committee. The chairman of the county committee, E. L. Pierce, was a Marked Tree businessman-planter. Prominent planter Charles Albright of Truman was vice chairman, and banker-planter Jake Schonberger of Marked Tree was secretary-treasurer. They supervised the distribution of relief supplies, and they looked to the local town committees to administer the program. In Marked Tree businessman John Brunner was the chairman of the town committee. J. G. Waskom, a Marked Tree attorney, was the vice chairman, and Harry Ritter, the second son of Ernest Ritter, was the secretary. Marked Tree was the focal point for the distribution of relief, and those who controlled the mechanism included of the town's most prominent citizens. The Red Cross's efforts, therefore, did not challenge the authority of those who traditionally held power within the delta.[17]

The extension of health services to flood victims, however, was delayed when the Red Cross nurse encountered resistance from local doctors who had been appointed "the local county health officers" during the emergency. According to Geraldine Green Orrell, the doctors resented the "interference" of health nurses who invaded their domain. The misunderstanding delayed the distribution of quinine intended to prevent an epidemic of malaria until Orrell "finally got the health officers to agree to giving prescriptions for those having chills, I got a druggist here (in Lepanto) to agree to handle the quinine, and I went to Marked Tree to the Red Cross nurse and got the quinine and gave it to the druggist."[18] When the refugee camp in Marked Tree closed in June, the agency determined to extend "nursing follow up work for 120 days." A nurse Laughlin was dispatched to the county but was replaced by yet another nurse after a month. By late summer the county health officers were in complete accord with the new nurse, Annie Gabriel, on her program to examine the children who reported for school in September.[19]

Despite resistance from the local doctors, the National Red Cross, in conjunction with the Rockefeller Foundation, established a county health unit in Poinsett in July. Through this agency hundreds of children were examined. The clinic reported upon its closing eighteen months later in December 1928 that its nurses had visited 3,300 children in fifty-three schools and had paid 800 home visits to 1,300 of these children. Twenty-eight hundred (85 percent) "had some physical defect," but the clinic reported a "total correction" of their problems in only 129 children.[20] Significantly, the cost of the operations was paid for by the children's families, and the names of the children operated on at the clinics were those of some of the more prosperous citizens in the delta.

The activities of the Red Cross shook the foundations of the plantation system in the delta, where for the first time sharecroppers were given aid by an agency rather than by their employers. But in some of the larger camps, the Red Cross avoided conflict with the delta planters by holding tenants and sharecroppers in the refugee camps until their respective planters signed them out. There is no evidence this occurred in Poinsett County, but it is clear that the agency perceived a narrow role for itself in the region. Rather than challenging planter authority, it demonstrated to the planters that aid from an outside agency need not represent a threat to their hegemony. Planters or those connected to them socially, politically, and/or economically formed and led the county Red Cross committees.[21]

County farm agent Ezell was perhaps overwhelmed by the responsibilities he shouldered during the flood. He helped coordinate both rescue and recovery operations, but by September 1927 he had had enough and departed. He was not replaced until mid-1928. The fifth farm agent since 1916 to serve Poinsett County, H. S. Hinson, arrived in June after the spring session of the county court finally voted to fund the extension program again. He remained in the county only five months, however, and his report reads like a reassessment of the extension program there. Highly critical of the ridge farmers, he suggested that "the only solution for the farm will have to come thru the boys and girls, due to the indifference and lack of cooperative spirit among most of the people."[22] He lauded the passage of the stock law and outlined an approach to be used in fencing and the development of permanent pastures, and he worked among the dairy farmers in immunizing their herds against tuberculosis.

Although Hinson focused more attention on the delta than had any previous agent, holding demonstrations on the use of fertilizers there, his report, like those of his predecessors, reflected a preoccupation with the ridge.[23] A. Raybon Sullivant, who replaced Hinson in 1929, departed from the previous agents and, although maintaining the headquarters in Harrisburg, oriented his appeal to the delta. Sullivant was the county's third farm agent in three years and the first since Dent to remain in the county for more than a year and a half. He stayed until the end of 1933 and was instrumental in securing the support of important delta businessmen and planters.

Of the four organizations in the county in 1929 that actively supported Sullivant's extension work, two were centered in the delta and a third included delta members. With businessman T. C. Brigance and businessman-planter L. V. Ritter, son of the now deceased Ernest Ritter and head of E. Ritter and Company, as president and secretary, the Marked Tree Rotary Club voted unanimously to support the work of the 4-H club "within the trade territory of Marked Tree." Sullivant reported that the Tyronza Business Men's Club was "in strong accord with the extension program and have cooperated in carrying out the county plan of work."[24] J. A. Emrich, the former business associate of the now deceased Louis Ritter, Ernest Ritter's brother, headed that club. Emrich had become one of the most powerful businessmen-planters in the delta, owning thousands of acres in southeastern Poinsett County, operating a huge commissary in Tyronza, and owning the bank there.[25] W. W. Warren, the county judge in 1922 who had supported the extension service, was secretary of the Tyronza Business Men's Club, and he too operated a plantation in the delta. The third organization supporting the farm agent, the Poinsett County Bankers' Association, represented all the banks in the county. J. L. Dean, the cashier of the Bank of Tyronza, was the association president. According to Sullivant, the bankers were primarily interested in promoting the development of agriculture in the county, and they devoted special attention to how agricultural problems could be "bettered by efficient extension methods." The Harrisburg Chamber of Commerce was the only organization supporting Sullivant that included members entirely from outside the delta. They were specifically interested in improving the dairy program on the ridge.[26]

Sullivant became the county agent at precisely the time that conditions in the county worsened dramatically. His association with bankers probably accounted for his participation in helping delta communities establish agricultural credit associations in the late 1920s. Arkansas led the farm belt in the creation of such organizations.[27] Like much of the extension program, however, the efforts of county farm agents to help farmers devise sources of credit were limited. The mortgage debt of Arkansas farmers increased by over 25 percent between 1900 and 1929.[28] The continuing decline in the price of cotton and other agricultural commodities made it increasingly difficult to pay off this mortgage debt and contributed to the record number of foreclosures that typified the 1920s and early 1930s.

Aside from working closely with delta planters and introducing agricultural credit associations in the delta, Sullivant preached diversification as a response to overproduction, but at the same time he introduced innovations such as the use of a nitrogen-based fertilizer to increase cotton yields, reporting in 1929 that the "most important thing the cotton farmers needed was increased yields." He also introduced a better-quality seed and made strides in persuading them to shift from short-staple to long-staple cotton. Sullivant's efforts to convince farmers in the delta to substitute another crop for cotton proved to be unsuccessful; although he encouraged them to plant more corn, "farmers in the St. Francis delta section think they can grow cotton and ship in the corn cheaper than they can grow it." On the other hand, the radish-growing enterprises of men like L. V. Ritter in the delta, seemed ideal to him because they did not interfere with cotton production. According to Sullivant, "The radishes were harvested in time to plant a crop of cotton on the same land without any material delay."[29]

Sullivant's activities in the prairie and on the ridge were similar. He focused on helping rice farmers increase their yields, and although he did convince some to rotate rice with soybeans, he stressed that in doing so they increased their rice yields. On the ridge, where farmers were struggling with soil infertility and erosion, he demonstrated better pruning and spraying methods in order to increase the production of peaches. He encouraged dairy farmers to buy better sires and lauded the countywide stock law that forced them to ask for advice on pastures.[30] Their new reliance on feed for their dairy herds was likely to increase the production of milk, even as it made them less self-sufficient and more vulnerable to fluctuations in the price of feed during a period when they were least able to absorb the costs.[31]

At the end of his first year in the county, Sullivant reported that "the cooperation and reception given extension work has thus far been unsurpassed in this county."[32] Indeed, never again would the county quorum court fail to appropriate funds for the county farm agent's salary. Although much of Sullivant's success resulted from his association with the county's most prominent planters and businessmen in the delta, it also reflected the desperate situation, which prompted planters and farmers all over the county to heed advice some of them had previously spurned. By the time Sullivant arrived, landowners had been unable to pay their drainage, road, and school taxes for two successive years. The price of cotton continued to decline and that of other commodities was sinking to new lows.

Sullivant devoted most of his efforts in 1930, however, to a severe drought which struck Poinsett County in the summer and ruined what had promised to be a fine crop. By fall the cotton crop had almost totally failed. Many planters were on the verge of bankruptcy and could not provide for their tenants and sharecroppers. The local Red Cross committees declared themselves completely unable to meet the needs and appealed to the national organization for additional funds. The thirty-seven men and three women who met with the Red

Cross investigator in Marked Tree in August reported that the "chapter couldn't raise $500 under present conditions."[33] The town's economy, dependent as it was upon the health of the farming sector, was faltering. In addition, a downturn in the forest products industry had led Chapman and Dewey to close two of its mills and put hundreds of men out of work. An August 1930 assessment of the situation in Marked Tree reported that "few families have any resources to carry them thru the winter" and added that the short crop of cotton would give few employment in the fields. The resources of local planters and merchants were severely taxed. They expected to harvest only 30 to 40 percent of the cotton crop and only 20 to 25 percent of the corn crop. The hay crop, the gardens, and the pastures had failed completely. Many farmers had begun "the sacrifice of stock," marketing cattle and hogs at below market prices because of the lack of feed. While only two hundred families had approached the local Red Cross committee, the investigator reported that nine hundred families needed help immediately. He added that "all of the farmers are hard hit and it's the exception to find one that isn't mortgaged to the hilt but the tenant farmers and sharecroppers will need help first."[34]

The approach adopted by local Red Cross chapters in the Arkansas delta, however, resulted in controversy. The leaders of the chapter in Marked Tree included John E. Buxton, a local planter-merchant, J. S. Schonberger, another planter, and Harry Ritter, an officer in E. Ritter and Company. What evolved over the winter of 1930–31 was a "works program" that tied the receipt of Red Cross rations to the willingness of the able-bodied to work. Red Cross officials were fully aware of this arrangement. One Red Cross representative wrote that "all seem to agree that a mere feeding program involving large numbers of drought sufferers would be unwise and would only tend to pauperize people and perpetuate the problems in these states." The local chapters made no attempt to conceal the works program. A Mississippi County chapter in February 1931 distributed a leaflet explicitly advertising jobs for those who wanted Red Cross rations. But an article in *Labor* that same month created a nationwide storm of protest, for the author declared that people were being forced to work for below-normal wages in return for Red Cross food. The journalist held that those in charge of local Red Cross chapters in the South used the agency to supply their own tenants and sharecroppers or those of their friends. He referred specifically to Poinsett County where he said planters required drought sufferers to work "clearing and grubbing" acreage for planters in return for a slip entitling them to one dollar's worth of Red Cross supplies for one day's work. He claimed planters normally paid two to three times that for the same work.[35]

In response to the ensuing controversy, the Red Cross asked local chapters to report on their activities. John Buxton, chairman of the Marked Tree chapter, defended the local works program, holding that it was only natural for certain leading citizens to be in charge. He pointed out that local donors to the Red Cross fund were straining their own resources to meet the needs of the drought sufferers and that it was only right that they should determine how the funds were to be dispersed. He denied the allegation that some prominent citizens had appropriated control of the local agency in order to further their own interests and insisted that the work being done consisted primarily of cleaning the town's drainage ditches and clearing church and city lots. He held that the program actually raised morale among most of the people who did not want a "hand out."[36]

Approaching the issue from the standpoint of class, Buxton argued that the works program eliminated "the professional bum" and "the street corner daily gatherings of the hundreds who had nothing to do and would group up and talk about how badly they were being treated and that some one had gotten more than they had and all the other things that idleness and discontent can breed." He went even further and implied that the underlying problem with the area's economy was the idleness of these people.[37]

E. M. Perry, the Marked Tree enumerator for the 1925 agricultural census and an honorary member of the A.F. and A.M., a fraternal organization, challenged Buxton's account. First, he argued that once it was clear that Red Cross supplies were forthcoming, certain "plantation owners, or managers, the larger merchants, the bankers, the organized politicians" met in order to discuss gaining control of those supplies. According to Perry, "One merchant in Lepanto said it cost him $58 to get the man he wanted on the committee as feed distributor." He declared that some poor farmers were refused feed for their livestock while "great loads were taken to some large plantations." The relief committee, meanwhile, put people to work not merely on community projects but on some of the plantations as well. He indicated that tenants and sharecroppers were given a slip for a day's work redeemable for Red Cross supplies but that many merchants shorted the orders. Merchants and planters, said Perry, conspired to reduce the debt of the planters by passing it along to the respective planter's tenants and sharecroppers. He concluded that rather than raising morale as Buxton claimed, the program crushed and subdued the spirit of the drought sufferers.[38]

The public outcry following the revelations in *Labor*, soon led the Red Cross to order the local chapters to suspend the works programs. Buxton of Marked Tree wrote to Red Cross officials indicating that his chapter had complied and bitterly complaining about the grave mistake they were making. He repeated his earlier assertion "that the effect of the dole on a person is terrific. They become lazy, dissatisfied and quarrelsome and very much opposed to resuming labor of any kind."[39] Clearly, the works program, in addition to paying off the debts of certain planters and satisfying certain merchants, was an attempt to minimize the dissent of the unemployed and to maintain their dependence upon local employers.

Although the Red Cross officials interfered with the way local relief was being administered, they did so only when bad publicity threatened to disrupt their own sources of revenue. They did not challenge the position of those who were in control of the community chapters, and community leader John Buxton did not blame the Red Cross for the problems encountered over the works program but directed his anger at those who had exposed the program to criticism and against the unemployed men who haunted the streets of Marked Tree. Although the works program was disbanded, the distribution of food and feed in the county remained in the hands of the local elite.[40]

County agent Sullivant was silent on this controversy, and while his annual reports for 1930 and 1931 reflect his attention to the problems caused by the drought—he placed a greater emphasis on food crops in 1931—he continued to pursue his preexisting program. He encouraged the use of fertilizers to increase cotton yields and pressed farmers to use soil-building crops such as soybeans and hairy vetch. With the help of businessmen from Marked Tree, including L. V. Ritter and J. D. DuBard, Sullivant worked to establish a local branch office of the Mid-South Cotton Growers Cooperative Association, an organization designed to promote cooperative marketing. Two new business organizations, the Truman Lions Club and the Lepanto Chamber of Commerce, now joined in with those which had previously supported the extension service. These organizations, as well as four others that had supported the agent's work since 1929, "publicly and privately endorsed" the extension service. Sullivant especially praised the efforts of delta bankers, noting that "their support and influence has been invaluable to extension work in Poinsett County."[41]

In response to the bank failures brought on in part by the drought of 1930, other sources of credit available to delta farmers in 1931 opened up. Three banks, including both the Harrisburg banks, closed after the drought of 1930–31, but a federal emergency loan program was organized and helped farmers to finance the next year's crop. The Federal Feed, Seed, and Food Loan program which Sullivant described as "almost literally a life saver," extended credit to farmers, and the Red Cross provided hundreds with seeds. In addition, three new agricultural credit associations were formed when the state made available funds through the Intermediate Credit Bank. Two agricultural credit associations, one of them dating back to 1929, already existed in the county. A Red Cross–sponsored agency, the Arkansas Farm Credit Association, which had begun extending loans after the flood of 1927, also continued to answer some of the credit needs of Poinsett County farmers. Early in the process 122 Poinsett County farmers received $20,250 in loans from the association. By May 1931, 602 farmers had received loans of over $130,000. The officers of the local Agricultural Credit Association included J. A. Emrich, the Tyronza banker-planter, and M. T. Dilatush, the president of the Marked Tree Farmers and Merchants Bank and Trust Company. The loans went almost exclusively to landowners. Farmers who had some form of collateral, especially real property, turned to these agencies and to the remaining three banks (two located in the delta and one in the prairie). The county farm agent reported that while he had helped only 13 farmers meet their credit needs in 1930, the next year he helped 585 farmers obtain credit.[42]

Sullivant clearly understood that the only way to have an effective program in a county like Poinsett was to align with the delta elite. His predecessors' difficulties in appealing to ridge farmers stemmed from the nature of the agricultural economy on the ridge, which was more resistant to the extension agent's program. The enterprises of the farmers there bore greater resemblance to subsistence farming than they did to the commercial plantations of the delta. The rejection to the stock law on the ridge sprang from the reluctance of farmers there to engage in costly improvements to their property and to purchase feed for their stock.

While it took until 1929 for a farm agent assigned to Poinsett County to recognize the importance of appealing to delta planters, Red Cross officials understood at once that they needed the support of community leaders in the delta in order to render aid to flood victims. Planters appropriated control over relief supplies, but their behavior during the drought created a controversy which led to restrictions being placed upon them. Those restrictions, however, did not shake their control over the relief supplies.

For twelve years before the implementation of New Deal programs, Poinsett County farmers faced a series of disasters that prepared them to accept a more innovative solution to the problems they encountered. During the disastrous 1920s foreclosures became commonplace. Paying taxes became almost impossible. The delta elite turned to the county farm agent and to the Red Cross in addressing the problems confronting them and broke down certain barriers that might have inhibited a willingness to cooperate with government agencies in the next decade. At the same time, as the controversy over the works program associated with Red Cross aid during the drought indicates, while planters were being schooled in how to appropriate control over government programs, tenants and sharecroppers were learning how to mount a challenge against what some considered unfair practices. They doubtless observed that negative national publicity helped expose the Red Cross works program and ensured its termination.

7

THE NEW DEAL AND
THE OLD PLANTATION

Four thousand people flocked to the delta town of Lepanto in October 1936 to witness the annual Terrapin Derby, an event sponsored by the local merchants and appreciated by those with a taste for gambling. "Roosevelt" and "Landon" were popular "in betting circles," but neither finished in the money. Although "New Deal" was another favorite, it was perhaps most appropriate that a turtle named "Hard Work" won the race.[1] The implementation of New Deal programs in Poinsett County necessitated concerted effort. Local leaders directed the administration of the agricultural programs and the distribution of relief. They determined each farmer's allotment of cotton and whether he was qualified to receive a production credit loan. They fashioned a relief program reminiscent of their experience with the distribution of Red Cross goods during the drought of 1930–31. Only the elderly, the infirm, and the mothers of young children whose husbands had deserted them received direct relief. Able-bodied men and women were expected to work in return for relief supplies that were often distributed through plantation commissaries or local merchants. "Hard work" was, indeed, the hallmark of the county's so-called New Deal.

The New Deal, moreover, favored the delta political elite. New Deal programs funneled more money into that region than it did into the prairie and ridge combined. The cotton allotment was by far the most important single source of federal funds, and since the delta produced most of the cotton in the county, most of those federal funds ended up in the hands of delta planters. The recovery of the delta plantation system stimulated a mild recovery in the delta towns and encouraged local businessmen to sponsor works programs for civic improvements. In order to take advantage of the federal works programs, communities had to provide a percentage of the funds required to finance improvements. The ridge and prairie were less able to accommodate that requirement than was the delta. The New Deal thus helped delta elites disproportionately and underwrote their continuing economic dominance in the county.[2]

The problems confronting Poinsett County upon Franklin Roosevelt's election in 1932 were enormous. The steady decline in agricultural prices during the 1920s coupled with the natural disasters that hit the county in 1927 and again in 1930–31 devastated the local economy. Almost no one was left untouched. Planters found themselves on the verge of bankruptcy and unable to supply their tenants and sharecroppers. Planters and small farmers alike faced foreclosure, and many of the landless tenants and sharecroppers feared starvation. Things were little better in town. The banks and businesses were dependent upon the health of the farm economy, and by 1932 both Harrisburg banks had closed. Only one bank was left in the prairie, and only one bank each was left in four of the principal delta towns. Merchants, too, faced bankruptcy, and like some bankers, many foreclosed on local farmers they had furnished for years in order to save themselves. Men thrown out of work in the countryside flocked into the towns and stood on the street corners with men who could no longer find work in the mills. Unemployment became a major concern, second only to the problems in the agricultural arena.[3]

Upon taking office in early 1933, Roosevelt sought to address the problems confronted by farmers and the unemployed, and county leaders responded enthusiastically. The cutting edge of the president's farm program was the Agricultural Adjustment Administration (AAA), an agency designed to raise the price of farm products by limiting production. But farmers needed immediate relief, on the one hand, and the incentive to participate in the production control programs, on the other. Thus, Roosevelt and his advisers devised a plan whereby farmers "rented" acres to the government in return for a cash payment and then pledged not to plant the rented acres in a specific commodity crop. Cotton was by far the leading commodity crop in Poinsett County, and the AAA program there concentrated primarily on the withdrawal of cotton acreage. Once the crop was in, farmers also received "parity" payments in order to bring the return on their investment up to an acceptable level. But the heart of the AAA was the crop reduction program.[4]

Cotton farmers overwhelmingly supported the crop reduction program; rice farmers in the prairie were less enthusiastic. They, too, had the option of renting acres to the government, but not every rice farmer chose to participate in the program. The AAA program for rice production had a price fixing component that created greater problems than the ones it sought to remedy. By 1938 rice farmers in Poinsett County were unhappy enough to vote down participation in the government's rice allotment program. Ridge farmers, meanwhile, were largely uninvited to the New Deal banquet. Only those few who grew cotton along the edge of the ridge participated substantially in the crop reduction program. The dairy, poultry, and orchard farmers of the ridge found themselves without a specific New Deal plan. A corn-hog campaign of modest proportions did have slight impact, but it was insignificant compared to the cotton program.[5]

Because the cotton program was the most important, influential, and financially rewarding plan in the county, and because the overwhelming majority of cotton producers were in the delta, the New Deal's impact on the county's agricultural recovery was felt disproportionately there. The delta elite, meanwhile, dominated the reduction control committees and assumed positions of even greater authority and power than before. They also organized and directed the production credit associations that were an important complement to the AAA program. Designed to provide production loans at low interest to farmers in need of short-term credit, the production credit associations were controlled by local bankers and businessmen, and only those who chose to participate in the reduction control program were allowed to join the associations (for a nominal fee).[6]

Long-term loans were rarely available to farmers or planters, at least not through any New Deal agencies. Those wishing to mortgage their land turned to the traditional local and nonlocal sources of credit to answer their needs. Banks, including the Federal Land Bank of St. Louis, provided some loans, but mortgage and insurance companies also marketed an increasing number of loans in the delta during this period. Except insofar as expanding the funds available through the Federal Land Bank, the Roosevelt administration did not address the long-term credit needs of farmers and planters. Only large farm companies that had incorporated were eligible for Reconstruction Finance Corporation (RFC) loans. None were received by Poinsett County enterprises. But Drainage District Seven, the ambitious drainage enterprise initiated in

1917 in the delta, received a $2 million loan which enabled it to avoid bankruptcy.[7]

Drainage District Seven was of considerable importance to farmers and planters alike in the delta, for its drainage enterprises, in conjunction with those of the St. Francis Levee District, were all that kept the delta from returning to swampland. But the depression in the agricultural sector in the 1920s, the flood of 1927, and the drought in 1930–31 ruined the local economy, and as a result farmers and planters routinely failed to pay their drainage taxes. By 1933 the drainage district was on the verge of bankruptcy, and only the Reconstruction Finance Corporation loan saved the enterprise. Again, delta elites played the most important roles. The drainage district commissioners who sought the RFC loan were delta men and included Curtis Dewey of Chapman and Dewey Land and Lumber Company. L. V. Ritter of E. Ritter and Company served as the trustee for the loan.[8] Ironically, the salvation of Drainage District Seven and the completion of a government-sponsored project to coordinate the drainage enterprises of the counties in the St. Francis River Basin helped to counteract one of the principal goals of the AAA. By 1940 an additional 105,000 acres were brought into cultivation in Poinsett County alone. Moreover, the AAA program had not substantially reduced the number of acres committed to cotton production. The crop reduction program, therefore, was a failure. But the New Deal had placed millions of dollars into the hands of the delta's planters and farmers and contributed to the recovery of the local economy. Roosevelt and the New Deal, consequently, continued to enjoy the approbation of the elite.[9]

The county agricultural extension agent was a central figure in the implementation of the crop reduction program. Indeed, the local farm agent's work increased considerably because of his expanded role in administering the New Deal agricultural programs. His staff was enlarged to help carry his additional responsibilities, and the state extension service instructed him to rely on community leaders to augment the programs. T. Roy Reid, the director of the Arkansas Extension Agency, was an enthusiastic supporter of Roosevelt's agricultural policies and only too willing to comply with the suggestions of the undersecretary of agriculture, Rexford G. Tugwell, in June 1933. Tugwell wrote Reid that "outstanding men in business and banking locally, and the recognized leaders among farmers must be encouraged to assist in the program. . . .

Those who believe in the program should be asked to assume the responsibilities that their positions in their communities place upon them."[10]

A. Raybon Sullivant, the county farm agent in 1933, drew his committeemen from among those men he had come into contact with in his work. When it became his responsibility to administer the plow-up campaign, he turned to the men who had assisted him with the Red Cross relief in 1931. Sullivant's annual County Plan of Work Committee, a feature of the farm agent's agenda, included several men who went on to become county or township committee members representing the New Deal agencies, including L. Hogue, L. J. Reynolds, J. H. Joiner, L. V. Ritter, J. A. Emrich, and T. J. Bennett. All of these men helped orchestrate the 1933 AAA plow-up campaign. At least two were listed as "demonstrators," farmers who grew particular crops under the guidance of the extension agent during 1933. Other demonstrators included B. F. Taylor, W. L. Cook, W. T. Walker, E. T. Byrn, H. Norcross, W. J. Roberts, R. A. Ziegenhorn, and H. F. Sloan. All were involved in the plow-up of 1933 and/or the cotton reduction campaigns of 1934 and 1935.[11]

R. L. McGill, who replaced Sullivant as Poinsett County's farm agent in 1934, inherited the relationships Sullivant had established with the county elite and appointed many of the same men to the agricultural reduction program's county and township committees. He was happy to allow the local committeemen to assume a large part of the responsibility for fashioning the county's crop reduction program. The committeemen's efforts "permitted the Agent more time" and freed him from "detail work." Farm agent McGill, new to the county in 1934, regarded his duties principally as administrative, and he generally relied on the committeemen to work out the specifics of the crop reduction programs.[12]

The choice of committeemen in Poinsett County's plow-up may help account for the remarkable rate of compliance achieved there. In 1933 farm agent Sullivant reported that 607 cotton producers in the county "signed contracts pledging twenty thousand six hundred and three acres to cotton for destruction." He calculated that county farmers "received $661,040.07 more than they would have received had there not been such a campaign or had they not participated." He lauded their enthusiasm for the program and said that they "displayed full faith in their government and those administering the plan in their county. Not a single cotton producer in this county had to be forced to comply with his contract."[13]

The county agricultural committee, which supervised the administration of the program, included some of the most influential men in the county. Two of them had served on Sullivant's County Plan of Work Committee, and all had participated in one or more of his demonstrations. Because very little cotton was grown in the prairie or on the ridge, four of the five members of that committee were prominent delta planters (table 7.1). L. V. Ritter of Marked Tree had weathered the depression of the 1920s and early 1930s well. Planter-banker-merchant J. A. Emrich of Tyronza had watched his bank go under in 1931, but he had paid all his depositors from his own personal finances and continued to operate the largest mercantile establishment in Tyronza. The only member from outside the delta was ridge farmer and banker J. H. Joiner, a longtime resident of Crowley's Ridge and a prominent citizen of White Hall, a community a few miles south of Harrisburg.[14]

The farm agent advised the county committee, which in turn directed the township committees. The agent selected three citizens from

Table 7.1. County committee for 1933 plow-up campaign

Committeeman	Residence	Acres owned in 1930
L. V. Ritter Marked Tree	Delta	5,001
H. F. Sloan Marked Tree	Delta	1,618
J. A. Emrich Tyronza	Delta	1,976
M. T. Byrn Truman	Delta	257
J. H. Joiner White Hall	Ridge	148

Sources: *Modern News*, June 30, 1933; Poinsett County Real Property Records, 1930, 1940, Poinsett County Courthouse, Harrisburg, Ark.

each of the county's eleven townships to organize the program on the township level. The committeemen typically approached farmers within their respective townships and urged them to sign contracts. With the farm agent's help, they determined how many acres of cotton each farmer would be required to destroy, drew up the contracts, and secured the farmer's signature. The township committee, working in conjunction with the farm agent and the county committee, then supervised the plow-up of the cotton. The township committees in the delta had the greatest number of contracts to handle. All of the members of the delta committees located in the extreme eastern sector of the delta were landowners, and at least three of the nine were prominent planters. Dr. B. F. Taylor of Marked Tree owned almost 2,000 acres within the delta and served on the Little River Township committee. Hiram Norcross, who owned 2,564 acres of valuable bottomland in 1930 and 6,221 acres in 1940, served on the Tyronza Township committee. He would achieve notoriety in 1934 when the Southern Tenant Farmers' Union charged him with evictions and with refusing to share crop subsidy payments with his tenants and sharecroppers. Another committeeman who was challenged by the STFU, planter H. H. Howington of Greenwood Township, owned 355 acres in 1930 and 2,359 acres by 1940.[15]

The configuration of the ridge committees, however, was somewhat different. Only one ridge committeeman owned more than 500 acres, even ridge landowner S. A. Clements, who owned 734 acres in 1930 and 851 acres in 1940, was no match for Norcross, Taylor, or Howington in the delta. The prairie committeemen more nearly resembled their delta counterparts. The real property records for the prairie in 1930 are not complete, but it is clear from those available that Reynolds, Hogue, C. O. Wofford, Chester Senteney, and R. A. Zeigenhorn were prominent prairie rice farmers. L. J. Reynolds was on the board of directors of the Arkansas Rice Growers Association.[16] Louis Hogue was one of the founders of the Weiner Bank and exercised great influence in the prairie. The 1940 records demonstrate that at least three of the nine men who served on the 1933 plow-up campaign owned over one thousand acres. The same was true in the delta. The main difference between the two sections, aside from the value of the land, was that two small landowners served in the priarie, and their acreage was less than half of that of the smallest landholder in the delta, W. T. Walker with 263 acres.[17]

Whatever differences may have existed among the county and township committeemen, they had considerable power and influence within their own sections. The farm agent relied upon them to put the plow-up campaign into operation. The acting chief of the AAA cotton section in Washington, E. A. Miller, directed the district agents to "contact county agents, county and local committeemen to ascertain . . . whether or not cotton has been immediately and completely destroyed."[18] It was assumed, therefore, that the committeemen would enforce compliance.

The reliance upon the county and township committeemen extended into the crop reduction program that took shape in 1934. Yet the legal office of the AAA was called upon almost immediately to settle disputes between county committees and farmers, and one such dispute arose in Mississippi County, bordering Poinsett to the east. The complaints of the officers of Lee Wilson & Company of Wilson in Mississippi County led the AAA's legal department to issue a memorandum in an effort to establish guidelines for AAA officials in dealing with similar situations. The document detailed the responsibility and authority of the county committeemen and revealed that they had great discretion in determining each farmer's allotment. Although they were to use the formula for determining allotments provided by the AAA rather than devising one of their own, they were required to "examine all data in the applications" and to "correct and adjust figures" if they believed them to be in error. The county committee, moreover, was to act as a "quasi-judicial board" if any producer complained about his allotment. The findings of the county committee were regarded as final. Jack L. Levy, an AAA official in the legal department, wrote that "an application that is approved by the county committee will necessarily have to be accepted by the State allotment board as representing the opinion of persons in the county who are best qualified to judge the equitableness of any claim."[19]

Poinsett County's committee, organized in 1934, closely resembled the committee that directed the 1933 plow-up campaign. The committee helped shape agricultural policy in the county for the next decade. Although three rather than five men served on the crop reduction committee, two of them had acted as county committeemen during the plow-up campaign, and two out of the three committeemen were delta planters.[20]

Many of the cotton reduction program township committeemen of 1934 had served on the 1933 plow-up campaign, and most of them would go on to be committeemen in 1935. The 1934 cotton reduction committee, however, was reduced in number, perhaps because there were fewer cases in certain sections. For example, the prairie committeemen numbered five rather than nine in both 1934 and 1935. The ridge committees had seven rather than nine members in those years. Finally, the number of members in the delta committees was reduced from 15 to 13.[21]

The new 1935 committeemen were not dissimilar from their predecessors. John T. McCalla replaced C. B. McGee of Harrisburg and differed in no significant respects. McCalla, like McGee, was a small landowner on the ridge. In 1930 he held 140 acres valued at $550; in 1940 he owned 157 acres worth $700. B. P. Biggs, also of the ridge, was replaced by J. W. Coker, who owned 59 acres in both 1930 and 1940 valued at $500 in both years. Significantly, the delta's committeemen served almost without interruption during the first years of the New Deal's agricultural programs. While G. C. Jernigan of Lepanto replaced Howington and Jess Wolfe replaced Joe Goodin in 1934, the two men who did not serve in 1935, Bartholomew and Bennett, were not replaced. The delta committeemen, therefore, presented a solid front to cotton farmers in that region for at least a three-year period.[22]

Some of the men involved in the Agricultural Adjustment Administration program in Poinsett County were also connected to the Production Credit Association (PCA), which began operation in Poinsett County in 1934. The PCA program was administered out of the Farm Credit Administration (FCA), and compliance with the AAA program was required. The county committees were charged with responsibility for ascertaining whether applicants to the loan program were in compliance, but a mechanism was also established whereby the Washington office of the FCA rechecked the list of applicants against a list of "cooperating farmers" supplied by the AAA. Officials of the FCA, however, were careful to establish with the AAA that only recipients of short-term production loans like those administered by the PCA would be required to comply with the crop reduction program. Borrowers of long-term loans administered by the Federal Land Bank were excluded from compliance.[23]

Another program that included prominent local men was the Farm Debt Adjustment Association (FDA), which actually began as a state agency. It did not itself make loans but was designed to assist farmers in developing a comprehensive program for recovering from the burden of debt they had incurred during the years before the New Deal programs provided cash and cheaper credit. County committees assisted farmers in refinancing existing debts at lower interest rates and often stepped in just in time to prevent foreclosures. Committees served "without remuneration for either their time or travel expenses" until the federal government's Resettlement Administration assumed administrative responsibility for the program. Until that time the FDA was administered by the state Agricultural Advisory Board.[24] The need for such an agency was obvious. The federal government was offering loans through the PCA and the Federal Land Bank. Emergency crop loans were also available through the AAA, and many banks and other lending agencies were recovering from the worst years of the depression and again marketing loans. The FDA helped farmers walk through the maze of lending opportunities and adjust their debts accordingly.

Some of the men who served on the county's PCA or FDA had helped organize the plow-up campaign and/or the cotton reduction program. J. A. Emrich, for example, one of the county committeemen in the plow-up campaign, became chairman of the Poinsett County Farm Debt Adjustment Committee.[25] L. V. Ritter, also a county committeeman in the plow-up campaign, helped found the county's Production Credit Association. Louis Hogue and J. H. Prestidge, both of whom served as township committeemen in the cotton reduction program, were on the board of directors of the PCA along with Ritter.[26] All four of these men had developed a relationship with the farm agent before the advent of the New Deal programs: Ritter, Emrich, and Hogue were all businessmen who were members of community organizations that supported the work of the extension agent, and Ritter, Emrich, and Prestidge were demonstrators mentioned in farm agent annual reports. A. B. Caplinger and H. P. Maddox, both of whom had served as county judge in the 1920s, were members of the FDA and PCA, respectively.[27]

Clearly, those individuals with the requisite connections were able to control or heavily influence the implementation of New Deal programs. Thus, bankers, merchants, planters, and politicians held the important positions on the county and township committees. Because the crop reduction program's rental payment per acre was based on the productivity of the individual farm, delta planters derived the greatest benefit from the programs. According to R. L. McGill, the extension service's regional supervisor in 1934, "numbers of farmers who in the past had business worth several thousands of dollars have stated that the Agricultural Adjustment Program for cotton had saved their farms from bankruptcy."[28]

County cotton farmers rented 29,000 acres to the government in 1934 for an average of $11 per acre. In 1935 they rented 25,934 acres to the government and sent a committee to a meeting in Little Rock to endorse the Agricultural Adjustment Administration's program. That committee included some of the most influential delta planters or their representatives. Curtis Dewey of Chapman and Dewey Company and C. A. Dawson, general manager of E. Ritter and Company, attended from Marked Tree. G. C. Jernigan, chairman of the county agricultural committee, and Dan Portis, a banker-merchant-planter, both of Lepanto, also attended.[29] Planters and farmers believed the AAA deserved the credit for the increase in the price of cotton.[30]

Data for the South as a whole demonstrate that the price of cotton rose again in 1936, and AAA officials attributed that rise in price to the programs they had enacted between 1933 and 1936. The programs had "held the harvested cotton acreage . . . to a level more than 25 percent below the acreage harvested" during the control period; they reduced the bales harvested by more than 20 percent. But the apparent success of the programs inflated the optimism of the AAA officials. They reduced the number of acres to be withdrawn in 1937 and set a goal of 32,750,000 total cotton acres for the year, expecting a yield of 13.5 million bales. But farmers, like the AAA officials, were overly optimistic and planted a total of 34,471,000 acres and produced 18,946,000 bales; it was the second largest crop in history.[31] Prices dropped precipitously.[32] The state of Arkansas and Poinsett County participated in this glut on the market even though growers were warned in early 1937 about the dangers of overproducing. In 1936 the state harvested 1,265,622 bales and in 1937, 1,808,237 bales. Poinsett County cotton production increased from 61,500 to 88,296 bales in those years. AAA officials reinstituted more stringent restrictions, and Poinsett County farmers agreed: in early 1938, 3,550 voted in favor of the crop control program that imposed new quota restrictions, and only 66 voted against it. By 1940 Poinsett county's cotton quota had been reduced to 56,919 bales, down from 61,500 in 1936 and down by more than 30,000 bales from a high of 88,296 in 1937.[33] Yet it was still almost 20,000 above the absolute low reached in the first year of the crop reduction program, when cotton farmers in the county produced 29,607 bales. Moreover, the 1940 figure was only some 3,000 bales less than the total amount of cotton grown in the county in 1930 before the New Deal programs came into effect. Much of the cotton grown in 1940 was planted on rich new land brought into cultivation as a result of the drainage enterprises made possible, in part, by the St. Francis Valley Floodway project, a massive drainage project funded by the federal government that made the valley relatively safe from flooding. Almost 100,000 acres were added to the county's "land in farms" between 1930 and 1940, mostly in the delta. The number of acres in farms had increased from 182,761 in 1930 to 277,788 in 1940, and the number of farms reporting the cultivation of cotton had increased from 3,474 in 1934 to 3,786 in 1939. This stood in contrast to the pattern in the state as a whole where the number of farms producing cotton had decreased from 183,595 in 1934 to 150,667 in 1939.[34]

While much of the new acreage in the delta was planted in cotton, acreage retired from cotton production was often put into soil-building crops such as vetch and especially soybeans. The county farm agent worked with some of the most prominent planters in the delta to demonstrate the soil-building possibilities of vetch and soybeans and the commercial possibilities of the latter. As early as 1935, even before the soil conservation program came into being, G. E. Dyerle of Lepanto served as a demonstrator. The farm agent reported that "on six acres of cotton following vetch an average of 505 pounds of lint per acre was harvested. On six acres joining on which no vetch was grown, 375 pounds of lint per acre was harvested." In 1935, 3,000 acres in the county were planted in vetch, a "substantial increase over 1934." In 1936 the agent highlighted the Norcross plantation, where Hiram Norcross planted 200 acres in vetch. He continued to experiment with vetch and soybeans as soil builders in 1938, turning under approximately 250 acres of vetch. He had a total of 1,300 acres in soybeans and turned under "between 300 and 400" acres.[35]

Norcross and other farmers in the delta came to appreciate the commercial possibilities of soybeans. Norcross, for example, was farming nearly 6,000 acres and had committed one-fifth of his acreage to that crop. Another delta planter, S. C. Chapin of Truman, committed a quarter of his farm to soybean production in 1938. He operated 3,200 acres that year with "1,250 acres in cotton, 1,000 in corn, 800 acres in soybeans and 150 acres in truck and various other crops." Chapin reported that he planned to experiment with different varieties of soybean seed in 1939 in order to increase his yield to that enjoyed by farmers in the soybean-growing sections of the country.[36]

The farm agent's efforts to introduce soil-building crops to delta planters were successful. However, it appears that some land retired from production through the crop reduction program and put into soil-building crops was then put back into cotton production. By 1940 the cotton bales per acre produced on delta plantations far surpassed that achieved in the county in 1930. This increased productivity contradicted the crop reduction program's goal of limiting production. In 1930, when county farmers had 61,965 acres in cotton, they produced 43,223 bales. In 1940, when they had 57,828 acres in cotton, they produced 56,919 bales. Yet the value of the 1940 cotton was below that of 1930, one of the worst years of the depression.[37] The crop reduction and parity subsidies introduced by New Deal administrators had to stay into effect in order to ensure cotton farmers an adequate return on production. In spite of the failure of the programs to raise prices significantly, county cotton farmers remained favorably disposed to the New Deal.[38]

The crop subsidies more than made up for the low price of cotton. Yet perhaps the most startling change that occurred in the county's crop production during the decade was the increase in the cultivation of corn and soybeans (graph 1). Some of the soybean crop was "turned under" to enrich the soil, and some of the corn was used for feed, but the census does not give specific figures. What is clear, however, is that although the prices of soybeans and cotton fluctuated in the 1930s, the price received per acre for soybeans was much lower than that for cotton during the years when farmers were moving into soybean production. The move into soybean production was, therefore, not a straightforward, rational move based strictly upon market-inspired economic incentives. In other words, had there been no government programs subsidizing cotton in the 1930s, it is unlikely that planters would have turned to the production of soybeans (graph 2).

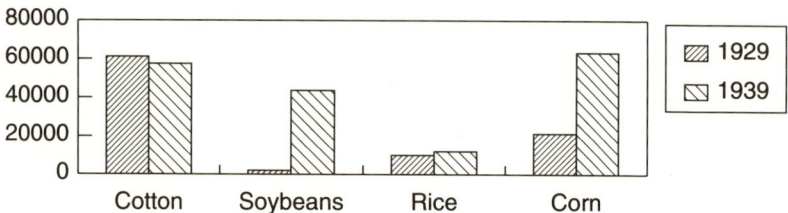

Graph 1. Acres in cotton, soybeans, rice, and corn in Poinsett County, 1930 and 1940. (*Sixteenth Census of the United States: 1940, Agriculture,* 4 [Washington, D.C., 1943]: 617–22)

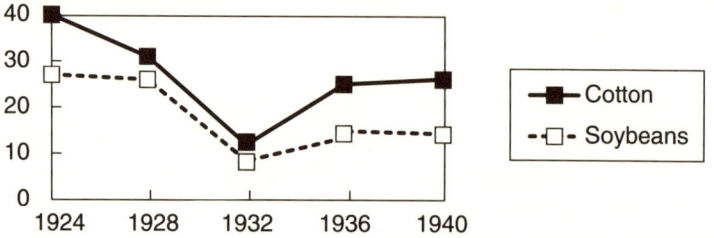

Graph 2. Gross profit in cotton and soybeans in Poinsett County, 1924–40. (U.S. Department of Commerce, *Historical Statistics of the United States: Colonial Times to 1970* 1 [Washington, D.C., 1975]: 513, 517)

Whether the soybeans and corn were harvested or turned under, the transition to less labor-intensive crops had important implications. Acreage in farms in Poinsett County continued to expand in the 1930s, and the adoption of soybean cultivation minimized the need for additional farm labor. What would prove even more significant, however, was the transformation of the farm labor force that accompanied the New Deal programs and the move toward the cultivation of less labor-intensive crops. Planters had traditionally depended upon tenants and sharecroppers to work the cotton fields during the period between planting and harvesting. The transition to soybeans rendered planters less dependent upon the tenancy system and more receptive to wage labor.[39]

Accompanying the shift into new crops was a new reliance upon tractors. Indeed, like the rice farmers in the prairie before them, the delta planters in time would learn to rely upon wage labor and mechanization. In spite of the fact that the land in farms increased by almost

100,000 acres, the number of farm operators increased by only 10.9 percent. Planters had begun to rely more on seasonal wage labor because of a transformation in the plantation labor force.[40] The use of seasonal wage labor was made more attractive because of the increase in the cultivation of crops that required less maintenance after planting, such as soybeans. The new reliance on tractors helped to alleviate some of the labor needs and bespoke the development of a more skilled farm laboring class.[41] The 1940 census added a category for tractors alone: 550 tractors were owned by the operators of 365 farms in the county. Some of those operators provided data regarding the age of the machines utilized. The average tractor was a 1936 model. In 1940 264 (75.8 percent) of 348 farms listed tractors of the 1936–40 model years; 45 (12.9 percent) used tractors of the 1931–35 period; and 39 (11.2 percent) used tractors produced before 1930. While prairie rice farmers may have purchased some tractors, few tractors would have been in use on the ridge. The cash that the New Deal funneled into the delta probably underwrote the growth in the use of tractors in that region and helped underwrite the transformation of the plantation.

Although New Deal programs rescued cotton planters and salvaged the economy in Poinsett County's delta, they wreaked havoc in the prairie and among rice farmers throughout the rice-growing region in the United States. According to historian Pete Daniel, cooperation among rice growers in the difficult 1920s had "saved them from the huge surpluses that drove cotton and tobacco prices below the cost of production."[42] But as conditions worsened in the early 1930s, Arkansas rice growers became discouraged, and by 1933 some had even begun to turn to stock raising as an alternative.[43] By the spring of 1933, "economic forces had already reduced rice acreage by 20%."[44] Even with this reduction in acreage, the price of rice dropped below the cost of production, and growers were induced to turn to the federal government for a solution.[45]

The program implemented by the federal government in 1933 for the 1934 crop year left both growers and millers dissatisfied. Instead of employing the processing tax used in the cotton reduction program, the rice reduction program fixed prices. Millers were saddled with the responsibility of enforcing the fixed price when purchasing the rough rice from the farmer and then when passing it on to the wholesaler. Rice millers initially accepted this responsibility, but when the AAA arbitrarily restricted the millers' margin of profit in the summer 1934, many refused to buy rice at all, and farmers found themselves in an even worse position.[46] Indeed, the old war between growers and millers which had surfaced in the first decade of the twentieth century was rekindled by the rice reduction program.[47]

By the end of 1934, the rice reduction program had virtually collapsed. Millers refused to buy the rice, the growers were unhappy, and both the millers' and the growers' associations began lobbying for a drastic change in the rice reduction program. What they wanted was a program identical to the one applied to cotton farmers: "a processing tax and adjustment payments to cooperating producers."[48] Such a program was worked out in late 1934 and put into effect during the 1935 crop year. But the damage had been done. Rice growers were "slow to cooperate" with the rice reduction program in 1935 because of the ill-feeling generated by the events of the previous year.[49]

Most rice growers would come to support the new program over the next several years, but those in Poinsett County balked again in 1938, voting not to participate in the rice reduction program of 1939. The problems in Poinsett County were reflected by the political and economic divisions within the county. The local farm agent paid little attention to the needs of rice farmers, focusing his energies on the delta and, to a lesser extent, upon the ridge. The geographical separation of the Poinsett County rice farmers from the principal rice-growing section in the state compounded their problems. Only two of the state's nine mills were in easy distance of the county's prairie, placing them at a distinct disadvantage, for they either had to accept the prices offered by the Jonesboro or Harrisburg mills or incur greater costs by transporting the rice to Stuttgart or DeWitt. The decision of the state extension service in 1926 to locate the rice experiment station in Stuttgart drew attention away from Poinsett County rice growers and at the same time may have encouraged the local agent to believe their needs were being addressed by the Stuttgart station.[50]

From the beginning rice farmers within the prairie complained about unjust treatment from those serving on the rice reduction township committees. One disgruntled Poinsett County farmer complained to Washington: "Not all acreage allotments for 1934 were just. Mr. L. Hogue, Weiner, Arkansas, boasts that because of his poor memory he had more rice in 1934 than in any year previous. Others (among them Charles W. Wharton, Fisher, Arkansas) were unjustly limited. The same local board recommended an 80 acre allotment for Mr. Schuster, Fisher, Arkansas, and some allotment for George McCracken and Earl Glenn, neither of whom were entitled to any allotment."[51]

The complainant blamed the township committeemen for the problem. A year later another farmer bypassed the local committee by writing to Washington requesting "an acreage allotment for her son and brother." The Washington office referred the matter back to the county committee.[52] Complaints continued to reach Washington from disgruntled Poinsett County farmers throughout the decade.[53] Yet the local farm agent rarely mentioned them in his annual reports. While a complete section of each report was devoted specifically to the cotton reduction program, no space was reserved for the rice program. The only reference made to rice between 1936 and 1940 was in 1936 when the farm agent wrote simply that rice growers needed to understand the value of soybeans and peas on idle rice lands.[54] By 1938, however, the rice grower's discontent prompted representatives from county farm bureaus and AAA committees in the rice counties to address their grievances. These representatives held a meeting in Brinkley and drafted a resolution calling for revisions in the rice reduction program. C. O. Wofford and G. C. Jernigan of the Poinsett County AAA township committee attended the meeting. Wofford was a landowning rice farmer and a member of the township committee. Jernigan was a prominent delta planter, a local farm bureau leader, and, like Wofford, an AAA township committeeman. The resolution drafted at Brinkley called for allocating the rice acreage as a percentage of the cropland, similar to the formula used to figure cotton acreage. By the end of the decade, the AAA revised the rice reduction program accordingly. It is likely that the rice farmers in the prairie finally became reconciled to the program once it more closely resembled that employed for cotton planters.[55]

While prairie farmers struggled with the implementation of a rice program they felt was unfair and ill-advised, ridge farmers found themselves almost completely ignored. Because few programs were directed to small producers of peaches, dairy products, poultry, and livestock, farmers on Crowley's Ridge were least affected by New Deal programs. There were special problems associated with these kinds of farming enterprises that made it difficult to fashion a meaningful program. The dairy and poultry industries were especially difficult to organize: "Both dairy and poultry products are produced in almost every part of the United States and by vast numbers of very small producing units. That type of industry organization does not lend itself well to coordinated action either economically or politically."[56]

The peach farmers on the ridge never attempted to form marketing agreements like those implemented elsewhere. Only the hog growers participated in any meaningful New Deal agricultural program, but they were few, and the total contracts signed in the county for the corn-hog program were always under fifty. Many of the corn and hog producers were in the delta rather than on the ridge.[57]

One of the greatest difficulties that Crowley's Ridge farmers faced in the 1920s and 1930s was the relative infertility of their soil. The rolling hills of the ridge were subject to erosion and not suitable for field crops. Although farm agent Sullivant classified Crowley's Ridge in 1933 as "primarily adapted to peach production," only 207 acres were devoted to peach orchards in that year. After the agents had spent almost a decade encouraging the development of dairy farming there, the industry was "barely holding its own." Only a handful of farmers shipped cream or milk. In fact, some dairy farmers were turning unproductive dairy cows to use as foundation cows to produce beef cattle. Similarly, the poultry flocks were decreasing in number, and fewer farmers were marketing poultry products.[58]

The farm agent reported that the ridge was poorly suited to anything but pastures and timber. Fertility was extremely low, and the land was "subject to serious erosion." W. T. Clark, who owned eighty acres on the ridge, had abandoned crop production altogether by 1938. He had turned much of his land to pasture and found "growing mules more profitable than growing cotton."[59]

In sum, the New Deal's agricultural policies affected the three sections of the county very differently. They gave delta planters the greatest advantage economically. While prairie farmers struggled with an ill-conceived rice program which, at least in the early years, created more problems than it solved, the ridge farmers were almost completely without assistance. Although the rice farmers eventually rose to the challenge and insisted upon more equitable treatment, many of the ridge farmers, caught in a losing battle with soil erosion and infertility, put more of their land to pasture and withdrew from the production of field crops.

The most significant transformation, however, was taking place in the delta where planters had gained control of the crop reduction program at the same time that they brought more total acreage into production. Rather than substantially reducing their cultivation of cotton, they only modestly cut back on production of that commodity, but they also moved enthusiastically into the cultivation of soybeans and corn. This new interest in crops that had previously received very little attention, together with the crop subsidies that put needed cash into the hands of producers in the delta, had a fundamental influence upon the organization of labor and underwrote the coming move to wage labor.

11. Sharecroppers "going home" from Blytheville, 1936. (Courtesy of the Arkansas History Commission)

12. Children chopping cotton near Marked Tree, ca. 1935. (Courtesy of the Arkansas History Commission)

13. Evicted from Dibble plantation, Parkin, 1936. (Courtesy of the Arkansas History Commission)

14. Ward A. Rodgers and H. L. Mitchell at Rodgers's trial. (Courtesy of the Southern Tenant Farmers' Union Papers, Southern Historical Collection, Library of the University of North Carolina at Chapel Hill)

15. Norman Thomas speaks at a sharecropper meeting. (Courtesy of the Southern Tenant Farmers' Union Papers, Southern Historical Collection, Library of the University of North Carolina at Chapel Hill)

16. Rev. E. B. ("Britt") McKinney, STFU vice president. (Courtesy of the Southern Tenant Farmers' Union Papers, Southern Historical Collection, Library of the University of North Carolina at Chapel Hill)

17. Arkansas farmworkers at a union meeting. (Courtesy of Louise Boyle)

18. Summer church, Phillips County, ca. 1935. (Courtesy of the Charlton Collection, Special Collections Division, University of Arkansas Libraries, Fayetteville)

8

MEETING ON THE "TURN ROW"

> It's seventy miles to Memphis
> It's a hundred and twenty to Wardell
> It's a thousand miles from here to civilization
> But only a few steps from here to hell
>
> —*The Story of a Union That Would Not Die*

Turmoil and unrest characterized conditions within Arkansas and other southern states in the early 1930s. Floods, droughts, and severe economic depression had driven many people—black and white—to desperation. While farmers in England, Arkansas, rioted over the distribution of drought relief supplies, some whites in Mississippi resorted to assassinating black railroad employees in hopes of taking their jobs. The exploitation of black levee workers in a project in Mississippi sponsored by the National Relief Administration (NRA) led the National Association for the Advancement of Colored People (NAACP) to demand an investigation. Other incidents reflected long-standing tensions between black sharecroppers and white planters: a black sharecropper in Union County, Arkansas, killed the planter for whom he worked, apparently as a result of an argument over a cow; blacks in Tallapoosa County, Alabama, organized a sharecroppers union and endured a violent reaction from planters. Conditions worsened in 1934 as planters evicted tenants and sharecroppers who were made superfluous by the New Deal's crop reduction program. A group of tenants and sharecroppers in Poinsett County responded to the evictions by founding the Southern Tenant Farmers' Union. As an interracial union it was an unheard-of phenomenon and, in the context of rising racial tensions in the thirties, a seemingly inexplicable one.[1]

But even as the interracial STFU began in Arkansas, the NAACP was at a crossroads that spoke to the issue of interracial cooperation. In July 1934 W. E. B. Du Bois, one of the organization's founders, severed his ties to the NAACP and resigned as editor of the *Crisis*, its news organ, largely because of the issue of self-segregation. Believing that separate black communities and institutions had an important place in the lives of black Americans, Du Bois broke with the organization's long-standing position on integration. His reservations originated in the early years of the NAACP when tensions first surfaced over what he felt to be the attitudes of white members of the association's board. Du Bois claimed that whites tended to control the meetings and "if Negroes attempt[ed] to dominate and conduct the committee, the whites become dissatisfied and gradually withdrew." Although Du Bois's departure from the *Crisis* two decades later coincided with the founding of the STFU, the interracial union received only nominal recognition from those who remained on the NAACP board. One might have expected more from the very people who continued to have faith in interracial cooperation despite Du Bois's skepticism. The NAACP annual meeting in June 1935 passed a resolution dealing with the problems faced by tenants and sharecroppers, but it failed to mention the STFU by name. This failure is all the more curious because one of the union's white organizers attended the meeting and had published a lengthy article about the STFU in that month's issue of the *Crisis*.[2]

The NAACP and the black press in general concentrated on other more pressing matters. The organization continued to advocate the passage of a federal antilynching law and to pursue the fight against peonage in the South, and it solicited funds for the defense of black Communist Angelo Herndon in Georgia and the "Scottsboro boys" in Alabama. The NAACP had long focused on the problems confronting blacks in the South; and it attacked the racially biased distribution of New Deal largess and remained aware of the problems facing black tenants and sharecroppers. Other black leaders demonstrated their continuing concern. In April 1934 an editorial of the black newspaper the *Cleveland Call and Post* noted that "conditions in the South today are so bad, that Negroes who came north and who are now on charity, are better off than many of their brothers who remained in the South and are working regularly." At the same time, the NAACP and black newspaper editors recognized that white landless farmers also faced enormous problems. In March 1934, four months before the founding of the STFU, the editor of the *Call and Post* revealed that reports "coming out of Arkansas" suggested that black farmers were not alone in suffering the consequences of the "inhuman and brutal system of exploitation" operating on southern plantations. Whites, who in other circumstances "permitted themselves to be used to keep the Negro down" and "who formed the mobs that readily lynched and burned Negroes," were now themselves "feeling the heel of oppression on their necks, the sting of the sheriff's whip and the force of the landlord's boot in the rear of their midsection."³

Indeed, the situation for white landless farmers in Poinsett County had deteriorated significantly. However, that alone does not explain the creation of the STFU, for the economic situation for blacks had declined even further. A complex constellation of factors combined to result in the formation of the STFU in Poinsett County, and ultimately the interracial union's survival there was directly challenged in the traditional way: some white tenants and sharecroppers would permit "themselves to be used . . . to keep the Negro down," and while some blacks and whites in Poinsett County's delta were uniting, delta and ridge elites were equally unified in their opposition to the STFU. When trouble first arose in Poinsett County's delta over the implementation of the crop reduction program in the spring of 1934, tenants and sharecroppers found few friends in Harrisburg. The speaker of the Arkansas General Assembly, Harrisburg's H. B. Thorn, "turned a deaf ear to the request of the tenants for a legislative investigation." Thorn could have initiated an inquiry, but he had long depended on votes from delta planters and refused to raise an issue that would threaten his base of support. It is likely, moreover, that he shared the views of a Harrisburg druggist who told an investigator that the country "ought to get rid of the 'onery low class tenants.'" Thurman A. Bettis of Harrisburg, the local Federal Emergency Relief Administration (FERA) director, also was unsympathetic and believed that those who organized the STFU in Tyronza were simply agitators who cared more for their organization than they did for the landless farmers they were supposed to represent. The response of ridge officials and citizens to the problems that delta planters confronted with tenants and sharecroppers reflected a consolidation of the county's elites in the face of pressure from below. No matter what else may have divided them, challenge from the "onery low class tenants" was not to be tolerated.⁴

While some local officials and citizens tended to blame the landless farmers or to minimize the extent of the problem, county agent R. L. McGill, who was responsible for administering the AAA program on the local level, simply wrote in his 1934 annual report that "due to some unavoidable conditions in some sections of the county, tenants and sharecroppers have been misinformed and are opposed to everything." Although McGill's reference to the difficulties in the delta is remarkably oblique given the storm of protest raised by landless farmers, his failure to address the problem more forthrightly is unsurprising. According to future congressman Brooks Hays—who toured the delta in early 1934 on a fact-finding mission for the FERA—McGill "was under [the] control of [the] politicians and landowners." When AAA representative E. A. Miller investigated allegations made against certain planters, he cleared them "of wholesale eviction charges," despite overwhelming evidence to the contrary. Both Miller and McGill understood that the success of the federal farm program depended upon the planters' cooperation, and they refused to acknowledge the problems arising from the implementation of the crop reduction program. Although some in the Department of Agriculture sought to address the grievances brought to the department's attention, most adopted the official line. Those who did not generally found their careers cut short. New Dealers like Jerome Frank of the legal division of the AAA attempted to represent the interests of tenants and sharecroppers and paid a high price: Frank lost his position with the AAA.[5]

The first stirrings of the landless farmers occurred in late 1933 when the federal work relief programs were implemented in the county. Many tenants and sharecroppers objected that the hiring practices of the Civil Works Administration (CWA) discriminated against them. A small delegation of these men asked two young socialists in the town of Tyronza, Harry L. Mitchell and Henry ("Clay") East, to accompany them to Harrisburg to ascertain why they were not being awarded any relief jobs. East was a town constable, and although he was from an old property-owning Poinsett County family, he and Mitchell had made themselves known as friends to the tenants and sharecroppers. Mitchell, who ran a dry-cleaning business in Tyronza, had worked as a sharecropper in Tennessee before following his father to Poinsett County in 1927 to work on shares for the Brackenseik brothers who owned a plantation about four miles from Tyronza. He was so disgusted with the housing they offered him that he decided to return to Tennessee, but his father, who operated a barbershop, convinced him to remain in Tyronza and helped establish him in the dry-cleaning business. Mitchell introduced East to socialist literature, and the two of them attended meetings in Memphis and established a socialist local in Tyronza.[6]

Rather than accompany the unemployed men to Harrisburg, Mitchell and East arranged for a meeting to take place in the Odd Fellows Hall in Tyronza. More than five hundred unemployed men filled the hall, stood on the steps, and milled in the street outside. They agreed to organize the Tyronza Unemployment League and elected officers; then the FERA director for Tyronza Township, Maxine East, appeared with the state director, Mississippi County planter W. R. Dyess. Maxine East, who was Clay's wife, was allowed to address the meeting, and she called upon those present to "join in repeating the 'Pledge of Allegiance to the Flag,' to show that they were good Americans." But she was interrupted by one unhappy man who said "We came about getting one of those relief jobs. If you have anything to tell us about the jobs, let's hear it. If not, you sit down, and let someone else talk who does have something to tell us what we want to hear." State FERA director Dyess then addressed the crowd and promised "that every man who showed up at the local office the next day would be hired by the CWA and paid five dollars a week for his work. He also promised to see to it that no more work was done on privately owned plantations, or at the homes of rich folks in Tyronza by the men on relief."[7]

The CWA, an arm of the FERA, was the agency that Cully Cobb of the AAA believed tenants and sharecroppers should turn to, but Arkansas proved to be "one of the most difficult states to organize." Governor Junius Marion Futrell thought "that poverty was the result of lack of individual initiative" and refused to provide matching funds to support the program. Only after Harry Hopkins "informed Futrell that Arkansas would lose not only funds for education and relief, but 'all' federal funds," did the governor capitulate. The conservatism at the top was reflected at the county level. T. A. Bettis, the Poinsett County administrator for the FERA, was a cotton planter who owned land on the edge of the ridge and sympathized with the needs of the delta cotton planters rather than with those of the tenants and sharecroppers, and because the state FERA faced continuing controversy and encountered two internal reorganizations within the first year of operation, county FERA directors like Bettis operated with little interference. They were rarely trained as social workers themselves, and their caseworkers typically had little or no training or experience. They were accused of having little sympathy for the poor whom they were supposed to be serving, and, indeed, many of those who were placed on the FERA rolls had some personal association with the local caseworkers or the local committees.[8]

Reminiscent of the Red Cross works program of 1930–31, the FERA program in Poinsett County was a works rather than a relief program. Those who were deemed qualified were put to work in a variety of enterprises. Marked Tree and other towns employed FERA workers to clean the streets and the gutters; planters used them to clear new land; Drainage District Seven utilized them to clear ditches. Only individuals regarded as legitimately "unemployable" were given direct relief, including the elderly, the disabled, and mothers of young children whose husbands had either died or deserted them. In its first year in operation, only 104,979 out of 287,444 recipients of relief in Arkansas received direct relief. In Poinsett County the figures were even more slanted toward work relief. Of the 4,341 individuals reached by the FERA in Poinsett County in 1933, only 350 received direct relief; the rest worked in a variety of jobs.

By early 1934, 750 men were working in one of the sixty-five CWA projects in the county. Pressured to provide more jobs, the CWA pledged to employ 50,000 more men in the state. In order to do so, administrators restricted the hours of each CWA worker, cutting average worker's time from thirty to fifteen hours per week and allowing road workers twenty hours per week. Yet tenants and sharecroppers found it difficult, if not impossible, to qualify for work relief. W. T. Brown, a cotton tenant near Harrisburg, complained in December 1933 that he had registered for work relief but had "not had any employment so far." He was concerned about a rumor which implied that because he had "raised a crop in 1933" he was not eligible for work in the CWA program. John Carlew, the secretary of the Unemployed Citizens League of Tyronza also wrote the CWA complaining about the allocation of jobs in the county's plantation belt, especially the policy relating to tenant farmers and sharecroppers.[9]

By February 1934 the CWA developed a policy and issued a decision which must have disappointed Brown and Carlew. All laborers who were living on plantations and who had "made a trade for next year's furnish" were officially eliminated from the CWA rolls. Eliminating farm laborers from the work relief rolls was necessary, in part, because funds "for carrying on the relief work in this county have been cut to about ⅓." The needs of an agricultural economy based on plantation labor were clear: "Such people have already been declared as unemployable by the government because they are not usually employed at this time of year and therefore are not eligible for CWA work. By the same ruling, they will have to be cared for by their landlord in the usual way instead of leaning on the arm of the federal government." This policy change was arranged partly by New Deal administrators who recognized the breakdown of the furnishing system in the plantation belt and wanted to ensure that planters continued to assume responsibility for their tenants and sharecroppers. It was also partly orchestrated by planters who feared a disruption in the plantation labor force, and did not want to compete with the federal government in the labor market. It ignored the fact that small farmers, whether landless or not, had traditionally worked off the farm in order to supplement their incomes.[10]

The policy constricted the opportunities open to tenants and sharecroppers and probably led some of them to become interested in socialism. Mitchell has claimed that the Unemployment League served as a conduit through which some passed into the socialist local. In fact, although Mitchell and East promoted the Socialist party by hosting such figures as J. C. Thompson, a veteran of the Southwestern Socialist Encampments which thrived in Arkansas in the early twentieth century, socialists had spoken to Poinsett County audiences at least two decades earlier. Socialism was not a new concept in the Old Southwest or to Poinsett County residents; J. Sam Jones, a socialist candidate for governor of Arkansas, spoke to a large audience at the Marked Tree Lodge Hall in 1908. Jones's attraction had its roots in his, and Thompson's, brand of grassroots socialism. When Mitchell and East invited Norman Thomas, the acknowledged leader of the Socialist party in the United States, to speak to their local, planters and businessmen attended the meeting along with tenants and sharecroppers. The elite did not attend simply to keep tabs on what the landless farmers were doing—they could have dispatched only a few of their number to accomplish that— but to give Thomas a hearing, albeit an unreceptive one, just as their predecessors had listened to earlier socialists speak in Marked Tree.[11]

Norman Thomas arrived in Poinsett County in February 1934, and the two young socialists took him into the plantation sector, introducing him to various tenants and sharecroppers. Among those Thomas met was Ed Boston who, together with his wife and seven children, was evicted from the H. H. Howington plantation near Tyronza soon after Thomas's visit. It was Thomas who suggested that Mitchell and East organize a union rather than simply seek solutions through the political process. East had just run unsuccessfully for county sheriff on a Socialist party ticket; perhaps Thomas recognized what must have been evident to Mitchell and East: no one who was not a Democrat and tied in with the local Democratic machine could hope to be elected to a countywide office. Thus, the formation of the STFU at the Sunnyside Schoolhouse in July 1934 grew out of a conversation with Norman Thomas over lunch at Clay East's home during his February visit.[12]

Local conditions also played a role in the founding of a union of farmers in Poinsett County, for the county was not without a history of such a union. The Homesteaders Union had united small white farmers against planters barely a decade before the founding of the STFU. Although that earlier union was a union of landowners—or men who had reason to believe they would soon achieve that status through having their homestead claims perfected—it was similar to the STFU in that it united a group of farmers against the forces of wealth and power in the county. Some of the homesteaders appeared to have been successful in achieving limited goals until the disasters of the 1920s. The Homesteaders Union owed much to the support of Lepanto's elite, just as the STFU had two small businessmen from Tyronza to thank for inspiration and guidance. And Mitchell and East were not the only businessmen in Poinsett County to display sympathy for the tenants and sharecroppers. One Marked Tree hardware store owner, H. J. Krier, together with a small farm owner, posted bond for union organizer Ward Rodgers when he was arrested. Almost certainly related to John Krier, a prominent businessman-planter and former mayor of Marked Tree with whom he was in partnership, H. J. had achieved local notoriety for a speech he gave in 1927 opposing a state anti-evolution law. Attorney C. T. Carpenter, although paid for his services, represented the STFU and in doing so made some powerful enemies and endured harassment and intimidation. Other small landowning farmers and some small businessmen in Tyronza, Marked Tree, and Lepanto also were sympathetic to the plight of the tenants and sharecroppers but did not wish to risk the ostracism that might come from active support of the STFU.[13]

Yet the fact that community sentiment existed favoring the sharecroppers and tenants undoubtedly played a role in the founding of the union in Poinsett County.

Another catalyst for the formation of the STFU was the actions of the manager of the Hiram Norcross plantation, Alex East, who demonstrated to the tenants and sharecroppers that those in control were not quite united on the issue of evictions. Alex East, who was Clay's uncle, had firsthand experience with economic misfortune. Long before Norcross arrived in the county, Alex East had acquired considerable property and standing, and between 1910 and 1920 he increased his holdings from 393 to 850 acres. Sometime toward the end of the 1920s, East entered an arrangement with Norcross to purchase 1,900 additional acres, and the Bank of Tyronza apparently handled the transaction. Meanwhile, John Emrich, the president of the bank and Norcross's partner in the Tyronza Supply Company, "lent Hiram Norcross the money out of the bank" to purchase his Fairview plantation "without any security. Norcross had the old man over a barrell," and the transaction "broke the bank." The Bank of Tyronza closed its doors in late 1930 at a time when Arkansas was beset by a chain of bank closures, but Emrich "pledged $300,000 worth of real and personal property as a guaranty that he will pay all depositors in full." Emrich fulfilled his promise to pay depositors, but borrowers like Alex East were not so lucky. East's inability to meet his obligation enabled Norcross to reassert his ownership of the 1,900 acres. In addition, probably because East used 560 acres of his own land as security for the purchase, he lost that too to Norcross. East lost an additional 170 acres and came out of the 1920s with only 120 acres intact. He then became Norcross's farms manager. Hence, some of the acres Alex East was managing were those he had once owned. Indeed, some of the tenants and sharecroppers on the Norcross plantation may have once worked for Alex.[14]

Described by his nephew as a hardworking man who "was always up before daybreak," East initially adopted an approach to the plow-up campaign used by many other planters across the South: rather than evict extraneous tenants and sharecroppers, he reduced the amount of acreage allotted to each sharecropper or tenant. The fact that they had been allotted acreage entitled them to an advance from the plantation commissary. According to Clay East, "A man was allowed $1 per month for each acre he was working and could buy that much groceries at the

Tyronza Supply Store that Emrich and Norcross owned." However, in order to be productive enough to pay off his account at the commissary, each farmer needed to operate a certain number of acres. By reducing the acreage allotments instead of the number of tenants and sharecroppers on the plantation, Alex East ignored the strictly economic equation. Although he was holding on to labor in accord with traditional wisdom, and thus his actions were not entirely altruistic, he was genuinely sympathetic to the tenants and sharecroppers. His sympathy, moreover, was not founded simply on paternalistic impulses. As a small farmer turned planter who had been reduced to farm manager, he knew the vicissitudes of the plantation system and had compassion for those threatened with displacement. Norcross, a St. Louis financier who had bought his plantation in the county in the late 1920s, saw things differently. When he examined his books and saw the small acreages allotted to tenants and sharecroppers, he decided to evict the extraneous tenants. Some other planters did likewise. What set Norcross apart was the number of families he evicted: twenty-three.[15]

While the actions of Norcross, the memory of the Homesteaders Union, and the appeal of socialism all figured in the formation of the STFU, some of the organizers were preachers, and meetings were often held in churches. Since the lives of the sharecroppers and tenants tended to revolve around their churches, the participation of preachers was not without consequence. The number of churches in the township had increased along with the rise of the plantation sector and the growth of the population. Of the fifteen churches holding services in Tyronza Township in the 1930s, nine were black churches and seven of those nine were Baptist. Half of the six white churches were Baptist. With fifteen churches for a population of 4,155 persons in 1940, there was one church for every 277 individuals.[16]

Recognizing the role that religion played in the lives of the sharecroppers and tenants, STFU troubadour John Handcox took familiar gospel tunes and wrote lyrics touting the union. Handcox's songs recognized not only the heartlessness of the plantation system but also the social distance between the planters and the sharecroppers/tenants. One song carried the line "I don't want to be like a planter in my heart." Another song referred to the planters "on their thrones" and called for sharecroppers and tenants to "join the union tonight."[17] The participation of preachers, the appeal of Handcox's songs, and the use of churches as meeting places demonstrate an intimate relationship to the community within which members of the STFU lived, but they also reflect the church's position as the dominant institution in their lives. The black church in particular was often a politicizing institution and provided a place where meetings could be held.[18] Although white religion rarely had the same politicizing impact on its members, both blacks and whites derived similar spiritual and community rewards from their respective churches.[19] The church served the purpose of answering their spiritual needs and providing them with a place of safety in a cruel and uncompromising world. While "churches across the region provided members with social and educational opportunities," they "were more than simply social institutions providing entertainment and education. The Delta was characterized by a strong spiritual pull that ran through most rural families. Frame churches dotted the countryside along the region's gravel roads. . . . People in the Arkansas Delta generally took their religion seriously." Religion provided the principal venue for the rituals that glued their society together: marriages, funerals, and baptisms. Coming together for these rituals necessarily included a social component which, independent of purely spiritual concerns, created a community of individuals who grew intimately knowledgeable about the lives and hardships of their neighbors.[20]

In Tyronza Township, as elsewhere throughout the South, women dominated church membership. They were typically behind the creation of both black and white churches (and schools) in frontier areas like Tyronza Township, and they played a crucial role in the STFU. When Myrtle Lawrence first learned of the STFU, she "thought it was another church getting started" and was well disposed toward the organization from the beginning.[21] Women were important to the STFU, in part, because of their greater literacy. Allowed to remain in school longer than male children, who often had to work the fields, women were more likely to be literate than their men. Female literacy always had a revolutionary potential. While southern rural women could use their literacy to counterbalance the power differential between themselves and their husbands, they could also use it to assist their husbands in demanding a fair settlement from planters. Black women especially could do what their men could not do: they could harangue the planter and get away with it. Within the STFU locals women often occupied positions as recording secretaries. According to a union report in 1935,

"The wives and daughters of Union men as a general rule have more education than the men, and so it is the women who must do the greater part of the educational work for which there is such a crying need."[22]

Women's role in southern rural culture was important both socially and economically, and their support of the STFU transcended their role as secretaries of locals. Bonded to each other in a multitude of supportive alliances, they glued their society together. In remote rural communities where midwives delivered most of the children, women took the responsibility of sitting up with one another and attending to each others' births. They also participated in all the rituals connected to engagements, marriages, deaths.[23] These relationships, moreover, played a significant role in "building . . . links between the local economy and farm social life." While the women "organized the primary social events that created a sense of community," they also bartered among themselves for foodstuffs, clothes, household goods, and even household livestock like chickens, geese, and turkeys. Thus, they performed a key function in the local subeconomy. The women of the STFU put their social and bartering skills together to raise funds for the union "by giving suppers, entertainments, etc." As fund-raisers, secretaries, and educators for the union, women were not simply ancillary agents. Their role in nurturing the community of sharecroppers and tenants was essential to the functioning of the society within which they lived.[24]

Men forged their own bonds of friendship. They attended church services in large numbers, to be sure, but the frontier had other attractions that provided opportunities for men to celebrate their lives together. Tyronza Township was one of the last few "hunters' paradises" in the state, and men not only supplemented their families' meager diets by "hogging catfish" and hunting game, they created lasting friendships that helped to sustain them. Much to their wives' chagrin, the men of Tyronza Township lived within a county notorious for bootlegging, honky-tonks, and gambling. They frequented backwoods blind tigers where they listened to roots blues or country ballads. There they fought over games of chance and women, and there feuds were born, and friendships were sealed.[25]

All of these activities helped maintain some semblance of place and security in the lives of both the men and women and, because their economic circumstances were so precarious, were all that much more important to them. "Between 1865 and 1940, virtually all rural black women belonged to families which just barely subsisted within the South's commercial agricultural economy, while the number of white women in a similar position rose at a steady rate."[26] The custom of moving regularly—for some families annually—from plantation to plantation created considerable instability in the lives of the sharecroppers, but most moved within a very narrow range and could maintain relationships with the same churches, schools, merchants, and friends. Because these relationships were often the only comforts they had, they were shocked and distraught when evictions threatened to divorce them entirely from their community. "The natural inclination is to stay where it is safe," so they resisted eviction not just because it threw them "outdoors" at a time when few employment opportunities awaited them elsewhere, but because it threatened the small islands of stability they had created for themselves.[27]

Living within a segregated society, blacks and whites created their own separate islands of stability. Although they operated independently of each other, they and their cultures intersected daily. They traded goods with one another, and they shared information about good and bad landlords, fair and unfair merchants.[28] Racism and segregation dictated that whites occupy the more dominant role, and even the least racist white held painfully ambivalent attitudes toward blacks.[29] Blacks recognized their precarious position in the social structure and regarded whites with suspicion and mistrust. Nevertheless, blacks often formed tenuous friendships with whites that were greatly complicated with contradictory feelings of fear and affection, hatred and admiration. Black farmer and union activist George Stith pointed out that "if you work together as a laborer, the white man mostly have the authority. No matter how good you were . . . they give him that privilege . . . he had to be real dumb not to get that privilege." But perhaps nothing more compellingly illustrates the day-to-day humiliation endured by blacks than the ritual of the separate water bucket or drinking cup. As Stith describes it, they had

> either two buckets or two drinking cups. For instance we used to chop cotton, and we had a water boy. Mostly it was a five gallon bucket of water. Some time we drank out of the same bucket. There was a tin cup for the blacks, and they had a white granite dipper for the whites. And we

dipped in the same bucket. He'd get his white granite dipper, and he'd dip it over and drink out of his, and I'd get my black cup and drink out of it. In some cases if it was a small group we drink out of the same cup. The whites got to drink first. When they got through drinking then we drink. It was sort of an unwritten law.

Despite the tortured racial pathology of their culture, they coexisted in a harsh environment. Although marked by incidents of nightriding and lynching, the relationship between blacks and whites was also cemented by the invisible bonds of shared hardship. George Stith reflected that although blacks and whites did not socialize together, they

> met on the turn row and we talked. We discussed our problems ... and talk about how tough it was and what ought to be done about it. And when the union came in it was the answer to our discussion. I didn't go over to visit your family and your family didn't come over to visit mine. But there was something common there about what we did. We knew something needed to be done. So when the union came along this was what we met on the turn row and talked about when I were joining you. We'd rest our mules, and we would go out on the turn row, and we'd set down [and say] "well you know this is terrible. You know I worked all the year and when I gather my crop they give me what they wanted. And I don't know what they owe me. I take what they say." We talk about it.

The relationships could turn violent in an instant and have disastrous results, especially for blacks, but something quite different was always possible and awaited only the right moment, the right set of circumstances. Their great impoverishment by the early 1930s and the evictions that followed the implementation of the crop reduction program provided both the moment and the circumstances.[30]

STFU members dramatically established their willingness to forgo racial animosities in late 1934 when "a band of plantation riding bosses" disrupted a union meeting of black sharecroppers and tenants in a church near Gilmore, a small Crittenden County town located a few miles from Marked Tree. After driving a white union organizer away, the plantation riding bosses severely beat a black union organizer, preacher C. H. Smith, and then jailed him. Smith was one of the founding members of the STFU, and both white and black members of the union began a determined campaign to secure his release. Some wanted to "storm the courthouse," but cooler heads prevailed. They hired an attorney who recognized the vulnerability of black men in a racist society and urged a careful strategy which stunned the white community into a temporary passivity. C. T. Carpenter, described by future Arkansas congressman Brooks Hays as a "quiet, conservative lawyer," instructed union officials to send only white members to attend court when Smith's case came up for hearing. Union members followed his advice, and the judge released Smith into the patrician attorney's custody. Smith then walked out of the courthouse with the white union members surrounding him. They went straight to a union meeting where both black and white farmers cheered the preacher, and "the few holdouts among the black sharecroppers joined the union. Black and white unity had carried the day."[31]

The white farmers who stood shoulder to shoulder with Smith defied the racist culture that had grown up around the plantation system in northeastern Arkansas. In so doing, they suggested that white and black landless farmers might finally surmount their differences and unite. When they walked out of that courtroom together, Smith and his STFU comrades issued a clarion call to all tenants and sharecroppers to reexamine their long-held hostility and redefine the nature of their struggle against the plantation system. From the first days of the plantation's existence in Poinsett County's delta, white tenants had struggled against the specter of proletarianization by attempting to drive blacks from the county and to force planters to accept share-tenancy rather than sharecropping arrangements. The shift in attitude toward blacks reflected an important transformation of the struggle.

The interracial nature of the STFU officially dated from its first meeting in July 1934 when a group of eleven white and seven black men gathered at the Sunnyside schoolhouse on the edge of the Hiram Norcross plantation to form a union of tenants and sharecroppers. Alvin Nunnally, who farmed for John Emrich, led the meeting, and one of the first issues to arise was what to do about black members. Should there be two separate unions, one for blacks and one for whites? Burt Williams, a white tenant farmer in Tyronza Township, rose to the question and, admitting that his own father had been a Ku Klux Klan member who had helped drive black Republicans from Crittenden County in the 1890s, insisted that black and white tenants and sharecroppers had to stand together. Isaac Shaw, a black sharecropper who had been a member of the ill-fated Progressive Farmers and Household Union of Elaine in 1919, agreed with Williams.[32] He had witnessed the violent suppression of the earlier all-black union amid charges that its members planned to murder white planters, and he feared that planters would again manipulate old animosities to drive a wedge between black and white tenants and sharecroppers. Both demonstrated a keen awareness of the divide-and-conquer strategy employed by planters in the past. After Shaw and Williams spoke to the need for one integrated union, the matter was settled.[33]

The men gathered at the Sunnyside schoolhouse that night had other pressing matters to consider. Some of them had been evicted from the Norcross plantation after the crop reduction program made them superfluous. Others hoped unionizing would give them the muscle necessary to force planters to share crop subsidy payments with them. In the previous year they had plowed up one-third of their crops, expecting to receive a proportional share of the government payments, but Norcross appropriated all the crop subsidy payments. Many other planters did likewise, leaving tenants and sharecroppers with less cotton and thus less income than they would have received had there been no New Deal program at all.[34] Those planters who appropriated the crop subsidy and those who evicted extraneous tenants and sharecroppers acted as though they were simply businessmen who could hire and fire as they chose. Tenants and sharecroppers saw themselves as independent farmers in copartnership with planters and rejected the planters' perspective.

Black and white sharecroppers and tenants could not fail to understand the implications of the planters' actions. For whites the ramifications were particularly stunning. The evictions and appropriations of crop reduction payments demonstrated that planters saw little real difference between whites and blacks, between tenants and sharecroppers. And, in fact, the economic circumstances of both blacks and whites had deteriorated sharply since the relatively prosperous period before the agricultural depression and natural disasters of the 1920s. A look at the category of property tax payers wherein tenants and sharecroppers would be found, that is, the first quartile (25 percent) of the personal property wealth in Tyronza Township—where the STFU was founded—demonstrates the economic devastation confronting both blacks and whites. The decline reported by whites, from $15,616 in

Table 8.1. Change in personal property tax filers in Tyronza Township, first quartile, 1920–34

	1920	1934	Difference	% Change
Whites	239	322	+83	+34.7
Blacks	196	159	−37	−18.9
Total	435	481	+46	+10.5
Missing cases*	10	4		

Source: Personal Property Tax Records, Poinsett County Courthouse, Harrisburg, Ark.
Note: The 1920 values have been deflated (on all tables listing quartiles). See Appendix for details.
*These property tax filers did not list their race.

1920 to $14,463 in 1934 (−7.4 percent), taken in conjunction with the fact that the number of white taxpayers had increased by 34.7 percent (table 8.1), represents a substantial decline in personal property wealth.[35] Blacks suffered an even greater decline, however, and the numbers reflected in the tax records only begin to reveal the distress facing poor blacks in the township. While the number of white property tax filers increased by over a third between 1920 and 1934, the number of black property tax filers declined by 18.9 percent. The decrease in black filers helps to explain the staggering decline in personal property wealth experienced by blacks: from $10,075 in 1920 to $4,916 (−51.2 percent) in 1934. This held serious implications for blacks. Many African Americans, the most impoverished of all farmers in the township, saw their holdings completely wiped out, and while they listed their names on the personal property roll, possibly in order to secure the poll tax receipt, they failed to list any property at all. Others simply did not file personal property taxes and thus lost the opportunity to secure the poll tax receipt.[36]

Although the number of black property tax filers in Tyronza Township decreased in this period, the population of black farm operators in Poinsett County increased slightly. In prosperous times the number of black property tax filers would have increased as well.[37] But these were not prosperous times, and a closer look at the 1920 personal property tax records demonstrates the vulnerability of black farmers in the first quartile of the population. While the average personal property reported by whites in 1920 was $65, the average for blacks was $51. Between 1920 and 1934, moreover, blacks suffered a more drastic percentage of decline in average wealth than did whites. In 1934 blacks reported an average of $30 and whites $44 in personal property. That amounts to a 41.2 percent decline for blacks and a 32.3 percent decline for whites.

Perhaps the most significant loss the tenants incurred was of workstock, which essentially reduced them to the status of sharecroppers. In 1920, 118 (27.1 percent) of the 435 individuals who made up the first quartile of the population in Tyronza Township reported owning mules. By 1934, only 37 (7.7 percent) of 481 taxpayers owned mules, representing a 68.6 percent decline between 1920 and 1934. Both black and white taxpayers reported this decrease, but blacks slid further down the ladder quicker than did whites (table 8.2). Between 1920 and 1934 the number of blacks owning mules had dropped from 53 to 7, an 86.8 percent drop, while the number of whites owning mules had dropped from 65 to 30, a 53.8 percent decline. This decline in the

Table 8.2. Mule Ownership by race in Tyronza Township, first quartile, 1920–34

	1920		1934	
	No.	%	No.	%
Whites who own mules	65	27.2	30	9.3
Whites who own no mules	174	72.8	292	90.7
All whites	239	100.0	322	100.0
Blacks who own mules	53	27.0	7	4.4
Blacks who own no mules	143	73.0	152	95.6
All blacks	196	100.0	159	100.0
All who own mules	118	27.1	37	7.7
All who own no mules	317	72.9	444	92.3
All filers	435	100.0	481	100.0

Source: Personal Property Tax Records, Poinsett County Courthouse, Harrisburg, Ark.

number of taxpayers reporting workstock suggests that fewer farmers of both races enjoyed the status of tenant farmer as opposed to sharecropper within Tyronza Township.[38]

One farmer whose status had been reduced was Robert B. Barnes. In 1910 he had owned his own farm, but by 1920 he had lost it and worked as a tenant farmer, possibly on the land he once owned. In 1920 he held $150 in personal property, but by 1934 he had lost his workstock, held only $35 in personal property, and had been evicted by Norcross from his farmstead. Barnes's bitter experience probably accounted for his attendance at the organizing meeting of the STFU. The future for Barnes and his neighboring tenants and sharecroppers looked dim indeed in 1934. And New Deal programs inadvertently made the situation worse for Barnes and other like him who were evicted from plantations at a time when little or no other employment was available.

The nature of the economic decline among the poorest whites and blacks in Tyronza Township provides a clue to the founding of an interracial union. Both had been hard hit by the depression and the natural disasters of the twenties, but blacks had been essentially emasculated. They were more greatly impoverished, fewer in number, and, since so many of them had failed to pay the poll taxes, less likely to be voting. Perhaps the white tenants and sharecroppers who founded or joined Poinsett County's interracial union no longer perceived black farmers as competition, no longer saw them as an economic or political threat. Certainly their founding of an interracial union and their rush to the rescue of a black comrade signaled a militant awareness of the fact that they shared a common enemy: the planter.

But landlords themselves did not escape the economic vicissitudes of the period. Situated in quartiles three and four were the largest landlords. Smaller landlords were located in the second quartile. Taxpayers in those quartiles experienced a sharp decline between 1920 and 1934. Only five taxpayers were located in the fourth quartile, and only twenty-one (twenty in 1934) were located in the third quartile. Together they represented 4.2 percent of the population in 1934 but held 50.2 percent of the wealth (table 8.3). And in one way at least, not much had changed for them in fourteen years: they had held 50.3 percent of the wealth in 1920. Yet the wealthiest 4.2 percent of the population had suffered a 26.2 percent decline in personal property wealth. The catastrophe suffered by landlords encouraged them to accept New Deal programs;

Table 8.3. Decline of personal property wealth and change in number of filers in Tyronza Township by quartile, 1920–34

	1920			1934			Difference	
	Amt.	Holders	%	Amt.	Holders	%	Amt.	%
First quartile	$26,330	445	81.5	$19,492	485	82.0	$6,838	−26.0
Second quartile	$26,142	75	13.7	$19,359	82	13.9	$6,783	−26.0
Third quartile	$25,258	21	3.8	$16,738	20	3.4	$8,520	−33.7
Fourth quartile	$27,899	5	.9	$22,500	5	.8	$5,399	−19.4
Total	$105,629	546	99.9	$78,089	592	100.1	$27,540	−26.0

Source: Personal Property Tax Records, Poinsett County Courthouse, Harrisburg, Ark.

in doing so, they relinquished their long-held fears of labor shortages and began to abandon the strategies they had devised to attempt to maintain a labor supply: the crop lien and the commissary system. Thus, rather than holding fast to labor, they evicted extraneous tenants and sharecroppers. Appropriating the crop subsidy payment can be regarded as merely an extension of the planters' prerogative, but when they introduced the federal government as an actor in the struggle between landlords and tenants over scarce resources, the planters altered the power relationship between themselves and their tenants. Tenants were quick to see something others before them had questioned: planters in Poinsett County were not omnipotent, and they perceived an opportunity to influence the new actor, the federal government. The introduction of the federal government into the complicated relationship between planters and tenants necessitated a renegotiation of that relationship. In the ensuing negotiations, however, planters would best the tenants and sharecroppers.

The businessmen-planters of Poinsett County were aided in their negotiations because the AAA wished to avoid endangering its program by alienating the planters. New Deal officials recognized the need to court the planter elite in order to facilitate the implementation of the AAA program; as a result, they refused to recognize the legitimate grievances aired by the STFU. Cully Cobb, who headed the cotton section of the Agricultural Adjustment Administration, was determined to pursue a course of action which would not endanger Roosevelt's fragile

New Deal coalition by alienating southern Democrats who represented the planter elite.[39]

Reporting to the FERA in 1934 after his mission to the delta, Brooks Hays, urged an investigative apparatus that would bypass the county agent.[40] He must certainly have been disappointed in the system erected by Cully Cobb. When sending investigators into local communities to look into charges of impropriety, Cobb was careful to provide them with instructions designed to create as little disharmony with the local elite as possible. The AAA issued a memorandum to the investigators briefing them carefully on the purposes of their mission: "It is very important that these problems be dealt with in such manner as not to unduly disturb the progress of the cotton adjustment program or disturb the relationships between landowners and their tenants." The investigators were further instructed on how to avoid offending planters:

> The work to be done must be carried on in close cooperation with those who have had charge of the cotton adjustment program in the states and counties and in close cooperation with the county cotton adjustment committees in the various counties. Nothing must be done which might cause them to feel that their actions are being questioned and nothing must happen which might create an impression that the committeemen and others have not been fair and just. Also, the impression must not be created that there has been widespread effort to "chisel," cheat, or be unfair.[41]

Rather than bypass the county agents, as Brooks Hays had suggested, the investigators were instructed to confer first with the state extension director before moving into the counties. They were reminded that it was crucial that their investigations take place with the "understanding and approval" of the county agents and that they were to "maintain close contact with them and keep them fully informed of the progress of your work. It will be very helpful if the director of extension or someone representing him will go with you to the first county visited and assist you in developing the procedure." Not only were the investigators to approach the extension agent first, they were to "go into the county without public announcement and . . . first go over the whole problem with the farm agent and the county committee." The county committee in Tyronza Township included Hiram Norcross, the planter against whom a great many complaints were lodged; the investigators were to confer first with the very people against whom complaints concerning the program had been filed. Only then were the complainants brought into the procedure. The investigators were instructed to "have one or more members of the county committee present at these conferences" and to hold these meetings in closed session. A tenant or sharecropper with a grievance would be faced with the farm agent, a member of the county committee, and the planter with whom he was in disagreement, as well as the AAA investigator, whose sympathies would most likely rest with the county committee.[42]

Planters who felt inclined to negotiate with the union found themselves branded as traitors and victimized by the local power structure. After indicating a willingness to negotiate with the union, planter C. H. Dibble of southern Cross County evicted STFU members from his plantation in late 1935, characterizing them as "undesirable" because they had created "agitation and unrest." He was prompted to serve the eviction notices after his banker pressured him to fire the union sharecroppers or risk foreclosure. In a show of unity among planters, the banker indicated to Dibble that "the ginner would not even gin his cotton the next fall if he signed a contract with the union."[43] Mitchell, who by this time had been driven from the county and was operating out of an office in Memphis, had been negotiating with Dibble before the planter's "change of heart" and sprang to the defense of the Cross County union members. He sent STFU attorney Herman Goldberger to represent the union members, but Goldberger was arrested and charged with barratry.[44] The case stimulated considerable controversy and drew the attention of Eleanor Roosevelt who requested Governor Futrell to investigate the matter. Futrell, who was hostile to the STFU, compared the incident to the Elaine race riot and suggested that planters in northeastern Arkansas "remember vividly the Elaine affair, where the Negroes, under the guidance of white men, had planned to dispatch several owners of land in that country."[45] By "remembering" the incident in Elaine as a plan by blacks to murder whites, Futrell was using racist and inflammatory language to inspire further resistance to the union. By suggesting that blacks in Elaine had been led by whites—a total fabrication on Futrell's part—he connected the STFU to an incident in Arkansas history that still echoed threateningly in the hearts and minds of white Arkansans.

Even as the AAA implemented an investigatory process weighted heavily in favor of the planters, it instituted regulations designed to respond to the grievances aired by the STFU, that were all too easily circumvented by planters. The new regulations, introduced in 1935, required that the crop subsidy payment be divided by the farm operators according to status, from landowner at the top to sharecropper at the bottom. The higher the status, the greater the share of the crop subsidy. Thus, many planters began to redefine their relationships to their tenants and sharecroppers. It was to the planters' decided advantage to define the tenants and sharecroppers as wage laborers, thus allowing the planters to appropriate all the crop subsidy payments. This induced planters to substitute wage labor and further weakened the tenancy and sharecropping system on the plantations. While tenancy remained viable for some time longer in Poinsett County than it did elsewhere, there was a dramatic increase in wage labor. In 1935, the first year such statistics were kept, 1,173 farm laborers worked on Poinsett County plantations. Except for a sharp drop during the war, the number of wage laborers steadily increased, reaching a height of 9,947 in 1955 before beginning to decline as mechanization spread.[46]

Like the extension agents and the Red Cross before it, the AAA understood that the goodwill of prominent planters and businessmen was the key to success on the county level. By the time the AAA ruled in 1935 that planters would have to maintain the same number of tenants and sharecroppers they were employing, planters had already pared down the number of tenants and sharecroppers. An addendum to the ruling was probably more important to the planters. While they could not reduce the number of tenants and sharecroppers any further, they did not have to retain any particular persons. In other words, they could evict activist tenants and sharecroppers and replace them with more compliant men. Planters used this indulgence to rid themselves of the troublesome union members.

Pressed hard by evictions, tenants and sharecroppers formed the STFU, but although planters viewed the organization as radical, in part because it was interracial, STFU leaders adopted an essentially conservative strategy.[47] They pursued the Norcross evictions through the courts, and they attempted to persuade the government to force planters to share crop subsidy payments with tenants and sharecroppers. The leadership of the organization, with few exceptions, forswore violence. However, one young FERA-worker-turned-union-organizer, Ward Rodgers, inspired an audience of sharecroppers and tenants to "throw their hats in the air" when he declared "I can lead a mob to lynch any planter in Poinsett County." Rodgers made his declaration at a meeting of union members in the town square of Marked Tree. As soon as he left the platform, he was arrested for "Anarchy, Blasphemy and Barratry," among other things.[48]

Although lynching planters was not what the union leadership had in mind, it fell in behind Rodgers, worked to secure his release, and paid for his defense. Suits, strikes, and nonviolence were the hallmarks of the STFU. But planters used other methods. Between August 1934 and April 1935, twenty-three separate acts of violence were directed against members of the STFU in Poinsett, Mississippi, Cross, and Crittenden counties. Planters, plantation riding bosses, and local law enforcement officers were implicated, but often masked men acting under the cover of darkness ambushed union members going to or from meetings. Mobs, sometimes led by planters or plantation riding bosses, attacked union meetings and in at least two cases mistakenly assaulted innocent groups. According to Clay East, Harry Ritter—who was the vice president of E. Ritter and Company, the youngest son of the founder of the company, and the son-in-law of John Emrich—led a party of men who planned, unsuccessfully, to ambush East. But more subtle forms of intimidation were also used. In March 1935, for example, two law officers escorted the newly elected [STFU] president William H. Stultz, a schoolteacher turned sharecropper, to an office at the Chapman and Dewey Land and Lumber Company in Marked Tree where A. C. Spellings, the company's farm manager, confronted him. During the "interview" threats were made against Stultz as well as H. L. Mitchell and another union organizer, Howard Kester. On the following day Clay East, cofounder of the union, barely escaped an angry mob which followed him to the office of attorney C. T. Carpenter. The leader of the mob was one of the law officers who had intimidated Stultz the day before. East was saved when Mayor Fox of Marked Tree was prevailed upon to intercede with the mob.[49]

While mobs sanctioned by law officers, authorized by planters, and led by plantation riding bosses sought to disrupt union meetings and impede the union, the sharecroppers and tenants persisted in their organizing efforts. Despite the opposition, they organized well enough to create a farm cooperative in Mississippi and to mount successful strikes in 1935 and 1936. But the civil suits brought by the STFU against specific landowners for evicting tenants failed. In the most celebrated case, the local chancery court in Poinsett County ruled against twenty-two tenants and sharecroppers who sued planter Hiram Norcross. They appealed to the state supreme court, but that body determined it had no jurisdiction. The U.S. Supreme Court refused to hear the case.[50]

The violence and intimidation made life in Poinsett County increasingly difficult for Mitchell and East. After a group of planter churchwomen instituted a boycott against them, Mitchell lost his business altogether, and East sold out and bought a store closer to Memphis. Both men moved out of Poinsett County, but they attempted to keep the Southern Tenant Farmers' Union alive from an office in Memphis, and Mitchell remained the duly-elected secretary of the union. They made trips to Washington, D.C., to keep the pressure on the AAA, and they created alliances with and solicited donations from northern and southern liberal institutions and individuals, including a collection of remarkable women. One such woman was Jennie Lee, a radical member of the Union Labor party in England and a member of the House of Commons. In February 1935 she spoke to an audience of five hundred tenants and sharecroppers from the back of a truck and then, along with two other women from Britain, led them in "singing 'The Union is a Marching, We Shall Not be Moved'—back through the streets of Marked Tree, across the railway tracks to the Negro Lodge Hall, where the march disbanded."[51] Mitchell later believed that such incidents as these marked the true beginning of the civil rights movement.

The union spread throughout the Arkansas delta and into other states, and one key to its appeal was its decision not to demand that all local affiliates adopt the interracial approach. The leaders recognized that conditions differed on the local level from one state to another as well as from one county to another. George Stith's local in Woodruff County, Arkansas, was not integrated, nor were locals in neighboring Crittenden and Cross counties.[52] Despite a history of whitecapping and KKK activity in Poinsett County, certain factors combined to make an integrated union possible there. Blacks and whites had long frequented the same saloons and honky-tonks in the delta, but more significant was the tendency on the part of Poinsett County planters like Norcross to evict both white and black tenants and sharecroppers. In some counties planters evicted only white farmers, who were more likely to be tenants rather than sharecroppers and thus held a greater claim to ownership of the crop. In evicting only white tenants, they did not run the risk of driving blacks and whites together in what might prove to be a dangerous alliance. At the same time some planters, like C. H. Dibble of Cross County, evicted only black sharecroppers for they were the only operators he employed. This did not encourage unity among blacks and whites in Cross County.

While Norcross and other planters evicted tenants and sharecroppers and New Deal administrators grew less and less interested in preventing evictions, internal disputes within the STFU began to damage the organization as much as the violence and intimidation visited upon it by the planters. The first such dispute concerned the thorny issue of race; in mid-1935 Poinsett County planters took advantage of tensions within the union and employed the time-tested strategy of divide and conquer for sowing dissent. Mitchell and other union members had been driven from the county, but Stultz, the president of the union, moved back to the county later in the year. He hired as his attorney Fred H. Stafford, the assistant prosecutor who had convicted Ward Rodgers, and "tried to turn it [the union] over to the planters," reorganizing it "along racial lines." Stafford filed suit against Mitchell, the union's secretary, "to turn over the union's seal and charter and other papers" to Stultz. Mitchell complied when the court ruled in Stultz's favor, but he characterized Stultz as "in a complete fog." Stultz later changed his mind and suggested "that he had been misled by anti-union and racist agents of the Arkansas cotton planters."[53]

Perhaps even more damaging to the union was the abdication of E. B. ("Britt") McKinney. McKinney farmed for absentee planter V. C. Walton and, according to local FERA director Bettis, was "one of the leaders of the negro race around Marked Tree." Bettis described McKinney as "a fairly good farmer" who "pays out of debt every year and has livestock and farming equipment sufficient to run a farm" and called him "quite an agitator" who "makes speeches very often at the mass meetings" and was even billed as the "principal speaker" at one meeting. Clay East remembered that McKinney once rented acreage from his father and ranked him "a top man" as a farmer. As vice president of the union, black preacher McKinney eventually came to believe that blacks were being "beaten, year after year, to produce rewards for whites." Referring to white union leaders as "new masters," he chastised them in a letter and was promptly expelled from the union and replaced by F. R. Betton, a black organizer from Woodruff County.[54] McKinney then unsuccessfully attempted to recruit other blacks to join him in a new group.

Although technically stripped of his association with the STFU, Mitchell continued touting the union cause from his office in Memphis and took over nominal leadership. But some of his own actions inadvertently hurt the union as much as had the Stultz and McKinney controversies. The second serious internal dispute to threaten the union resulted when Mitchell orchestrated an ill-fated alliance with the CIO. When that union insisted that STFU members send dues directly to the Washington office of the CIO, Mitchell and others bolted. Added to the Stultz and McKinney crises and the challenge from the CIO was a breach with Commonwealth College in Arkansas which had originally lent considerable support to the union. Together these controversies seriously weakened the union.[55]

Even as STFU organizers and members were being driven from Poinsett County, planters were appropriating control over the new works programs implemented after the CWA was disbanded in 1935. Although the works programs that followed—the Public Works Administration (PWA) and the Works Progress Administration (WPA)— were organized differently, planters exercised even more influence over them than they did the old CWA. The elite within Poinsett County had almost complete control over the way the new programs functioned on the local level. They solicited the bids for the jobs and awarded the contracts. What they got from the New Deal agencies responsible for the programs was funding and what they were forced to adhere to was the wage rate, which, thanks to southern pressure, was below the national norm. And much to the chagrin of the NAACP, blacks received less than whites for the same work.[56] Civic leaders within each community submitted proposals to the state Public Works Administration and the federal Works Progress Administration for specific projects. A given community's success in attracting PWA or WPA funds depended on its ability to provide matching funds. Most of the PWA and WPA projects served the delta because the agricultural programs had funneled more money into the delta and helped the local town economies there to recover. The town governments and the businessmen and merchants in the towns had the funds to sponsor improvements.[57]

Both the scope and the magnitude of the projects in the delta were much greater than those in the prairie and on the ridge. The prairie town of Weiner applied for funds to build a school and make improvements to its city water system in 1937 and 1938 while Harrisburg citizens also applied for funds to build a new sewer system in 1937. By this time, the two communities could manage to provide the required matching funds up to one-third of the cost. Economic recovery was lagging behind in the prairie and ridge, and local contributions were simply not as readily available as they were in the delta.[58] In a community such as Marked Tree, where the recovery was taking shape because of the additional resources available to farmers and planters, more local funds were available for expenditure on local projects. In 1937, for example, Marked Tree came up with some matching money and received funds for community sanitation and for the construction of a recreation center, bath house, and wading pool. In that same year the town paved its streets with WPA funds; citizens pledged $10,000 for the project. In 1938 a new street drainage system was sponsored by a citizens group with half the cost to be borne by donations from the town's residents. Harrisburg and Weiner, meanwhile, participated more modestly in the WPA programs. Harrisburg streets were paved in 1938 with WPA matching funds, and Weiner's streets were graveled.[59]

Most of the funds expended in the county went to road construction, and most of the roads were located in the delta. The largest road project approved in 1935 was the "graveling and rebuilding of the Boat Run Road leading from Marked Tree to the Poinsett County/Cross County line." That ten-mile stretch of improvements cost $14,731. Another delta project that year involved the graveling and building of a road leading from Lepanto to Dyess colony, a resettlement community just across the Mississippi County line, for $3,777. Deckerville improved eight miles of road at a cost of $17,850, and Truman improved eight miles of road at a cost of $12,666. The total cost on all the delta projects combined was $49,025; community contributions amounted to $26,669. Ridge projects included improvements to Harrisburg's highways 14 and 18 for $13,834 and its Bay Village farm-to-market secondary road for $14,688 for a total of $28,522; community contributions were $8,398. Prairie projects were the least expensive. Weiner improved its farm-to-market secondary road for $5,188 with a total sponsor contribution of $1,054.[60] This trend continued throughout the 1930s. By 1938 the WPA had spent $333,556 in Poinsett County with the government contributing $242,130 and communities contributing

$91,425.⁶¹ The expenditures for roads and highways alone was $178,778. The average community contribution was 18.2 percent. The total funds expended on WPA projects in the state was $43,300,700 with community contributions of $7,829,738 and the federal funds amounting to $35,471,031. Fifty-one percent was devoted to the construction of rural roads.[62]

The benefits accruing to the delta elite were perhaps most dramatically evident in the construction of the St. Francis Levee Project, which reclaimed tens of thousands of acres of Poinsett County land. Federal funds partially financed the drainage of hundreds of thousands of acres in Arkansas and, coordinated with similar projects in Missouri, created a complex system of ditches which secured the St. Francis Valley from serious flooding once and for all. The St. Francis Levee District, headquartered in West Memphis, Arkansas, responsible for a complicated series of levees designed to protect five counties in northeastern Arkansas, including Poinsett, from serious flooding, negotiated and received a $451,000 grant from the PWA in April 1934 and hired hundreds of men to do the work. Drainage District Seven, located in Marked Tree, negotiated a $1,675,750 loan with the RFC in early 1934 to refund its bonds. The district had negotiated with its creditors, and the RFC loan enabled it to avoid liquidation. In 1935 an additional loan of $200,000 was negotiated from the RFC to facilitate repairs to the levees. In 1936 a PWA grant in the amount of $103,091 provided further funds for repairing the levees. After the flood of 1937, Congress authorized the Corps of Engineers to rebuild the levees in the entire St. Francis Valley with a view to coordinating the efforts of the counties within the states of Arkansas and Missouri. Drainage District Seven negotiated a final RFC loan in order to participate in this effort, and its levees were again refurbished in 1939 and 1940.[63]

Certainly these drainage efforts resulted in some benefit to white tenants who kept a place on Poinsett County plantations. The ability to work for a wage to supplement their incomes, something farmers had traditionally done even before the New Deal, would have appreciably improved their circumstances. But planters had firm control over the hiring and firing of workers, and both the PWA and the WPA systematically suspended men during harvest. Once the harvest season was over, the men were eligible for reemployment.[64] In this manner the works programs came to supplement the needs of the delta planters and did not interfere with the expansion of their enterprises.

At the same time that planters were successfully co-opting works programs, they controlled the two resettlement communities placed in Poinsett County's delta. The New Deal administrators eventually reacted to the problems posed by the abundance of displaced farm laborers by creating the Resettlement Administration (RA), which later became the Farm Security Administration (FSA). Brooks Hays worked with Rexford Tugwell in the RA from 1934 to 1941 and must have been disappointed by how planters were able to co-opt the program. The Arkansas RA purchased land and created two resettlement communities in Poinsett County that ultimately functioned as government-run plantations. The headquarters of the Poinsett County RA, which grew out of the rural rehabilitation program of the FERA, were moved from Harrisburg to Marked Tree in 1936 because both resettlement communities were located in the delta. T. A. Bettis, who had been the county administrator for the FERA, headed the RA and the FSA and moved to Marked Tree to oversee implementation of the programs.

The RA/FSA was designed to put displaced tenants and sharecroppers on homesteads, but the southern committeemen who implemented the program on the local level were accused of discriminating against tenants and sharecroppers and accepting only former small landholders as RA/FSA clients.[65] Only former landowners were regarded as good risks, and thus few tenants managed to find a place in RA projects. No sharecroppers were deemed eligible. But in Poinsett County some tenants did locate on at least one of the two resettlement communities. No STFU members found a home there, however; given the way the RA communities were organized and run, it is unlikely that STFU members would have found the environment to their liking.[66]

The county agricultural committee took applications for RA projects and made recommendations to the district office, which usually accepted its evaluations without question and forwarded them to the state office which then passed them on to Washington. Once the clients were approved, the county committee, working in conjunction with the county farm agent, devised a "program of work" for the RA clients. It was the county agent's responsibility to work directly with the RA clients in implementing that plan. The county committee exercised greatest influence in selecting the clients, while the farm agent typically had the most contact with them. It was a carefully orchestrated project that was designed to fit into, rather than challenge, the existing plantation system in the delta. The clients functioned as closely supervised tenant farmers and encountered some of the same problems faced by other sharecroppers and tenants in the county.[67]

The Arkansas Rural Rehabilitation Corporation (ARRC), founded in spring 1935 to oversee the administration of the resettlement communities, had taken options on thousands of acres all over the state by that fall. A 2,328.38-acre resettlement community located near the delta town of Truman on the northern border of Poinsett County was one of the first to be organized. It came to be called Campbell Farms because most of the land had been purchased from Dr. G. O. Campbell of Truman for approximately $90,000. During 1935, the RA project's first year, the ARRC leased the land from Campbell and farmed it with "rehabilitation clients on the share crop (half and half) basis."[68] At least some of the RA clients situated on the farm had been inherited from Campbell rather than carefully selected from a pool of eligible farmers. Almost all of the land was already developed and was easily put into cultivation. During the crop year of 1936, after the land had been purchased from Campbell, forty-four clients operated as tenants on 35-acre tracts in Campbell Farms.[69]

The second resettlement community to be organized in Poinsett County was what came to be called the Northern Ohio Farm, near Marked Tree. Much of that 4,111-acre tract was purchased from the Northern Ohio Cooperage Company in 1936 at a cost of $122,000, and was undeveloped land. Rural resettlement clients "received the use of the land for one or two years in return for placing it under cultivation."[70] Unlike the Campbell farm where some tenant houses already existed, almost all the houses and outbuildings required to make the Northern Ohio Farm a functioning plantation had to be constructed. Approximately thirty families lived on and worked the project during the 1937 crop year. A loan request submitted to the ARRC in May 1938 called for the construction of fifty additional units. Each individual farm was to measure approximately forty acres, and each farmer was to be provided with workstock, milk cows, brood sows, and chickens.[71] The county farm agent advised the tenants about how to implement the "plan of work," and the local RA supervisor, T. A. Bettis, served in effect as the plantation boss.

Complaints began to surface from both projects and were reminiscent of the kinds of complaints made by tenants and sharecroppers generally during the New Deal. Felix Dickens, a tenant of the Campbell

Farm, hired an attorney in 1937 because the local Resettlement Administration had failed to pay him his AAA payments of 1935 and 1936. Investigation revealed that Dickens's account at Campbell's store had been settled with his 1935 AAA check without his knowledge or approval. Even though the farm functioned as an ARRC project in 1935, Campbell had signed the AAA crop reduction agreement and appropriated the farm's subsidy payments as partial satisfaction of debts owed to him by the RA clients. Dickens's problem with the 1936 payment was more easily remedied. The 1936 soil conservation payment had yet to be sent out but would be delivered directly to Dickens. Once the ARRC purchased the land from Campbell, the AAA contracts were negotiated with each client.[72]

The problem that surfaced at the Northern Ohio Farm was more complex and involved every tenant there. In March 1937 tenant Alvin Inman wrote to President Roosevelt about his pending eviction from the resettlement community near Marked Tree. He attached a petition signed by his fellow RA clients supporting his claim that the local RA director, Bettis, "wants him to move at once without any cause or excuse any more than a personal grudge" and testifying that Inman was a good neighbor and good member of the community. All twenty-nine of his fellow clients signed it. J. O. Walker of the Resettlement Administration in Washington wrote to Inman in March 1938 informing him that the matter had been referred to T. Roy Reid, the regional director of the RA. It is likely that Reid referred the matter back to the local agricultural committee, which would have had little sympathy with Inman; there is no evidence that his eviction was forestalled.[73]

When the Farm Security Administration took over the Resettlement Administration projects in 1938, a new agenda was outlined. The FSA officials wanted to allow the clients within the communities to purchase the land they were farming. Although the descendants of some Northern Ohio and Campbell Farms residents speak reverently of their "original forties," the FSA records turned up only four completed ownership contracts. Neither Alvin Inman nor Felix Dickens was among them. Regardless of the precise number of residents who came to own their "forties," it is clear that some of them sold out to E. T. Burton and other prominent delta businessmen-planters. The pull of the war industries during the early forties encouraged some farmers to move out of agriculture entirely. With the outbreak of World War II, draft boards tended to draft the FSA clients first, for local opposition to the FSA program was considerable. The program was the brainchild of the Bureau of Agricultural Economics, an agency widely believed to have a substantial reform agenda, and what set the bureau apart from other government agencies was its effort to bypass the planter elite and work directly with the client. Finally, more conservative forces within the Department of Agriculture attacked and destroyed the bureau. No program that did not have the sanction of those in power both on the local and federal level had much hope of survival. As a result, no program that sought to address the grievances of poor farmers could long exist.[74]

In every possible way the Poinsett County elite appropriated control over New Deal programs and integrated them into their operations. The immediate results were not always efficacious for the planters, but even in the worst situations, planters eventually turned things to their advantage. Although their evictions and appropriations of the crop subsidy payments led to the formation of the Southern Tenant Farmers' Union and to unfavorable publicity that infuriated and embarrassed planters, the Department of Agriculture capitulated to planter demands and did little to assist tenants and sharecroppers. Ultimately, planters assumed unquestioned authority over the implementation of the AAA program on the local level. Although required to share crop subsidy payments with tenants and sharecroppers, planters merely redefined their relationships with those who worked their plantations so that they had to share less. This redefinition of relationships played an important role in the demise of the tenancy and sharecropping system and the growth of wage labor. Even though the Department of Agriculture insisted that planters cease reducing the total number of tenants and sharecroppers on a given plantation, the tendency of planters to redefine these relationships to their advantage was not seriously challenged.

Planters, meanwhile, controlled access to the works programs and ultimately influenced wages, establishing a differential for southern as opposed to northern workers. This set a pattern of control over wage labor that is unique to the South. The works programs also established a precedent for government participation in solving the labor needs of planters in the South. Although the World War II and post–World War II labor programs were designed to assist all agriculturalists throughout the nation, southern planters, who were rapidly learning the ramifications of wage labor, were especially receptive to government labor programs during and after World War II.

The problems confronting tenants and sharecroppers in Poinsett County were shared by their counterparts all over the South. Conditions at the local level explain how the challenge to the planters arose in Poinsett County and why it took the form of an interracial union. A complicated constellation of factors combined to make the emergence of the Southern Tenant Farmers' Union possible in Poinsett County. The tradition of southwestern socialism together with the existence of a socialist local and two activists of talent and perseverance played a role. The memory of the Homesteaders Union, which had defied the rich and powerful in a neighboring township just a decade or so earlier, may have remained sufficiently vivid to influence the founding of the STFU. Some small farmers and a few small merchants in key delta towns supported both the earlier Homesteaders Union and the STFU. Their support was important and reflected the existence of individuals who recognized the place of tenants and sharecroppers within the local culture.

While a close examination of the situation on the local level helps to explain the emergence of a union, a similar analysis throws light on why that union would welcome both black and white members. The declining percentage of blacks in the county population and their even greater impoverishment encouraged whites to cease considering them a threat or as economic competition. At the same time, planters began to evict blacks and whites without regard to race. Thus, planter actions were as responsible for the emergence of an interracial union as the decline in the black population. Planters would reintroduce the old divide-and-conquer strategy to good effect in Poinsett County, at least for a time, but union president Stultz's compromise, wherein he sacrificed the principle of interracial membership, did not alter the official union's position.

The willingness of black and white tenants and sharecroppers to move beyond their profound suspicion of one another suggests that despite the significant shortcomings of the tenancy and sharecropping system, they believed there was something in it worth defending. What they were defending was not their own exploitation under the system but a way of life and a place within a culture that they cherished. Although their lives were marked by considerable mobility, most tenants and sharecroppers, after immigrating to a particular region, stayed within a fairly tight radius and grew familiar with the culture they settled within. They attended churches, shopped with merchants, sent their children to schools, and formed relationships with other people that they were reluctant to give up. These relationships provided a point of reference that gave their lives purpose. Their customs and rituals grounded them in a place and a situation that made their lives meaningful. The evictions divorced them not merely from a job but from a community and a way of life. They resisted the demise of tenancy and sharecropping not because they wanted to maintain a pernicious and exploitative system but because they wanted to reform it. The arrival of the federal government on the local scene made reform seem possible. Instead, the federal government made further exploitation likely and the destruction of the tenancy and sharecropping system a reality.

9

NO WAY FOR A MAN TO LIVE

When Herbert E. Kennedy, an African-American farmer, took his daughter and her children away from a Poinsett County delta farm in the summer of 1952, he incurred the wrath not only of his son-in-law but also of the man who was in charge of the farming operation upon which Kennedy was a tenant. V. O. Isbell, the Marked Tree farm operator for E. Ritter and Company, had reason to be troubled by the loss of a tenant at the height of the growing season. The dramatic exodus of tenants and sharecroppers from Poinsett County delta farms following World War II added to Isbell's alarm. Although mechanization was underway on the Ritter and other delta farms, labor needs were so critical that Isbell decided to follow Kennedy and bring him back.

Kennedy fled to a farm near Elaine, where he joined his twin brother, Hubert, who worked a tenant farm for a Phillips County planter. Isbell and Kennedy's two sons-in-law went after Herbert Kennedy and found him working a field late at night. There Kennedy "saw the lights of the pickup coming across the field . . . [the] next thing I knew Mr. Isbell came up and grabbed me by the collar and hit me with a hammer, knocking me down." Isbell and Kennedy's sons-in-law then forced Kennedy into the truck, collected his daughter and grandchildren, and began the drive back to Marked Tree. Hubert Kennedy, however, alerted Marianna authorities, and when the truck reached that city, the Marianna police arrested Isbell and the other two men on a charge of kidnapping. The assistant prosecuting attorney reportedly even considered filing peonage charges against Isbell. The disposition of the case remains unknown, for no records of it have survived.[1]

Perhaps it is no coincidence that the year following Isbell's arrest, L. V. Ritter, president of E. Ritter and Company, hired a college-educated farm manager to take over the management of at least part of his vast farming operation. Jean Thatcher, who graduated from the University of Arkansas College of Agriculture in 1948, was frank with his new employer about how he wanted to do business: "When I went to work for the company I told Mr. Ritter it would have to be on the up and up, and he said that was just exactly the way he wanted it. I had lots of opposition in the company, but I stayed with it, and we got it organized and Ritter always backed me up." Thatcher was in charge of estimating the plantation's labor needs and procuring labor when necessary. He engaged in farm planning and developed a modern system of production records which allowed him to project costs. Ritter essentially gave him a free hand to modernize the operation. Although the company had already purchased three one-row cotton pickers, it was during Thatcher's tenure that the Ritter plantation made the important transition from a labor-intensive to a capital-intensive enterprise. Key to that transition was the demise of tenancy and sharecropping. By the time Thatcher retired in 1986, no tenants or sharecroppers worked for E. Ritter and Company. Instead, the company employed a handful of wage laborers.[2]

Tenants and sharecroppers had no place on the neoplantations that emerged in the postwar period. With the advent of mechanization, the old labor-intensive system became obsolete. A massive exodus from the county's plantations began during World War II and accelerated in tandem with the rise of the neoplantation. The county's population dropped 31.7 percent in two decades: from 39,311 in 1950 to 26,822 in 1970. While the population in the county's delta countryside declined 62.5 percent, from 18,516 to 6,934, the towns there increased in population only modestly, and merchants in those towns suffered significantly. Efforts to attract and maintain small industries met with only modest success.[3] Between 1949 and 1969 the number of farm operators decreased from 4,523 to 1,042, and the number of tenants from 3,206 to 380. All the while, farmers and planters became more dependent upon the use of chemicals and fertilizers, tractors, cotton pickers, and combines. In 1949 only 1,635 (36.1 percent) out of 4,523 farms reported using tractors. By 1969, 890 (85.4 percent) out of 1,042 utilized farm machinery. In 1949 Poinsett County farms had 4,682 mules and horses; by 1969 only 244 horses grazed in the county's fields, and mules had become so scarce that the census bureau ceased reporting their numbers.

Black and white landless farmers, who had been resisting being reduced to wage labor since the Civil War, found it increasingly difficult to secure independent farmsteads in the post–World War II period. Planters had gained the upper hand during the New Deal when they convinced federal officials to accept their definition of the status of tenants and sharecroppers—as a form of wage laborer rather than as independent farmers—and would build on that advantage during and after World War II. Indeed, the war represented another watershed. During that national crisis the federal government assumed a role in securing and maintaining an adequate supply of farm labor on plantations, and it continued to play that role. By increasing the supply of foreign laborers, it reduced the cost of wage labor, and this encouraged planters to cease using tenants and sharecroppers.

The extension service played a crucial role in addressing planters' labor needs, and the activities of the Poinsett County farm agent demonstrate how planters were schooled in the ramifications of wage labor. As early as February 1942, the county agent called a meeting and supervised the formation of a farm labor committee of Poinsett County planters and farmers. The committee was a direct result of House Resolution 96, which put county agents in charge of the recruitment and placement of farm labor in order to deal with the emergency facing American farmers across the country.[4] The Poinsett County situation was complicated by the county's "fast development" resulting from the drainage projects engineered in part by the federal government in the previous decade and leading to "a marked increase in the cultivated acreage of cotton, corn, soybeans, alfalfa and other crops." The year 1942 passed without a disruption in harvesting the crop, but in 1943 a crisis loomed; by August the farms of Poinsett County had "a demand for farm labor that was 125% of normal while the available labor was 57% of normal supply." Having advocated a "work, fight, or go to jail" policy since early 1942, local planters' fears of a labor shortage increased significantly in the months preceding the harvest season in 1943.[5]

The Poinsett County Farm Labor Committee, which by 1943 included C. A. Dawson of E. Ritter and Company, considered recruiting additional labor from hill counties, but a severe drought lengthened the harvest season and relieved some of the pressure in harvesting the county's 1943 crop.[6] According to a Department of Agriculture report prepared for a Senate postwar commission hearing, "The record-breaking agricultural production of 1943 was achieved with the smallest average number of persons working on farms in the whole 35-year period for which farm employment estimates are available." Women, children, retired persons, and townspeople assisted in the longer than normal harvest of 1943, but this labor source had been stretched to the limit. The county farm labor organization set to work to deal with the upcoming 1944 labor crisis, and its efforts would not be unrewarded, for the agricultural agent praised the farm labor program that year as "the outstanding program of the extension service in Poinsett County."[7]

The major way the county organization devised to deal with the next harvest was their successful campaign to get two German prisoner of war camps placed in the county in 1944. L. V. Ritter of Marked Tree and banker Louis Hogue of Weiner played the key roles in securing the camps. Ritter, along with a number of other prominent delta planters and businessmen, formed an association of farmers "to deal with the government in securing the camp," and the War Department approved the placement of a camp near Marked Tree in April.[8] The local farm labor committee secured the use of property belonging to the Chapman and Dewey Lumber Company located three miles from town. An abandoned barn on the property would serve as a mess hall for the anticipated 350 prisoners. The committee also secured materials to build a separate mess hall for the guards, expected to number 75.[9] Although army officers were in charge of locating the sites and supervising the construction of the buildings, local planters and farmers contributed the materials, and local workers built the buildings and the fences surrounding the compound.

The initial camp in the county, however, was located in Harrisburg and prisoners from that camp assisted in harvesting the largest rice crop in the history of the county.[10] The first prisoners to arrive in Poinsett County, from a contingent of some thirty-seven hundred captured in North Africa, some of them near El Alamein, were placed in the Harrisburg camp in early 1944. Local farmers quickly learned that they were "very good workers but you cannot push them." One Saturday, for example, "they refused to work because the canteen had not been established." They all had at least the equivalent of eighth-grade educations, and many had completed high school, so that they were far better educated than the typical Poinsett County resident. By mid-1944 the Marked Tree facility had been completed, and some of the POWs in Harrisburg were moved there and began to work on the cotton plantations. By early the next year, when an estimated twenty thousand German POWs were placed in camps in Arkansas, the Marked Tree camp was filled to capacity. The German POWs proved critical to easing the labor shortage.[11] Remaining nearly ten months after the German surrender, they were not repatriated until March 1946. Significantly, the program was designed so as not to interfere with prevailing wage rates; while planters were prohibited from paying less than the prevailing rate, they were not forced to pay more than the prevailing wage rate. Yet, by its very existence in a labor-scarce economy, the POW labor program depressed farm wage rates and may have encouraged some landless farmers to move to employment in war industries.[12]

Although Poinsett County planters met their labor needs by using German prisoners of war, other agriculturalists relied on Mexican laborers. The use of Mexican laborers by American farmers was officially recognized by a treaty signed in 1942.[13] Renegotiating that treaty every succeeding year, the Mexican government attempted to, and often succeeded in, guaranteeing the rights of its nationals and assuring minimum standards in both pay and living conditions for them. Between 1942 and 1949 an estimated 400,000 Mexicans were brought into the United States to fill the growing labor needs of farmers throughout the nation.[14] The extension service provided those counties with the most acute labor problems with a farm labor clerk to assist the county farm agent in coordinating the labor program. Poinsett County's farm labor clerk was employed on a full-time basis for the first time during 1946.[15] Once the German POWs left the county that March, planters eagerly accepted the clerk's services. The cotton planters were particularly receptive to the program, for they faced the most serious labor problem. Rice farmers in the county's prairie had adopted new and improved mechanical harvesters, but a marketable and cost-effective cotton picker was just off the drawing boards, and only one cotton plantation, owned by the Chapman and Dewey Farms Company, had purchased a cotton picker by 1946. Planters, according to the provisions of the farm labor program, petitioned the farm labor clerk in the spring, estimating their labor needs. Forty meetings were held during 1946 attended by 882 farmers. The farm labor clerk estimated that planters would need 1,200 cotton pickers.[16] The county agent, the farm labor clerk, and cotton farmers on the farm labor committee attended a meeting in Jonesboro on August 6, 1946, where the use of Mexican laborers was discussed. In the end 600 Mexicans were brought into the delta section of the county to help with the cotton harvest, and an additional 300 laborers were brought in, largely from Memphis, to assist as well.[17]

The existence of the Mexican labor program, like the POW program during World War II, allowed planters some control over the prevailing wage rates. Farm labor groups, such as the STFU, protested the use of these foreign workers but to no avail. H. L. Mitchell at first attempted to persuade the government to provide its own citizens with the same treatment accorded Mexican nationals: a guaranteed wage and minimum living conditions. Unsuccessful in this endeavor, he shifted strategy and protested the importation of Mexican labor, arguing that there existed "a surplus of labor here." By 1944 STFU members picketed Arkansas cotton plantations in protest against the use of German and Italian POWs. The War Manpower Commission had brushed aside the STFU's wage demands, labeling them "too high," and increased the number of POWs and Mexicans in Arkansas.[18]

Mitchell also was alarmed by another trend which had accelerated during the war. He complained that due to mechanization "on the large scale cotton plantations, sharecropping was rapidly changing into a system of wage labor." The STFU was not the only organization concerned about the threat that mechanization posed. Farmers Union representatives from twelve Arkansas counties dispatched a resolution to the state's congressional delegation in 1945 lamenting the fact that some farm families engaged in cotton production would be "displaced by the mechanical cotton picker, flame cultivator, mechanical cotton chopper, four-row cultivator and other labor-saving devices." The union charged that "economic planning for the postwar period has . . . been completely dominated by power company officials, plantation owners and other selfish interests whose view of common people is not far advanced from the viewpoint of the slave owner. Labor and family farmers alike are considered objects for economic exploitation."[19]

The road to mechanization was a bumpy one, for the keys to complete mechanization were a marketable mechanical cotton harvester and changes in the ginning system that would be needed to process the machine-picked cotton. Planters had a sizable investment in mules, implements, and gins, and some of them were resistant to change. Even though the figures reflect a dramatic transformation that occurred in a relatively short period of time, many planters and farmers turned to mechanization cautiously. Accompanying a rise in the number of tractors in Poinsett County, for example, was a well-publicized debate entitled "Which Is the Most Valuable to the Farmer, Mules or Tractors?" In 1942 the Marked Tree Rotary Club sponsored this debate between Ed Pittman and Charlie Causey, two prominent local planters. Pittman defended "Old Beck," and Causey "went to bat for the iron steed." The very use of the term "iron steed" to describe the tractor by the man who was defending it illustrates the ambivalence even its proponents felt about mechanization. By using that term, Causey provided the tractor with a link to the past and connected it to something alive and familiar.[20]

Although some Poinsett County farmers were reluctant to put Old Beck out to pasture, others demonstrated keen interest in using tractors and the development of a marketable mechanical cotton picker. Equally enthusiastic about the coming changes was the Department of Agriculture's Post War Planning Commission. Delta planters, who dominated the Poinsett County Post War Planning Committee established in April 1944, were no more sympathetic to the needs of displaced farmers than were planters elsewhere and paid no heed to the pleas of the Farmers Union or the STFU. Planters and larger farmers dominated the local postwar planning committees and fed information to the Commission in Washington. In Poinsett County the chairman of the Post War Planning Committee was A. Carlson of Truman, works manager for Poinsett Lumber and Manufacturing Company (a subsidiary of Singer Company). E. P. Burton, a prominent Marked Tree planter, shared the vice-chairmanship with George Mouton, the proprietor of the Mouton Rice Mill in Harrisburg.[21] The needs of their own enterprises influenced their vision of the future, and when they formulated policy, it was those interests the local group had in mind. The Post War Planning Commission experts based their decision to discourage a "back to the land" movement after the war on their view that the family farm was a thing of the past, that large-scale, mechanized, scientific agriculture was the wave of the future. They predicted, and accepted as natural, the greater concentration of land ownership and thus focused on the need to maximize profits through greater efficiency and economies of scale. One such expert harkened back to what the application of Frederick Taylor's ideas had done in the industrial sector earlier in the century and recommended that his methods be employed on the farm.[22]

The Post War Planning Commission predicted the continuation of the labor shortage that emerged during the war; therefore, it supported the extension of the Mexican labor program in the postwar period. Labor organizations tried to counteract the view. George Stith, a black STFU organizer in the 1930s who became vice president of the National Farm Labor Union, an affiliate of the American Federation of Labor, insisted in 1950 that a sufficient number of domestic farm laborers were available and suggested that planter reliance on Mexican laborers was connected to keeping labor costs down: "I have seen truck loads of American citizens riding the roads looking for cotton to pick. When they ask for work the answer was 'Sorry boys, would like to have you but I got these Mexicans, got to take care of them you know!' When others refused to pick cotton for the low wages of $1.50 to $1.75 per hundred pounds they were told 'You don't have to take it. I can get Mexicans even cheaper than that.'"[23]

In June 1950, just weeks before the North Koreans crossed the thirty-eighth parallel, President Harry S. Truman appointed a presidential Commission on Migratory Labor. In a series of hearings held around the country between July 31 and September 16, no fewer than sixty-six farmworkers and labor representatives testified before the commission. Their testimony echoed Stith's, and apparently the commission found them more convincing than the seventy farm employers and officials of farm organizations who argued that a serious labor shortage existed. The commission submitted its report to the president on March 1, 1951, and argued that "in the present emergency, first reliance should be placed on using our domestic labor force more effectively. No special measures should be adopted to increase the number of alien contract laborers beyond the number admitted in 1950. Future efforts should be directed toward supplying agricultural labor needs with our own workers and eliminating dependence on foreign labor."[24]

But the Korean War had renewed fears of war-related labor shortages and played a role in encouraging the government to continue its involvement in the supply of labor on southern plantations. Truman's Department of Agriculture ignored the presidential commission's report.

Arguing that there was an insufficient supply of domestic farm labor, it increased the number of Mexican nationals brought into the country. Even as sharecroppers found themselves increasingly reduced to the status of wage laborers, Mexican nationals dominated the agricultural labor market.

Accompanying the erosion of the position of tenants and sharecroppers on the plantations was the deterioration of the economic base of many small-town merchants. According to Stith, some of those merchants, reminiscent of men like H. J. Krier of Marked Tree during the 1930s, sympathized with the tenants and sharecroppers. They recognized that the depopulation of the countryside threatened their livelihoods and that the seasonal presence of Mexican nationals offered little remedy. Although the Mexicans eagerly traded in the local mercantile establishments, and many merchants found it necessary to hire bilingual clerks to handle the traffic, most of the Mexicans lined up at the express office every Saturday and sent as much, if not more, of their pay back to their families in Mexico.[25]

Even though the displacement and depopulation of the countryside threatened the very existence of the towns in the Poinsett County delta, three of its prominent businessmen-planters, J. C. Baird, Jr., J. A. White, and Henry Craft, testified before the presidential commission arguing in favor of the importation of the Mexican nationals and thus, whether knowingly or not, supported a policy which would prove detrimental to the local economy.[26] Indeed, Poinsett County planters intensified their use of Mexican nationals during the cotton harvest seasons in the 1950s. From the beginning Arkansas planters had been particularly energetic in their recruitment of this labor supply, to the point that planters in Tennessee and Mississippi complained loudly in 1948 when it seemed that Arkansas planters were getting an unfair share of the supply.[27] By the mid-fifties Poinsett County planters like E. Ritter and Company were employing the labor of eight hundred Mexicans on its twenty-two-thousand-acre operation.[28] Between three and five thousand Mexican nationals per year were brought into Poinsett County in the 1950s and early 1960s. Those Mexican workers played a crucial role in helping planters to maintain their operations while they transformed their methods of producing and harvesting the crop.

Even as the federal government pursued a labor policy that seriously eroded the position of tenants and sharecroppers, it expanded the number of black county agents and offered a program to black farmers that was clearly out of step with its support of mechanization, wage labor, and scientific agriculture. While German POWs were helping to ease the serious labor shortage confronting Poinsett County planters and farmers during World War II, black sharecroppers and tenants found a new ally in the extension service. The appointment of Dr. F. D. Patterson of Tuskegee Institute as a special assistant to the secretary of agriculture in 1942 resulted in the expansion of the number of black agents but ensured that it would continue to adhere to the Tuskegee model of economic development. Patterson's appointment resulted from the need to increase production of agricultural products at a time when unprecedented labor shortages confronted farmers. But when the extension service assigned black agents to counties with heavy black populations during and immediately following the war, they outlined a program designed to ameliorate the harsh circumstances most tenants and sharecroppers endured rather than one intended to prepare them for a leap into capital-intensive agriculture.[29] Thus, black tenants and sharecroppers were encouraged to participate in the "live-at-home" program and produce more food crops for home consumption. Only a modest program was designed to encourage their transition into landowners. The black county agent merely made his clients aware of federal loan programs, but given the cost of land and the pegging of the purchase price of a particular farm to its productive capacity, the likelihood of black tenants and sharecroppers making the leap to farm ownership in Poinsett County was extremely slim.[30]

The first black extension service representative assigned specifically to Poinsett County was a home demonstration agent, Lena Eddington, who, like white home demonstration agents, worked with farm women. Eddington's narrative report for the year 1944 highlighted her efforts to introduce nutritional improvements and at the same time underscored a serious problem prevalent on the county's plantations. Convinced that the rejection of young black men for military service was largely due to poor nutrition, she assisted 350 families in improving their diets and instructed another 185 families in better food preparation. The "Movable School Agent" first introduced the "live-at-home" program to Poinsett County black families. Eddington enthusiastically supported it, reporting that 116 black families were in the program in 1944 and that it had "improved the standard of living of many families in this county."³¹ The home demonstration agents who succeeded Eddington continued in these activities in subsequent years while visits from the black district agent sufficed to reach black farm men until the arrival of the county's first black farm agent, L. J. Jackson, a graduate of Arkansas A. M. & N College with a degree in agriculture, in May 1946. The program adopted in 1946 encouraged black farm men to grow more food crops for home consumption, to increase and improve poultry flocks, dairy cows, and hogs, and to make modest repairs to equipment and buildings. The agent recognized the financial limitations of his clients. Fully 81.4 percent of the black farm operators were sharecroppers, and most of the rest were tenants. Only 3.7 percent were farm owners. He recommended that future work with black farmers include a "better live-at-home program" and suggested that farmers "repair homes where they will be convenient," and "watch the up keep of out houses in order to better observe sanitation and health."³² Jackson left the county on September 30, and a new agent replaced him in January 1947. The declining number of black farmers in the county led to the removal of the county's last black agent in 1957. After the Civil Rights Act of 1964, the Extension Service integrated its offices and no longer operated a separate black agent program. Black agents typically became assistant agents under the supervision of the white county agents. In the process of implementing the provisions of the act, the extension service also reevaluated the entire operation and determined that not enough blacks remained in Poinsett County to warrant the assignment of a black assistant agent.³³

It was no coincidence that when the number of farm operators in Poinsett County declined in the postwar period, black farmers, who were almost exclusively tenants and sharecroppers, declined at a much higher rate than white farmers. They did not have the means to invest in the machinery and chemicals needed to participate in the postwar revolution in agriculture, and planters' ability to influence the cost of labor exacerbated their situation. Between 1940 and 1965 black tenants and sharecroppers declined 91.9 percent, from 642 to 52, and white tenants and sharecroppers declined 76.1 percent, from 2,536 to 605.³⁴ As elsewhere in the South, some white tenants were related to the landowners for whom they worked and thus enjoyed advantages not extended to black tenants. And although the white county agent worked principally with farm owners, he occasionally extended advice to white tenants. White tenants operated larger farm units than did black tenants and were not so impoverished. Finally, the failure of a black business class to develop in the delta towns meant that blacks had to trade exclusively with white merchants.[35]

By the time the Student Non-Violent Coordinating Committee (SNCC) began actively working in the Arkansas delta to register blacks to vote in 1963, the black population of Poinsett County had declined to the point that the county was not targeted by that organization. No mention of Poinsett County is made in SNCC's extensive archives. Mississippi, Crittenden, and Cross counties, which form the eastern and southern borders of Poinsett County, all witnessed considerable SNCC activity. Poinsett County also received no attention from the Arkansas Council on Human Relations, an organization founded to help facilitate peaceful integration. And, finally, none of the reports in Governor Orval Faubus's papers from white law enforcement officials in the delta on surveillance of black organizations mention Poinsett County.[36]

During the period when the extension service was willing to play a role in the affairs of black farmers, it wanted to be careful about upsetting prevailing racial mores. It told its staff to avoid all association with the NAACP or other black organizations advocating social change. One black employee of the Poinsett County farm agent's office, however, had to be reprimanded for an alleged connection to the NAACP. Mrs. Irma Kimbell, the secretary in the black agent's office, was given special instructions after rumors surfaced concerning her involvement with the NAACP. The black district agent, T. R. Betton, defended Kimbell by ascertaining from her immediate supervisor that "as far as he knows she was not connected with this organization; nor is she connected with any controversial organizations. There is no NAACP organization in this county." Betton, who may have been the son of F. R. Betton, a former vice president of the STFU, also denied the rumor that Kimbell had participated in a sit-in demonstration. Nevertheless, the "recommendations" to Kimbell indicate that she had expressed opinions that fed the rumors concerning her activities. She was instructed to "be more careful in making statements to her associates and friends of a controversial nature. . . . Refrain from being in company in public with anyone who is a suspect of controversy. . . . Do everything possible to leave a favorable impression with the public that is healthy for Extension."[37]

The fact that Kimbell's supervisor found it necessary to counsel her concerning her alleged civil rights activities suggests that at least one black woman voiced opinions which challenged prevailing racial mores and that at least some blacks in the county may have found what she had to say appealing. Certainly, Marked Tree's black community demonstrated greater presence in the postwar period. The men founded an American Legion Post in 1950 and the women a Garden Club in 1951. The black community also succeeded in getting the black school expanded from eight grades to twelve in 1945.[38]

The white community, meanwhile, reacted with relative calm to the *Brown v. Board of Education* decision in 1954 and to subsequent developments regarding desegregation. Editorials in the *Marked Tree Tribune* criticized extremists on both sides and, watching with alarm the developments in Little Rock in 1957, called for a "new note of responsibility" among leaders in the state and the nation.[39] The editorial writer of the newspaper, Dorothy Stuck, fully supported restoring voting rights to black citizens, but was not enthusiastic about desegregation. She regarded desegregation statutes as infringements upon individual liberties, but at the same time she decried racial prejudice: "There is no reason in prejudice—no sense of responsibility. Prejudice is the manifestation of an inner weakness that turns away from the self-discipline it takes to be reasonable and just. Prejudice rests on the unsteady ground of rationalization that seeks only to justify—never rectify."[40] Stuck was only able to express such moderate sentiments because at least some of the white community was receptive to them. In 1965 the Marked Tree schools began to integrate without incident, and other Poinsett County towns did likewise. The only exception, ironically, was Tyronza which as late as 1964 had no desegregation plan in place.[41] Stuck wrote in 1965 that "our Board has worked hard to comply with a law that goes against the grain of the vast majority of citizens of both races in our area." She urged compliance and argued that "students of both races deserve an opportunity to meet this challenge without the club foot of prejudice."[42] The sentiments expressed in this editorial, published in May 1965, stand in stark contrast to those expressed nearly sixty years earlier by the first editor and publisher of the Marked Tree newspaper.

By the time delta schools desegregated, the black population in Poinsett County had declined dramatically, and the transformation of the plantation system in Poinsett County, and in the South as a whole, was all but complete. The tractor, the cotton picker, and the combine had taken the place of the mule and the hand hoe. The land tenure system had gone through a metamorphosis which saw the almost complete disappearance of sharecroppers, who were the most vulnerable of farm operators, and a drastic reduction in the number of tenants on farms. By 1965, for the first time since before the emergence of the plantation system in Poinsett County, the number of owners and part owners there exceeded the number of tenants. In 1900 owners held an average of under fifty acres; by 1965, 89 percent of the acres in farms were held by individuals who farmed in excess of five hundred acres, and half of those acres were held in farming operations in excess of a thousand acres.

L. V. Ritter presided over a vast farming operation and served as president of both E. Ritter and Company and the St. Francis Valley Company, co-owning the latter operation with a group of Memphis investors.[43] The two companies together held twenty-two thousand acres in farmland, including several thousand acres purchased after Chapman and Dewey ceased doing business in 1948. Although both companies had been mechanizing throughout the 1950s, the trend accelerated in the 1960s, especially after the cessation of the Mexican Labor program in 1964. According to Jean Thatcher, in 1960 E. Ritter and Company absorbed the St. Francis Valley Company, and the entire operation employed approximately seventy-five farm laborers. By 1975, after full mechanization, that number had dropped to about twenty workers. Similarly, when the Ritter Company had its five gins in operation, it hired close to sixty employees; after it constructed a large new gin in 1975, it closed its five old gins and hired only twelve workers to run the new one. Significantly, it was in 1975 that the company began to offer hospitalization and a retirement plan.[44]

Meanwhile, black farmers in Poinsett County found fewer opportunities open to them. By the time Thatcher retired in 1986, E. Ritter and Company employed no black farmers; according to him, blacks "stopped working for us when we mechanized."[45] This development was apparently not by design, however, and the kind of enterprise into which E. Ritter and Company evolved would not have been attractive to the tenants and sharecroppers who founded the Southern Tenant Farmers' Union. Those tenants and sharecroppers were fighting for the right to hold on to farmsteads and operate independent of supervision. Although E. Ritter and Company provided a retirement plan and hospitalization for its farmworkers and gave them greater job security by hiring them year-round instead of seasonally, their employees were wage laborers rather than semi-independent farmers. Because they operated expensive machinery, those who worked for the neoplantations were closely supervised, and probably few of them had any visions of farm ownership.

When Jean Thatcher left E. Ritter and Company in 1986, the tenants and sharecroppers who helped to transform the delta in the early part of the century had been replaced by wage laborers. But most of the planters and farm owners in the delta hired wage laborers seasonally, for only large enterprises like the Ritter Company could engage in the sophisticated year-round operation described by Thatcher. Thus, the labor system that arose in the delta in the postwar period came to resemble that which had existed from the beginning in the prairie. Rice farmers there had long used a form of wage labor. The similarities between the two sections do not stop there. Along with the transformation in the delta came an increase in the production of soybeans and rice, the latter a crop traditionally grown only in the prairie. Rice farmers in the prairie, meanwhile, adopted the use of fertilizers and began to cultivate acreage in cotton and soybeans. Today both sections are flat, relatively treeless, and completely drained. They produce the same mix of crops and rely on expensive equipment and chemicals.

While the prairie and delta have come to resemble each other, at least superficially, the ridge has maintained its own distinctive economy. Broken by rolling hills and streams, ridge land was not conducive to large-scale production, and thus farmers there continued to rely on livestock and orchards for their livelihoods. It is likely that Edwin Palmer, who described his trip through the delta to Harrisburg in 1881, would find the area surrounding the town much more recognizable than the swamps. Harrisburg is itself about as prosperous today as it was then. The courthouse still stands, and a local man, Steve Ryan, completed a sixteen-year term as county judge in 1993. Echoing the old battle over the courthouse in 1917, Judge Ryan recently led a successful ridge effort to thwart the designs of certain delta elites to locate a new million-dollar county jail in the delta. The new jail will be built in Harrisburg. Economic power continues to have its political limits.

In uniting to form an interracial union in Poinsett County, black and white plantation laborers overcame the legacy of racism. This realization of their common interests was one of the most important things they left behind when they moved from the rural South to the urban ghettos in the post–World War II period.[46] At the same time not only were culturally distinct communities eradicated, but hopes for land ownership and independence were dashed. Landless southern farmers who for generations had avoided proletarianization now joined others in trekking to the cities and rising to the factory whistles rather than the sun. The industrial jobs in cities provided them with the means to purchase items like radios, refrigerators, and even automobiles, and these things seduced their children, whose attachment to the rural way of life and adherence to their parents' dreams were tenuous at best.[47] The transition to the city was most difficult for those who had spent much of their lives in the countryside and had been nurtured and sustained in that culture. Although they welcomed the rewards that jobs in the city could provide them, they reminisced about that inexplicable something they left behind: "It seemed dreadfully unnatural to them to stand on their front porch and be able to talk to somebody else standing on *his* front porch. It sometimes happened in the county that a man could *see* another house from his front porch, but not often. In the city, though, they were forever cheek to jowl. They felt like animals in a pen. It was, they said, no way for a man to live."[48]

APPENDIX

NOTES

BIBLIOGRAPHY

INDEX

APPENDIX
QUANTITATIVE METHODS

Much of the evidence for *Contested Ground* is derived from quantifiable sources. My use of such material in the tables and in the text included analyses of the federal manuscript census; the Poinsett County real property tax records; the Poinsett County personal property tax records; the Poinsett County mortgage records, and the Arkansas Slave Narratives.

My use of the federal manuscript census, schedule of population, was limited by the seventy-year privacy act, which make records unavailable after 1920. I selected four townships in the county for the years 1860, 1870, 1880, 1900, 1910, and 1920. A township is a geopolitical entity, a voting area defined by geographical boundaries. I chose townships that included the principal town in the particular section: Marked Tree, the principal town of the delta, is within Little River Township; Harrisburg, the county seat and the principal town on the ridge, is in Bolivar Township; and Weiner, the principal town of the prairie, is in West Prairie Township. Because I wanted to focus on the development of the delta, and because Little River Township was very small, I took an additional delta township, Tyronza Township, adjacent to and southeast of Little River Township. Between 1860 and 1870 the boundaries of the county and of the townships within the county changed. Bolivar township was much smaller after the partitioning. The prairie and the delta, meanwhile, were so sparsely populated in this period that they were simply marked as uninhabited and no names were enumerated.

I entered all available information on every individual. In order to keep track of the households, I assigned an identification number to

Table A.1. Cases in selected Poinsett County townships from census of population

Township	1860	1870	1880	1900	1910	1920
Bolivar	1,302	865	1,000	1,590	2,387	3,040
East Bolivar						917
West Prairie			178	524	477	965
Little River and Tyronza			142	1,833	4,510	6,197
Total	1,302	865	1,320	3,947	7,374	11,119

Source: Federal Manuscript Census, Arkansas, Poinsett County, Schedule of Population, 1860, 1870, 1880, 1900, 1910, 1920.

Note: East Bolivar was not officially recognized by the census bureau as a township separate from Bolivar, but it was treated as such by the census taker.

each one. But the unit of analysis was the individual. Table A.1 shows the number of individual cases. The information entered included sex, race, age, occupation, nativity, and position within the household (whether head of household, wife, son, daughter, boarder, etc.), so that I could link records from one census year to another. For example, Vernon and Catherine Thrower appeared on the 1860 census with a two-year-old son, William, and an infant daughter. I recorded all four Throwers as individual cases. The 1870 census included Vernon Thrower with Catherine and their two children plus a third child, a four-year-old son named James. Because they were all ten years older than they had been in 1860, and they listed the same sex, race, and nativity as they had before, I gave them the same identification numbers I had given them in 1860 and assigned the new young son his own identification number. By 1880 Vernon and Catherine Thrower were no longer in the record. William Thrower, however, appeared as the head of his own household. He had an eighteen-year-old wife and a fourteen-year-old brother, James Thrower, living with him. Because William and James listed the same sex, race, and nativity as they had in 1870, and their ages tallied with the time lapse, I gave William and his brother the identification numbers they had been carrying and assigned William's wife a new identification number.

In the process of matching from one census to another, I made certain allowances for the specified ages, for ages were sometimes reported incorrectly by a year or two, one way or the other, and I occasionally accepted as a match a case where all the information was consistent but the nativity. But if I was in serious doubt about a certain individual, I would assume she or he was not the same person and assign her or him a new identification number. The error involved in this kind of record linkage was minimized by my decision to apply the four standards of comparison (sex, race, age, and nativity). The use of the family unit itself as an indicator was often useful. Those individuals named Jones and Smith presented, of course, the greatest difficulty. I often resorted to a close examination of the individual's family. I found that I could eliminate a great many with those surnames, that is, assign them new identification numbers, keeping them in the record for purposes of analysis independent of the record linkage.

Two problems with the 1910 manuscript census for Marked Tree prevail. First, six pages were badly microfilmed so that 194 names were illegible; the paper manuscript of the census was destroyed after the microfilm was made. Second, the census taker who handled the town of Marked Tree and the township of Little River confused the boundary lines. Fortunately, even though the census sheets are not clearly marked, the total for Little River, including Marked Tree, reported in the census almost precisely to the figures derived for the two (minus the illegible entries mentioned above). It is significant that there were very few sawmill workers on the unclearly marked sheets so these were probably outlying areas.

In addition to linking these records from one census year to another, beginning in 1860 and ending in 1920, I also linked the schedule of population in 1860 to the schedule of slaves in 1860 in order to determine the number of slaves each individual owned. The schedule of slaves was arranged like the schedule of population, by township.

I also linked the schedule of population with the schedule of agriculture of 1880. Again, I had to assume that the William Thrower who appeared on the schedule of population in Bolivar Township was the same William Thrower who appeared on the schedule of agriculture in Bolivar Township, for I had nothing more than the name and the township within which the individual resided to draw upon. In the process of linking these two records, I eliminated not only a great many Smiths but other people with common surnames.

The schedule of agriculture for 1880 provided valuable information regarding the farm economy in the county. Selecting the same townships

I had used in the schedule of population, I entered each individual. If he did not match with a name in the schedule of population, I assigned him a new identification number. The schedule of agriculture detailed not only the kinds of crops grown, the acreage planted, and the bushels or pounds, etc. produced but also included a category for land tenure. Unfortunately, the schedules of agriculture for 1900, 1910, and 1920 had been destroyed by the National Archives. In 1880 there were 241 individual cases in the ridge area, 38 in the prairie, and 25 in the delta, giving a total of 304 from the manuscript census of agriculture.

Another source I used for 1880 was the Poinsett County real property record, arranged by section, township, and range. In this case "township" is a cartographical entity to be taken in conjunction with "section" and "range" and is entirely different from the geopolitical township used by the census enumerators. In any case, the county tax assessor enters each parcel of land beginning with the first parcel located in the southwestern prairie and working to the last parcel located in the northeastern delta. In order to determine how much acreage a given individual owned in the county, it was necessary to consider all the parcels and sort the record by the name of each individual. I had only the name of the individual as a basis for determining whether I had a match once I sorted the file alphabetically. This presented certain obvious problems. I chose to assume I had a match only if the name was not a very common name. If a John Davis owned several different parcels that were not listed contiguously on the record, I assigned each John Davis a separate identification number. Fortunately, this problem occurred relatively infrequently in the real property records. Either the names were sufficiently distinct or the parcels were contiguous and the names were entered in such a way as to leave little doubt that the individual was the same, as in

Name	Section	Township	Range	Acreage
John Davis	28	10	7	40
"	29	10	7	40
"	30	10	7	40
"	31	10	7	40

I then added the acreages and valuations together and came up with a total acreage and valuation for each individual landowner which I entered in new columns in the record.

The real property record supplied data regarding the acreage owned, its tax valuation, its location according to section, township, and range, and the school district within which the acreage was located. For the years after 1900, the numbers of the road and drainage districts were provided. I chose to enter all the information listed above for the years 1880, 1900, 1910, 1920, 1930, and 1940. Poinsett County real property records were not available before 1878. Table A.2 shows the number of cases from these records.

I then linked the names on the real property records with those on the schedules of population and agriculture in 1880. Although there was a complete run of real property records from 1880 to 1940, there was no manuscript census of agriculture for the twentieth century or schedule of population for the years beyond 1920. But certain county records proved extremely useful. The mortgage records were available from 1880 to the present, and the personal property tax records were available for every year after 1917; of these, I used the mortgage records for 1909–10 and the personal property records for 1920, 1933, and 1934. I chose to focus on the mortgage record for 1910 because I suspected that the plantation system had been established in the first decade of the twentieth century. I would have looked to the personal property tax records in that year had they been available. Although there was no schedule of agriculture for 1900, 1910, and 1920, the schedule of population in 1900, 1910, and 1920 provided information about farm ownership, for there was a category stipulating whether a farmer owned or rented his farm or home. Most farmers listed themselves living on farms rather than living in homes. I selected only those farmers who listed whether they owed or rented their farms,

Table A.2. Cases from Poinsett County real property records

	1900	1910	1920	1930	1940
Ridge	572	692	843	608	721
Prairie	437	505	598	309	500
Delta	193	303	631	808	576
Total	1,202	1,500	2,072	1,725	1,797

Source: Real Property Records, 1900, 1910, 1920, 1930, 1940, Poinsett County Courthouse, Harrisburg, Ark.

for example, to construct table 2.3. Another category detailing "weeks/months-out-of-work" in 1900 and 1910 also proved useful.

The Poinsett County mortgage records were of use in filling the gap left by the loss of the manuscript schedule of agriculture; in fact, the mortgage records provided valuable information that would not have been available on the schedule. The mortgage records were arranged according to date of filing. From 270 mortgages for the years 1909 and 1910, I selected from the documents not only the names of the lenders and borrowers, but the amounts of and lengths of the indebtednesses, the interest rates charged, the nature of the specific property being pledged (whether land or personal property, household goods or livestock, etc.), I also noted whether the borrowers pledged to grow cotton and whether they agreed to market the cotton through the lenders.

Because the mortgage records did not indicate where in the county the particular borrower resided, I collected only those from the record that I could locate on the manuscript of population for 1910. I then looked more closely at the lenders and was able to identify 201 and established their locations. I was then able to trace the credit network in the county. Table A.3 showed the number of matched cases.

Linking the schedule of population with the Poinsett County mortgage records presented the usual problems. Yet I often had not only the name of the man mortgaging his property but also his wife's name, for many of the mortgages included the wife in the agreement. This facilitated a match between the mortage records and the schedule of popula-

Table A.3. Matched cases of borrowers in mortgage records and manuscript census of population in Poinsett County, 1909–10

	Real property	Personal property	All mortgages
Ridge	60	31	91
Prairie	33	35	68
Delta	51	70	121
Total	144	136	280

Sources: Real Property Records, Personal Property Records, and Mortgage Records, 1909 and 1910, Poinsett County Courthouse, Harrisburg, Ark.

tion, and it accounts for the relatively high number of documents I was able to utilize.

For the year 1920 I turned to the Poinsett County personal property records. Unlike the manuscript of population, which is arranged by township and the real property records arranged by parcel location, the personal property records are organized according to school district. On a map of the county listing school districts, I superimposed a section/township/range map. I could then code each entry from the personal property tax records for its school district.

The personal property tax records noted the number of farm animals and their value, the amount each person claimed in jewelry, musical instruments, and household goods as well as the dollar amount held in merchants' supplies, manufactured articles, automobiles, and the total tax paid by each property holder. Although the record does not reflect the sex or age of the property holder, it does indicate the race of the individual tax payer. Merchants, sawmills, and cotton gins could be identified because of categories listing merchant stock and manufacturing articles.

I linked the personal property tax records of 1920 to the real property records for that year by comparing the location in the county and the name. I derived a subset of 560 individuals who appeared on both sets of records; 240 from the ridge, 124 from the prairie, and 196 from the delta. The subset was made up of the most prominent citizens in the county, for it included only those individuals who owned real property. The subset was second only in importance in this study to the match between the mortgage records and the schedule of population of 1910.

Matching the 1920 personal property tax records to the 1920 population census provided valuable insight into the standard of of the three sections of the county. Again, the linking required conversion from school district to township. Having identified the township within which the personal property holders resided, I compared the list of names and made a substantial number of matches.

To examine the delta township where the Homesteaders Union was founded, Greenwood Township, I recorded all the information from the population censuses of 1900, 1910, and 1920 and matched it to the real property and personal property records where possible. I also linked names of individuals from that township to the real and personal property tax records.

In order to facilitate an analysis of Tyronza Township in 1934 when the Southern Tenant Farmers' Union was founded, I identified its school district and selected only those individuals from the personal property tax records within that school district. I then subdivided it into quartiles (four parts), took the 1920 data for the township, subdivided them into quartiles and deflated the values to conform to 1934 dollars. I also divided the 1920 and 1934 personal property tax records into two additional categories: individuals and companies. I then linked the names generated in these lists to the real property records and to the census records.

Finally, the Arkansas Slave Narratives provided a valuable source of information concerning the movement of freedmen and their families into Arkansas in the post–Civil War period. For all 690 individuals, I compiled and analyzed the following information: sex, age, occupation, marital status, number of children, political activity, economic condition, nativity, and the place from which they immigrated to Arkansas. Some people also included information on when they immigrated to Arkansas, the place to which they came in Arkansas, or whether a labor agent was involved.

NOTES

Abbreviations and Short Citations

See Bibliography for repositories.

AAA	Agricultural Adjustment Administration
Abbott Papers	Abbott Family Papers
ACHR	Arkansas Council on Human Relations
Agriculture	Department of Agriculture
ACES	Alabama Cooperative Extension Service
ASCS	Agricultural Stabilization and Conservation Service
BAE	Bureau of Agricultural Economics
Campbell Papers	Thomas M. Campbell Papers
CCC	Commodity Credit Corporation
Court Records	Poinsett County Court Records
CWA	Civil Works Administration
Davis Papers	Jeff Davis Papers
East Papers	Henry Clay East Papers
Emrich Papers	John Emrich Papers
Extension Service	University of Arkansas (System), Cooperative Extension Service Records
Faubus Papers	Orval Eugene Faubus Papers
FCA	Farm Credit Administration
Federal Extension	Agricultural Extension Service Records, Federal
FERA	Federal Emergency Relief Administration
FHA	Farmers Home Administration
FSA	Farm Security Administration
Hays Papers	Lawrence Brooks Hays Papers
Hearings of the Committee	Hearings Held Before the Committee on the Public Lands

Justice Department	Department of Justice
Labor	Department of Labor
Mayor's Court	Mayor's Court, Harrisburg, Ark.
Minutes of the Meetings	Minutes of the Meetings of the Board of Commissioners of Drainage District 7
Mortgage Records	Poinsett County Mortgage Records
MS Census	Federal Manuscript Census
NA	National Archives
Palmer Papers	Edwin Palmer Papers
Personal Property	Poinsett County Personal Property Tax Records
PWA	Public Works Administration
RA	Resettlement Administration
Real Property	Poinsett County Real Property Tax Records
Record of the Proceedings	Record of the Proceedings of the Council of the Incorporated Town of Marked Tree, Ark.
Red Cross	Red Cross Records
RFC	Reconstruction Finance Corporation
Rice Millers Papers	Rice Millers Association Papers
Robinson Papers	Senator Joseph Taylor Robinson Papers
Roosevelt Papers	Theodore Roosevelt Papers (on microfilm)
Sanborn Map	Sanborn Map Company
SCS	Soil Conservation Service
SNCC Papers	Student Non-Violent Coordinating Committee Papers
STFU Papers	Southern Tenant Farmers' Union Papers
Thornburgh Collection	George Thornburgh Collection
Treiber Papers	Jacob Treiber Papers
Truman Papers	Harry S. Truman Papers
USPHS	U.S. Public Health Service
WPA	Works Progress Administration
WPA Church Records	Arkansas Historical Records Survey, Church Records
WPA Place Files	Works Progress Administration Place Files, Arkansas

Introduction

1. On progressivism, see Moyers, "Arkansas Progressivism"; Grantham, *Southern Progressivism*; Dittmer, *Black Georgia*; Link, *Paradox of Southern Progressivism*. On prohibition, see Hunt, "History of the Prohibition Movement"; Boyer, *Urban Masses and Moral Order*; Gusfield, *Symbolic Crusade*; Buenker, *Urban Liberalism and Progressive Reform*; Duis, *Saloon*; Clark, *Dry Years*; West, *Sa-*

loon. For both see Timberlake, *Prohibition*. For race relations, see especially, Woodward, *Strange Career*, and Rabinowitz, "From Exclusion to Segregation."

2. Daniel, *Breaking the Land*; Kirby, *Rural Worlds Lost*; and Cobb, *Most Southern Place on Earth*.

3. Kirby, *Rural Worlds Lost*, xv; Daniel, *Breaking the Land*, xi.; Cobb, *Most Southern Place on Earth*, x; Gavin Wright, *Old South, New South*, 8.

4. Cobb, *Most Southern Place on Earth*; Brandfon, *Cotton Kingdom of the New South*. See also the three articles by Woodruff: "African-American Struggles for Citizenship in the Arkansas and Mississippi Deltas in the Age of Jim Crow"; "Mississippi Delta Planters and Debates over Mechanization, Labor, and Civil Rights in the 1940s"; and "Pick or Fight."

5. For explorations of conflict between contending groups and how it influences outcomes, see Royce, *Origins of Southern Sharecropping*, 17–22. Royce alerted me to several other perspectives on this phenomenon: Skocpol, *States and Social Revolution*, 17–18; Polanyi, *Great Transformation*, 36; Moore, *Injustice*, 376–81; Dawley, *Class and Community*, 196, 240; Thompson, *Making of the English Working Class*, 194; and Bendix, *Nation-Building and Citizenship*, 55–56.

6. For the origins of sharecropping, see Litwack, *Been in a Storm So Long*, 447–48; Royce, *Origins of Southern Sharecropping*, 1–3; Woodman, "Sequel to Slavery," 523, 550; Meier and Rudwick, *From Plantation to Ghetto*, 171–72; Mandle, *Roots of Black Poverty*, 25; and Shlomowitz, "Origins of Southern Sharecropping," 557. For the coercive and restrictive aspects of sharecropping and tenancy, see Ayers, *Promise of the New South*; Foner, *Reconstruction*; Mandle, *Roots of Black Poverty*; Ransom and Sutch, *One Kind of Freedom*; Reidy, *From Slavery to Agrarian Capitalism*; Wiener, "Class Structure and Economic Development in the American South"; and Woodman, "Post Civil War Agriculture and the Law." For alternative views of the nature of the southern economy and the mobility of blacks in the post–Civil War era, see Higgs, *Competition and Coercion*; Reid, "Sharecropping as an Understandable Market Response"; DeCanio, *Agriculture in the Postbellum South*; Cohen, *At Freedom's Edge*; and Wright, *Old South, New South*.

7. Woodman, *New South, New Law*.

8. Royce, *Origins of Southern Sharecropping*, 186.

9. Quoted in Royce, *Origins of Southern Sharecropping*, 187. See Woodman, "Post Civil War Southern Agriculture and the Law," 327n.13.

10. Stith interview.

11. Thompson, *Making of the English Working Class*.

12. Ibid., 11.

13. On the rise of the Southern Tenant Farmers Union, see Grubbs, *Cry from the Cotton*; Mertz, *New Deal Policy and Southern Rural Poverty*; Conrad, *Forgotten Farmers*; and Dunbar, *Against the Grain*. All four, however, assume that the plantations of Poinsett County were large business plantations similar in scale to the Delta and Pine Land Company, and that planters there had unquestioned power and authority.

1. "A Wild and Sickly Country"

1. Palmer, "Field Notes," Oct. 13, 1882, Palmer Papers.
2. Foti, "River's Gifts and Curses," 35.
3. Court Records, July 6, 22, 1886, Jan. 5, 1890.
4. Balch, "Story of Richwood Township," 379; Hubbell, "Always a Simple Feast," 187.
5. Hubbell, "Always a Simple Feast," 191.
6. Palmer, "Field Notes," Oct. 13, 1882.
7. MS Census, Agriculture, 1880.
8. Palmer, "Field Notes," Oct. 13, 1882; Goodspeed, *Biographical and Historical Memoirs*, 571.
9. Palmer, "Field Notes," Oct. 18, 1882.
10. In 1879 only 7,712 out of approximately 486,000 acres, or 1.6 percent of the county's total area, were under cultivation. *Tenth Census, Agriculture, 1880*, 51, 55–57.
11. Bank of Harrisburg, *Pictorial History*, 1.
12. Lynch, "Thru the Years," 1.
13. Rawick, *American Slave* 8: 163; Thompson, *Arkansas and Reconstruction*, 9.
14. Mathews, "History of Poinsett County," 1, WPA Place Files.
15. Wheeler, "People's Party in Arkansas," 29.
16. Harrison and Kollmorgen, "Land Reclamation in Arkansas," 416; Ellenburg, "Reconstruction in Arkansas," 105–7.
17. His slaves had increased by 17 since 1850. MS Census, Population, 1850, 1860; MS Census, Slaves, 1860; Walz, "Arkansas Slaveholdings," 56.
18. Chowning, *History of Cross County*, 3–5.
19. Goodspeed, *Biographical and Historical Memoirs*, 7, 572. Harris's father, William, had been the first county judge, and Benjamin Harris owned the site where Harrisburg was located.
20. Chowning, *History of Cross County*, 8.
21. Ibid.
22. Determining exactly how many of its 428,620 acres of swampland remained in Poinsett County after the creation of Cross County is problematic. Harrison and Kollmorgen, "Land Reclamation in Arkansas," 379.
23. Ellenberg, "Reconstruction in Arkansas," 11, 112; Harrison and Kellmorgen, "Land Reclamation in Arkansas," 416. See also Wood, "Development of Arkansas Railroads"; Tucker, *Arkansas*; Hacker, *Course of American Economic Growth and Development*, 116; North, *Economic Growth of the United States*, 206; Woodward, *Origins of the New South*; and Wright, *Old South, New South*, 39.
24. Wood, "Development of Arkansas Railroads," 158.
25. Lynch, *Then and Now*, 2, 6.
26. Fully 64.5 percent of the heads of household in the prairie in 1900 were northern born. MS Census, Population, 1900.
27. See Hahn, *Roots of Southern Populism*.

28. Mathews, "Industries of Poinsett County," 2.

29. Cart interview, April 1, 1983; Schloze interview, April 1, 1983; Senteney interview, Oct. 25, 1986; Ziegenhorn interview, April 2, 1983.

30. In 1900, 78.7 percent of the population in two large delta townships were southern born; by 1910 this had risen to 82.9 percent. MS Census, Population, 1900, 1910.

31. Dawson, *One Hundred Years of Progress*, 2–3.

32. *Tri City Tribune* (historical edition), May 14, 1992, 10b.

33. Ibid., (heritage edition), May 12, 1993, 4; Dewey interview, Dec. 28, 1985.

34. These figures include the unskilled category. In Marked Tree 43 of the 45 individuals in this category listed themselves as day laborers. MS Census, Population, 1900.

35. Ibid.

36. Davis, *Trials of the Earth*.

37. MS Census, Population, 1900; *Tri City Tribune* (heritage edition), May 12, 1993, 4.

2. From Railroad Camp to Plantation Town

1. *Marked Tree Tribune*, May 17, 1918.

2. Whayne, "Significance of Race, Class, and Family in the Battle for Prohibition in Small Town Arkansas," 138. See also Boyer, *Urban Masses*; Gusfield, *Symbolic Crusade*; Timberlake, *Prohibition*; Buenker, *Urban Liberalism*; Duis, *Saloon*; Clark, *Dry Years*; and West, *Saloon*.

3. On progressivism in Arkansas see Moyers, "Arkansas Progressivism." Ray Arsenault's *Wild Ass of the Ozarks* pays some attention to the issues of progressivism, disfranchisement, and prohibition in Arkansas, but only one study of prohibition itself exists, Hunt, "History of the Prohibition Movement in Arkansas." On progressivism in the South, see Grantham, *Southern Progressivism*; Link, *Paradox of Southern Progressivism*; Kirby, *Darkness at the Dawning*; as well as Dittmer, *Black Georgia in the Progressive Era*; Tindall, *Emergence of the New South*, 18–20; and Kousser *Shaping of Southern Politics*, 229. Prohibition, however, was a class rather than race issue. See Whayne, "Significance of Race, Class, and Family in the Battle for Prohibition in Small Town Arkansas."

4. West, "Life on the Urban Frontier," 4–15.

5. Smith, *Sawmill*, 22, 37; Davis, *Trials of the Earth*; Dawson interview.

6. *Marked Tree Gazette*, March 30, 1906.

7. In 1900, for example, 121 (34.4 percent) individuals out of a total population of 352 lived as boarders. By 1910, only 347 (17.2 percent) out of a population of 2013 lived as boarders. MS Census, Population, 1900, 1910.

8. County Court, Aug. 15, 1900, 321.

9. Dawson interview; Thatcher interview.

10. Real Property, 1910; *Arkansas Gazette*, May 5, 1921; *Commercial Appeal*, May 5, 1921.

11. Dewey interview.

12. Smith, *Sawmill*, 4.

13. St. Francis Levee District, *History of the District*, 289; Minutes of the Meeting, March 28, 1921–Feb. 13, 1939; Portis interview.

14. Smith, *Sawmill*, 37.

15. There were 94 adult black men in Marked Tree in 1900 and 87 adult white men. MS Census, Population, 1900, 1910.

16. Dewey interview.

17. MS Census, Population, 1900.

18. Jensen, *Lumber and Labor*, 75.

19. Tokes, *Labor and Lumber*, 130; "Red Cedar Shingles and Shakes," 115.

20. Smith, *Sawmill*, 25.

21. *Marked Tree Gazette*, Jan. 19, 1906; Jensen, *Lumber and Labor*, 26–27.

22. Dewey interview.

23. Ibid.; Dawson interview; Smith, *Sawmill*, 42, 47, 78, 69, 91, 96, 100, 161. The Lee Wilson plantation in nearby Mississippi County also issued doodlums (Wilson interview).

24. Dewey interview.

25. MS Census, Population, 1900; Dawson interview. Ernest Ritter and J. R. Wiggenton operated mercantile establishments in Marked Tree in 1900. Chapman and Dewey, however, also operated a commissary.

26. MS Census of Population, 1900. Note that these are "adults" rather than heads of households.

27. *Marked Tree Gazette*, Jan. 13, 20, March 17, 1911.

28. Ritter, "Marked Tree from 1883–1936: Pt. 2," 27.

29. *Marked Tree Gazette*, May 1, Dec. 11, 1908, Oct. 15, 1909, June 30, Sept. 8, 1911, May 17, 1912.

30. Whayne, "Significance of Race, Class, and Family in the Battle for Prohibition in Small Town Arkansas," 143; MS Census, Population, 1900, 1910, 1920. For the imposition of Jim Crow in Arkansas, see Graves, *Town and Country*.

31. MS Census, Population, 1900, 1910.

32. The 1891 election law, a "major instrumentality of Negro disfranchisement" in Arkansas, "established a residence requirement for voting of one year in the state, six months in the county, and one month in the precinct or ward." The census figures indicate that most millworkers were not native Arkansans, so the one-year residence requirement would have excluded them from the polls. Graves, "Negro Disfranchisement in Arkansas," 216; Graves, *Town and Country*; MS Census, Population, 1900.

33. When Ritter incorporated E. Ritter and Company in 1907, W. B. Miller and N. J. Hazel served as officers of the corporation along with M. W. Hazel. The latter was the town's first recorded marshal. Arnold interview; Record of the Proceedings, May 20, 1903, 13.

34. Record of the Proceedings, Oct. 14, 1902, 200, April 7, 1903, 210. J. F. Fuller, who brought the first blacks into the Poinsett County delta in 1892 to work his lumber mill, was also on the first council.

35. Ibid., Aug. 26, 1897, 11–22.

36. Ibid., Sept. 7, 1897, 35, Nov. 24, 1897, 57. They did not, however, sell the convict labor at this point but allowed the prisoners to work off their fines at the rate of $1.00 per day.

37. Ibid., July 12, 1900, 125–26.

38. *Marked Tree Gazette*, July 21, 1905, March 10, 1908.

39. Ibid., March 30, 1906, Sept. 27, 1907.

40. Ibid., Aug. 2, 1907, June 26, 1908.

41. Ibid., May 22, 1908, Dec. 24, Feb. 12, 1909.

42. Ibid., Nov. 12, 1909.

43. MS Census, Population, 1910. For the problems with the 1910 census of population, see the Appendix.

44. In 1900 there were actually 181 adult males and only 178 listed occupations. In 1910 there were actually 774 adult males and only 735 listed occupations. MS Census, Population, 1900, 1910.

45. Thornburgh Collection.

46. Moyers, "Arkansas Progressivism," 219; Hunt, "History of the Prohibition Movement."

47. *Marked Tree Gazette*, Jan. 20, July 14, 1911, Jan. 5, Aug. 16, Sept. 6, 13, 20, Dec. 6, 13, 1912, March 14, 1913; *Modern News*, Jan. 31, 1913.

48. *Marked Tree Gazette*, Sept. 20, 1912; West, *Saloon*, 51–72; MS Census, Population, 1900; *Polk's Arkansas State Gazetteeer*, 1906–7; Marked Tree, Ark., Sanborn Map, 1908.

49. MS Census, 1910; *Marked Tree Gazette*, Oct. 13, 1911.

50. The three merchants were John B. Claunch, W. H. Powell, and B. F. Taylor, who operated a drugstore. MS Census, Population, 1910.

51. *Marked Tree Gazette*, Oct. 13, 1911; *Marked Tree Tribune*, July 7, 1911; Arkansas State Bank Department, Liquidation Record, undated.

52. *Marked Tree Tribune*, April 5, Aug. 2, 30, 1912.

53. Although commonly referred to as the Grandfather Clause amendment, it actually proposed to establish educational requirements for suffrage. It "contained a grandfather clause intended to exempt illiterate whites from its provisions." Graves, "Negro Disfranchisement," 223n; *Marked Tree Gazette*, Sept. 13, 1912.

54. Ibid., Jan. 31, 1913. In an article entitled "The Liquor Fight Reopened," men like Ernest Ritter, C. T. Carpenter, Almonta Smith, B. F. Taylor, and J. W. Griffin, all prominent Marked Tree businessmen, challenged the 1912 election and charged that it had been a fraudulent election.

55. *Osceola Times*, Feb. 27, 1914.

56. *Marked Tree Gazette*, Dec. 6, 13, 1912, Jan. 31, Feb. 14, March 14, April 4, May 16, 1913; *Modern News*, Jan. 9, 1914.

57. *Polk's Arkansas State Gazetteer, 1906–1907*, 461; *Polk's Arkansas State Gazetteer, 1912–1913*, 406.
58. Record of the Proceedings, 35; MS Census, Population, 1900; Real Property, 1910, 1920; Personal Property, 1920.
59. *Marked Tree Gazette*, Nov. 22, 1907.
60. WPA Church Records.
61. Dawson interview. Ritter spent most of his time in Marked Tree.
62. *Marked Tree Gazette*, June 12, 1908, May 5, 1911. Before 1907 black children attended school in the "colored Methodist Church." In July 1907, two lots were purchased for the purpose of erecting a 30-by-40-foot building "because the number of scholars keep increasing." There were approximately 129 students enrolled in the black school during its first term with "an average daily attendance of 73." Dr. S. L. Mitchum was the chairman of the black school (ibid., Aug. 2, 1907, May 5, 1911). For white schools, see *Marked Tree Tribune*, Nov. 18, Aug. 21, 1908. There were 361 white children of school age in Marked Tree in 1910 (MS Census, Population, 1910).
63. *Marked Tree Gazette*, Aug. 21, 1908.
64. Dewey interview; Wilson interview.
65. In other words, only 45 of the 368 farmers who reported whether they owned or rented their farms in 1910 claimed to own the farms they worked. MS Census, Population, 1910.
66. Mortgage Records, 1909–1910.
67. Freeman interview; Scholtz interview; Ziegenhorn interview.
68. MS Census, Population, 1900, 1910.
69. Court Records, 1901–2.
70. Ibid.
71. *Lepanto News Record*, Sept. 9, 1938.
72. MS Census, Population, 1910; Mortgage Records, 1909–10.
73. In this discussion chattel mortgages refers to those mortgages that involved personal rather than real property.
74. *Acts of Arkansas*, 1868, 245, 1875, 85, 1885, 225–26, March 7, 1893, 75–76, March 27, 1893, 171–72, April 7, 1893, 228–29.
75. Mortgage Records, 1909–10; MS Census, Population, 1910.
76. Mortgage Records, 1909–10. The Bank of Tyronza, like the Bank of Marked Tree, filed real property mortgages only. While the Marked Tree bank held ten, the Tyronza bank filed only three.

3. Laboring in the "American Congo"

1. Unsigned letter to the President, July 22, 1898, Justice Department Records.
2. *Helena World*, Feb. 23, 1898.
3. Undated *Arkansas Gazette* newspaper clipping in Trieber Papers; *Forrest City Times*, April 1, 1904.

4. Undated *Helena World* newspaper clipping; Trieber Papers.

5. Edward Ayers indicates that the "weakness of legal institutions received even more of the blame for upcountry vigilantism than it did for low-country lynchings." Ayers, *Vengeance and Justice*, 260.

6. See Holmes, "Whitecapping," 167–68, "Whitecapping in Georgia"; "Moonshiners and Whitecaps in Alabama"; and "Whitecapping in Mississippi"; Brown, "American Vigilante Tradition"; Vanderwood, *Night Riders of Reelfoot Lake*; Brown, *Strain of Violence*, 24; Ayers, *Vengeance and Justice*, 260–62; Schlesinger, "Las Gorras Blancas". See also Nolan, *Vigilantes on the Middle Border*; and Crozier, *White Caps*. For the most recent study, see Waldrep, *Night Riders*.

7. *Commercial Appeal*, March 13, 1903; WPA Church Records, Cross County.

8. Harrel, *Social Sources of Division in the Disciples of Christ*, 200.

9. *Commercial Appeal*, March 13, 1903.

10. *Forrest City Times*, Dec. 4, 1903; Heaney, "Jacob Trieber," 442.

11. *Commercial Appeal*, March 13, 1903.

12. Ibid.

13. Heaney, "Jacob Trieber," 442.

14. *Commercial Appeal*, March 13, Aug. 23, 1903; *Forrest City Times*, March 20, 27, May 22, Aug. 28, Oct. 16, 1903; *Lee County Courier*, Oct. 17, 1903; *Modern News*, Oct. 17, 1903; *Arkansas Gazette*, Oct. 18, 1903; William G. Whipple (U. S. Attorney for the Eastern District of Arkansas) to Attorney General, Oct. 16, 1903, March 22, 1904, Justice Department Records; U. S. District Court, Eastern District of Arkansas, Helena, E39 Criminal Record Book, Oct. 9, 1903, 486–92, March 17, 18, 21, 1904, 509, 515–24, 530.

15. *Arkansas Gazette*, Oct. 18, 1903.

16. See St. Francis Levee Board, *History of the District*, 324–25, *Historical Report of the Secretary of State*; and Donovan, *Governors of Arkansas*.

17. Hodges v. U.S. 203 U.S. 1, 27 S.Ct. 6; *Osceola Times*, March 4, 1909; Undated *Helena World* newspaper clipping in Trieber Papers. For lynching across the South, see Cutler, *Lynch Law*; NAACP, *Thirty Years of Lynching*; Zagrando, *NAACP Crusade*; Raper, *Tragedy of Lynching*; Hall, *Revolt against Chivalry*; Brundage, *Lynching in the New South*; White, *Rope and Faggot*; Shapiro, *White Violence and Black Response*; Tolnay and Beck, *Festival of Violence*; and Lewis, "Mob Justice."

18. Cross County Court Records, True Bills, 1908.

19. *Osceola Times*, March 4, 1909; *Acts of Arkansas*, 1909, 315; *Marked Tree Gazette*, Feb. 2, Jan. 12, 1917.

20. *Osceola Times*, Sept. 17, 1908, Oct. 14, 1909, Jan. 11, 1912, March 19, 1915, March 19, 26, April 28, 1916, and April 22, 1921.

21. *Arkansas Digest* 1982: 117–18; Heaney, "Judge Jacob Trieber," 449.

22. Woodward, *Origins of the New South*; Brooks, "Use of the Civil Rights Acts of 1866 and 1871"; Heaney, "Busing, Timetables, Goals, and Ratios"; Colbert, "Challenging the Challenge."

23. Heaney, "Jacob Trieber"; LeMaster, *Corner of the Tapestry*, 166–67; XV Papers, Minutes, Feb. 4, 1904, 7.

24. Trieber to Roosevelt, February 27, 1905, Roosevelt Papers, ser. 1, reel 53.

25. Pickens, "The American Congo," 426–28.

26. Industrial Commission *Report*, xviii; Rawick, *American Slave* 9: pt. 4, 248.

27. Whipple to the Attorney General, Jan. 20, 24, Feb. 2, 7, 21, March 16, April 18, 1905; Lock McDaniel to Attorney General, Jan. 27, 1905; Attorney General to Whipple, Feb. 3, 25, March 1, 1905, Justice Department Records; *Marked Tree Gazette*, May 14, 1912.

28. See Royce, *Origins of Southern Sharecropping*, 198, for a discussion of the "partnership" of planters and sharecroppers.

29. Kelley, *Hammer and Hoe*, 101; Scott, *Weapons of the Weak*, 29.

30. See *Osceola Times*, Feb. 1, 1902; July 11, 18, 1903, Feb. 27, Aug. 20, Dec. 24, 1904, May 6, 1905; Dec. 1, 1906, Sept. 5, 1907, March 25, 1909, June 26, Oct. 23, 1914, March 3, Dec. 27, 1915, and Nov. 15, 1918. For merchant and tenant disputes, see ibid., Dec. 31, 1904, and June 27, 1907; *Forrest City Times*, June 19, 26, 1903. For an analysis of this confrontation between black working-class individuals and whites in the urban South, see Kelley, "'We Are Not What We Seem.'"

31. *Osceola Times*, March 10, 1910.

32. Ibid., July 18, 1903.

33. Pickens, "American Congo," 426–28. See also Lewis, "Mob Justice in the 'American Congo.'"

34. *Osceola Times*, Jan. 1, 1904.

35. Pickens, "American Congo," 427.

36. Ibid., 426–28.

37. Kelley, *Hammer and Hoe*, 103.

38. Industrial Commission, *Report*, xix. Gavin Wright finds, however, that "sharecropping was a high turnover system and indebtedness did not prevent blacks or whites from changing landlords frequently" (Wright, *Old South, New South*, 12). Roger Ransom and Richard Sutch, like Gavin Wright, emphasize the power that blacks were able to wield in the labor market but find that the free market actually worked against them. They conclude that racism forced blacks into the lowest-paid and least skilled positions (Ransom and Sutch, *One Kind of Freedom*, 13). Pete Daniel, meanwhile, stresses the emergence of debt peonage in the South in the late nineteenth century and finds that it was an especially acute problem for black sharecroppers. He finds that freedom of movement from one plantation to another was true for many rural laborers in the South but argues that some were locked in a vicious cycle of debt that rendered them virtual peons (Daniel, *Shadow of Slavery*).

39. Rawick, *American Slave* 9: 489.

40. Wiener, "Class Structure and Economic Development"; Higgs, "Comments"; Mandle, *Roots of Black Poverty*; and Daniel, *Breaking the Land*. See also Woodman, "Sequel to Slavery."

41. See Royce, *Origins of Southern Sharecropping*, for a good explication of the question of the so-called failed efforts of tenants and sharecroppers.
42. Stith interview.
43. Wright, *Old South, New South*.
44. See Whayne, *Shadows over Sunnyside*, for an experiment with Italian labor in southeast Arkansas.
45. Rawick, *American Slave* 8: 62, 9: 91.
46. See Robert Darnton's essay on the use of metaphor in French peasant tales, in which references to food frequently appear and illustrate the place that hunger occupied in the lives of peasants. Darnton, *The Great Cat Massacre*.
47. Rawick, *American Slave* 8: 58, 127, 212, 257.
48. The remaining occupations break down as: lumbering and railroad related, 115 (44.9 percent); service, 16 (6.3 percent); artisan and skilled, 15 (5.9 percent); other, 4 (1.6 percent); merchant, 2 (.6 percent); white collar, 2 (.6 percent). Only 21 individuals (8.2 percent) listed professional occupations, and mostly ministers and teachers.
49. Mary Grace Quackenbos, "Report on general conditions of Delta cotton plantations," Jan. 10, 1908, Justice Department Records; Whayne, *Shadows Over Sunnyside*.
50. For the best description of black resistance to malaria (due to the sickle-cell trait), see Wood, *Black Majority*.
51. Dewey interview, 1985.
52. *Marked Tree Gazette*, Dec. 24, 1909, May 8, 1908.
53. Ibid., Aug. 2, 1907.
54. Going faced trial in June 1909, but the "case was postponed indefinitely," *Modern News*, June 25, 1909. No evidence of its reaching a second hearing exists.
55. Industrial Commission *Report*, 496; Terris, *Goldberger on Pellagra*, 51–52, 287; *Marked Tree Gazette*, April 26, 1912; Report of C. M. Hubbard, April 23, 1912, 2, Red Cross, Box 51, File 805.6. See also Dabney, "1912 Flood on the Lower Mississippi."
56. Winters, "Tenant Farming in Iowa," 132.
57. On paternalism Genovese, *Roll Jordan Roll*, 110–11; Fox-Genovese and Genovese, *Fruits of Merchant Capitalism*, 398–99; Roark, *Masters Without Slaves*, 144–45; Mandle, *Roots of Black Poverty*, 30–31, 35; Powell, *New Masters*, 31; Williamson, *Crucible of Race*, 83; and Billings, *Planters*, 102–109.
58. Woodman "Sequel to Slavery," 553. Harold Woodman finds that the sharecropper, unlike the tenant, had no legal right to the "growing crop."
59. Royce, *Origins of Southern Sharecropping*, 186.
60. *Twelfth Census, Agriculture, 1900*, 268, 430; *Thirteenth Census, Agriculture, 1910*, 123, 802; Industrial Commission, *Report*, 497; Real Property Records, 1900, 1910; Spillman, "Types of Farming in the United States," 355. The average family size on the Poinsett County delta farms was 4.2. MS Census, Population, 1910.
61. MS Census, Population, 1920; Personal Property, 1920.

62. Royce, *Origins of Southern Sharecropping*, 186; MS Census, Population, 1920; Personal Property, 1920.

63. *Fourteenth Census, Agriculture, 1920*, 566, and *Fifteenth Census, Agriculture, 1930*, 1159. If a strict definition of sharecropper as one who owns no mules, horses, or implements is ascribed to, then a discrepancy between the 1920 printed census of agriculture and the county personal property tax records exists. According to the former, only 15.1 percent of landless Poinsett County blacks worked as tenant farmers, but according to the latter, nearly 25 percent had mules and/or horses.

64. Personal Property, 1920.

65. Woodman, "Post Civil War Southern Agriculture and the Law"; *Marked Tree Gazette*, June 12, July 3, 1908. One should not overestimate the status that Mitchum enjoyed in the community, however, for the lawyer who represented him in the hearing was the author of an editorial which had lashed out against racial amalgamation just a month before the incident (ibid., May 1, 1908).

66. An editorial indicates that as late as 1918 some confusion continued to exist as to whose lien was superior, the landlord's or the merchant's. *Osceola Times*, Nov. 15, 1918.

67. Woodman, "Post Civil War Agriculture and the Law," 321. See also Solberg, "The Legal Aspects of Farm Tenancy in Arkansas."

68. The Mitchum farm sold for $10,000 to E. M. Duren, a white farmer from Adamsville, Tennessee. *Marked Tree Gazette*, Nov. 22, 1912.

69. Trieber to Roosevelt, Feb. 27, 1905, Roosevelt Papers, ser. 1, reel 53.

70. Significantly, none of them appeared in the Mayor's Court in Harrisburg charged with any offense in the years 1901 through 1904. The only person connected to the *Hodges* case who appeared in the Mayor's Court in those years was L. C. Going, one of their attorneys who was charged with assault and battery, found guilty, and fined $4.80 on Jan. 5, 1902. Mayor's Court Docket.

71. The indictments may have used nicknames for two of the others. I could find no Wash McKinney in the manuscript census, for example, but the McKinney family had settled the Ridge many years earlier and several McKinney families appeared on the census. Rube Parker may have been Rufus Parker.

72. MS Census, Population, 1860, 1870, 1880, 1900, 1910, and 1920.

73. See Arsenault, *Wild Ass of the Ozarks*; Graves, *Town and Country*; and Niswonger, *Arkansas Democratic Politics*.

74. According to Edward Ayers, kinship relationships linked whitecappers and vigilantes in the Georgia upcountry. Ayers, *Vengeance and Justice*, 264.

75. Cross County Historical Society Archives. Two additional Hall brothers were charged in the affair but could not be located in the manuscript census.

76. Weeks, *From Madison to Marked Tree*, 49.

77. Daniel, *Breaking the Land*, 7–9; Holmes, "Whitecapping," 167–68; Holmes, "Labor Agents and the Georgia Exodus," 437–38; Industrial Commission *Report*, 492; *Marked Tree Tribune*, Sept. 8, 1908, Feb. 2, 1912.

78. Bratton to Terrell, July 5, 1905, Justice Department Records; White, "'Massacring Whites,'" 715.

79. Printed document with no author, no date, no place of publication in Trieber Papers.
80. Ibid. Estimates of the death toll range from 21 to over 100 (Smith, "Black Organization," 5; White, "'Massacring Whites,'" 715; Rogers, "Elaine Race Riot," 143–50; Desmarais, "Military Intelligence Reports"; Cortner, *Mob Intent on Death*, 30). For a provocative analysis of the situation in Elaine, see Woodruff, "African-American Struggles for Citizenship in the Arkansas and Mississippi Deltas," 41–43.
81. White, "'Massacring Whites,'" 715–16.

4. Sunk Lands and Lost Hopes

1. *Modern News*, Aug. 8, 1913; A. W. Cheatham to Secretary of the Interior, June, Aug. 21, 1913, GLO; *Marked Tree Gazette*, June 4, 1915; Della Abbott to Mary Ann Dick, Feb. 18, 1916, Abbott Papers.
2. Hannah Embrey to the Secretary of Interior, Dec. 26, 1917, GLO; *Osceola Times*, Nov. 23, 1917. See also, Ottoson, *Land Use Policy*; Carstensen, *Public Lands*.
3. St. Francis Levee Board, *History of the District*, 235.
4. *Marked Tree Gazette*, Aug. 2, 1907, Jan. 15, 1909, Oct. 4, Jan. 26, 1912; *Osceola Times*, March 10, 1910, Aug. 21, 1914; *Arkansas Gazette*, July 29, 1911; St. Francis Levee Board, *History of the District*, 234–36.
5. *Marked Tree Gazette*, Feb. 2, 1912.
6. St. Francis Levee Board, *History of the District*, 1.
7. "Chronological Statement of Proceedings Affecting Arkansas 'Sunk Lands,'" 1, GLO.
8. St. Francis Levee Board, *History of the District*, 234–35.
9. Harrison and Kollmorgen, "Land Reclamation in Arkansas," 414. See also, Harrison, "Clearing Land in the Mississippi Alluvial Valley"; "Formative Years of the Yazoo-Mississippi Delta Levee District"; and "Early State Flood-Control Legislation in the Mississippi Alluvial Valley."
10. "Chronological Statement of Proceedings Affecting Arkansas 'Sunk Lands,'" 1–2, and Secretary of the Interior to Land Commissioner, Nov. 17, 1902, GLO.
11. *Hearings of the Committee*.
12. First Assistant Secretary A. A. Jones to Commissioner of the General Land Office, Sept. 24, 1913, GLO.
13. R. E. L. Johnson to Secretary of the Interior, April 22, 1914, H. C. Hall to Secretary of the Interior, Feb. 17, 1911, GLO; *Marked Tree Gazette*, June 2, 1911.
14. R. E. L. Johnson to Secretary of the Interior, April 22, May 22, 1914, GLO.
15. Lee Wilson & Company v. U. S., 245 U.S. 24 (1917).
16. *House Reports*, vol. 2, 67th Cong., 1st sess., report 298.

17. *Osceola Times*, Feb. 17, 24, 1910.
18. *Lee County Courier*, March 25, April 1, 8, 15, 22, 29, May 6, 1905; Arsenault, *Wild Ass of the Ozarks*, 238; St. Francis Levee Board, *History of the District*, 287.
19. Ernest Ritter bought 2,000 acres and an option on another 8,000 acres on April 4, 1906 (Deposition of Ernest Ritter, Chapman and Dewey Land Company v. St. Francis Levee District, Justice Department, Appellate Case File 22839, 150–51). The St. Francis Valley Land Company purchased 39,370.26 acres located within Mississippi County in 1906 and began reselling them at a profit (*Osceola Times*, Feb. 17, 24, 1910). According to one account, the St. Francis Valley Land Company paid $100,000 to the district for the 40,000 acres (ibid., June 22, 1911).
20. W. B. Miller to Jeff Davis, March 4, 1910, Davis Papers. See also Arsenault, *Wild Ass of the Ozarks*, 238.
21. Levi Cooke testified on Feb. 15, 1910, that he represented "Mr. Blutenthal and Mr. Heilbronner, of Memphis, Tenn., and other purchasers." In fact, Blumenthal, Heilbronner, and Ritter were partners in the purchase of the land; on Feb. 17, 1910, A. J. Frieburg, an attorney from Cincinnati, spoke on behalf of "Max Heilbronner; E. Ritter; Herman Blumenthal." *Hearings of the Committee*, 4, 39.
22. Ibid., 4, 38–39, 79–149; Fred Dennett, Commissioner, Department of Interior, to Messrs. Murphy, Coleman and Lewis, Attorneys at Law, Little Rock, May 31, 1910, GLO.
23. *Hearings of the Committee*, 11, 279–395.
24. Ibid., 392–93.
25. Jernigan and Chambers, *History of Lepanto*, 57.
26. Ruben was twenty years old in 1900, James was twenty-five, and John was thirteen.
27. J. T. Lee lists himself as a merchant on the 1910 census. Ibid., 82.
28. Ibid., 3, 17, 4; *Hearings of the Committee*, 392.
29. Jernigan and Chambers, *History of Lepanto*, 4.
30. Ibid.
31. *Modern News*, Aug. 8, 1913, Sept. 25, 1914; *Marked Tree Gazette*, June 4, 1915.
32. MS Census, Population, 1920; Real Property, 1920. Finn held no real property in Poinsett County in 1910.
33. *Modern News*, Aug. 8, 1913.
34. Della Abbott to Mary Ann Dick, Feb. 18, 1916, Abbott Papers. See *Marked Tree Gazette*, Feb. 2, 1912, for reference to a black man named Johnston fleecing blacks by charging them $10 to locate homesteads for them.
35. *Modern News*, Aug. 8, 1913; MS Census, Population, 1900, 1910.
36. Names of settlers were compiled from *Modern News*, Aug. 8, 1913, the *Marked Tree Gazette*, Sept. 25, 1914, and the *Hearings of the Committee* and then compared to the MS Census, Population, 1900, 1910, and 1920.
37. *Modern News*, Aug. 8, 1913; Jernigan and Chambers, *History of Lepanto*, 43–44; Abbott interview.

38. Personal Property, 1920.
39. This letter is dated April 17, 1921, and appears in House Report no. 298 accompanying H.R. 1318 and in Senate Report no. 908 accompanying H.R. 6863. *House Reports*, vol. 2, 67th Cong., 1st Sess., no. 298; *Senate Reports*, vol. 2, 67th Cong., 2d Sess., no. 908.
40. *Congressional Record*, 67th Cong., 1st Sess., House, 4463; *Journal of the Senate*, 65th Cong., 3d Sess., 164.
41. *Journal of the House*, 67th Cong., 1st Sess., 245; *Congressional Record*, 67th Cong., 1st Sess., 6516–18.
42. *Osceola Times*, May 6, 1905; St. Francis Levee Board, *History of the District*, 287.
43. Real Property, 1910, 1920, 1930. The analysis of Ritter's holdings only covers the acreage in Poinsett County.
44. Abbott interview; Real Property, 1930; Jernigan and Chambers, *A History of Lepanto*, 44.
45. The remaining percentages were either government land or unclaimed land.
46. Court Records, Sept. 11, Oct. 5, 10, 1905, May 8, 1906.
47. T. F. Wilborn to General Land Office, April 6, 1910, GLO.
48. Harrison and Kollmorgen, "Socio-Economic History of Cypress Creek Drainage District," 25, 28, 33; Simonson, "Origin of Drainage Projects," 265.
49. Edrington, *History of Mississippi County, Arkansas*, 76, 80–81.
50. Court Records, Sept. 8, 1906, Jan. 7, 1907, July 8, 1908.
51. Ibid.
52. Ibid.
53. Ibid., Oct. 4, 1904.
54. The court was to be reimbursed once the taxes were imposed. Ibid., Jan. 9, April 5, Oct. 10, 1905.
55. Ibid. There is no evidence that he formally removed his name from the list of original petitioners (to form the district). Certainly his objection registered his disapproval.
56. The timing of his opposition is also curious. Ritter had not objected in January 1905 when the engineer's original report stipulated the cost per acre necessary to charge in order to pay for the drainage project. A few months after his opposition surfaced, he raised no objection when the court approved the district, and no mention of the appeal he was granted appears in the county court record. Ibid., July 9, 1906.
57. Ibid., July 23, 1906.
58. *Marked Tree Gazette*, Feb. 9, 1917
59. Ibid., Aug. 13, 1915; *Marked Tree Tribune*, Jan. 19, Feb. 9, March 2, 1917.
60. St. Francis Levee Board, *History of the District*, 274.
61. Court Records, Jan. 7, Feb. 25, July 5, 15, 1918.
62. *Marked Tree Gazette*, Oct. 19, 1915.
63. *Modern News*, Nov. 14, 1916; Real Property, 1910, 1920.
64. Court Records, Jan. 7, Feb. 25, 1918.

65. Embrey to Secretary of the Interior, Dec. 17, 1917, GLO; Minutes of the Meeting, Jan. 4, 1918, 45.

66. Court Records, Oct. 4, 1904, July 8, 1908. Drainage District One covered only a small part of the sunk lands, and Drainage District Six was restricted to the southeastern corner of the county, outside the parameters of the sunk lands.

67. *Acts of Arkansas*, March 10, 1921, 494–95; Minutes of the Meeting, March 21, 1921, 278, June 5, 1931, 115.

5. Different Roads to Development

1. *Marked Tree Tribune*, May 11, 1917.
2. Ibid., Oct. 18, 1907, June 5, July 31, 1908, July 6, 1917; Court Records, July 30, 1917, 305–7, Sept. 7, 1917, 324–25, Oct. 26, 1917.
3. *Modern News*, July 14, Aug. 4, 1922.
4. Ibid., Nov. 10, 1911, June 21, July 19, 1912. By 1920 the Illinois Farming Company listed neither real nor personal property, but the Kansas City and Memphis Farm Company listed 7,192 acres, and Chapman and Dewey, its parent company, listed 22,759 acres and $41,100 in personal property. Real Property and Personal Property, 1920.
5. *Thirteenth Census, Agriculture, 1910*, 110; and *Fourteenth Census, Agriculture, 1920*, 1152.
6. *Marked Tree Tribune*, Aug. 25, Sept. 15, 1916.
7. Ibid., Aug. 19, Nov. 22, 1912, Jan. 22, Feb. 12, 1915.
8. Ibid., Nov. 2, 1917, Dec. 14, Dec. 20, 1918.
9. Real Property and Personal Property, 1920. The townships were Little River and Tyronza; the landholders in the sample were those who were listed on the real property records and who could be matched with taxpayers listed on the personal property records.
10. Bolivar Township was used, and the individuals selected were derived by matching taxpayers on the real property and the personal property records.
11. *Modern News*, Feb. 14, 21, March 4, April 4, Oct. 31, 1913.
12. Real Property and Personal Property, 1920. Individuals held 86 percent of the ridge land but only 16.8 percent of the delta land, according to a match of the real and personal property records.
13. *Modern News*, Feb. 6, 1920. Land along the ridge in 1920 was selling for $100 to $150 per acre while land in the delta was bringing from $250 to $300 per acre.
14. MS Census, Population, 1920.
15. "Historical Appraisal of Extension Work in Poinsett County," Extension Service.
16. *Modern News*, March 7, 1913, June 9, 1916.
17. Ibid., July 28, 1916.
18. Ibid., June 30, Oct. 20, 27, 1916.

19. Personal Property, 1920. Only 42.7 percent of the prairie taxpayers and 28.5 percent of the delta taxpayers raised hogs in 1920.
20. *Modern News*, Jan. 24, June 13, 20, July 4, 18, Oct. 3, 1919, Oct. 22, 1920.
21. Ibid., Oct. 22, 1920.
22. Ibid., Oct. 29, Nov. 5, 1920. Earle is a community about thirty miles southeast of Marked Tree.
23. Ibid., Dec. 10, 1920.
24. *Fourteenth Census, Agriculture, 1920*, 327–29.
25. Unlike personal property taxpayers in the delta and the prairie who rarely entered anything in the category "materials and manufactured articles," certain rice farmers regularly recorded property there. For example, rice farmers Fred Ruesewald, William Koester, Max Schisler, and Harry, T. E. and A. E. Ziegenhorn, to name but a few, entered property in excess (collectively) of $2,000 in that category.
26. Bank of Weiner, *Second Annual Rice and Amusement Carnival*, 37.
27. MS Census, Population, 1900, 1910. Most of the farmers who settled the prairie were from Illinois and Indiana, although there was also a small but prominent group of Germans.
28. Real Property and Personal Property, 1920. These figures are taken from the subset generated by the match between these records.
29. Bank of Weiner, *Second Annual Rice and Amusement Carnival*, 37; *Modern News*, May 23, April 10, 1914.
30. Stuttgart is a community in the heart of the Arkansas rice belt about 150 miles southwest of Weiner.
31. *Modern News*, Jan. 13, July 25, 1913, Nov. 3, 1916, March 5, 1920; Real Property, 1920.
32. *Modern News*, Jan. 28, 1916.
33. *Marked Tree Gazette*, Sept. 10, 1915; *Modern News*, Jan. 21, 1916.
34. *Marked Tree Tribune Thirty-fifth Anniversary Number*, Aug. 3, 1939, reprinting article from the *Marked Tree Gazette* dated March 14, 1919; *Marked Tree Gazette*, March 16, 1915.
35. *Marked Tree Tribune Thirty-fifth Anniversary Number*, 1939, reprinting March 14, 21, May 2, 1919.
36. Ibid., June 5, 1908.
37. Court Records, Aug. 13, 1900, 312–14, Aug. 11, 1910, 70–74; MS Census, Population, 1900, 1910.
38. *Marked Tree Gazette*, Oct. 18, 1907, June 5, 31, 1908. Nettleton is a town in Craighead County, north of Poinsett.
39. *Modern News*, Jan. 15, 29, Feb. 5, 1904.
40. Ibid., Jan. 8, 15, 29, Feb. 5, 1904.
41. *Marked Tree Gazette*, March 27, 1908.
42. Record of the Proceedings, 50.
43. Harrison and Kollmorgen, "Land Reclamation in Arkansas," 416–17.
44. St. Francis Levee Board, *History of the District*, 274.

45. Arsenault, *Wild Ass of the Ozarks*, 13; Ledbetter, "Jeff Davis," 18–19, 29–31; Segraves, "Arkansas Politics," 147–48; Niswonger, "Study in Southern Demagoguery," 123–24.

46. Harrison and Kollmorgen, "Socio-Economic History of Cypress Creek Drainage District," 25–26; Simonson, "Origin of Drainage Projects in Mississippi County," 265–66; Edrington, *History of Mississippi County*, 79–82.

47. St. Francis Levee Board, *History of the District*, 287, 290.

48. *Modern News*, July 29, 1910.

49. Ibid., March 10, 1908.

50. *Marked Tree Gazette*, March 10, 1908.

51. Ibid., March 20, 27, 1908.

52. *Modern News*, Jan. 15, 1904.

53. M. W. Hazel served from 1909 until at least 1916; J. A. Emrich served from 1907 to 1910.

54. *Marked Tree Gazette*, March 15, April 5, May 31, Aug. 2, 30, 1912.

55. Voting returns are not available on the township level. Occasionally a newspaper would break them down to that level, and thus one can achieve a sense of what was occurring.

56. MS Census, Population, 1860, 1880, 1900, 1910.

57. See *Modern News*, Sept. 15, 1922, for insight into the relationship between the ridge and Willis Township.

58. *Marked Tree Gazette*, July 29, 1910, March 24, July 28, Aug. 4, 1911, May 31, June 7, July 12, 1912, April 23, 1915.

59. Ibid., Dec. 31, 1909.

60. *Marked Tree Tribune*, Sept. 14, 1917.

61. Ibid., Dec. 15, 1916, Sept. 7, 21, 1917.

62. Ibid., Aug. 10, Oct. 26, 1917.

63. *Marked Tree Tribune*, Dec. 5, 1930.

64. *Marked Tree Gazette*, April 7, 1916; *Marked Tree Tribune*, Aug. 10, 1917.

65. *Modern News*, March 10, 1916. H. B. Thorn was secretary of the CDCC during the 1916 election controversy.

66. *Marked Tree Gazette*, Nov. 12, 1915.

67. There is some confusion over when the personal property tax records disappeared; those prior to 1917 are not in the Harrisburg County Court House archives.

68. *Marked Tree Tribune*, May 25, 1917.

69. Ibid., May 18, 25, July 6, 1917.

70. Ibid., Aug. 3, 17, Oct. 12, Dec. 7, 1917.

71. Court Records, Jan. 7, 1918, 348–49. In fact, this record details the report of the engineers who had been appointed in late 1917 to draft plans for the drainage district.

72. *Modern News*, May 19, July 14, 28, Aug. 4, 11, 18, Sept. 29, 1922.

73. Ibid., Aug. 4, 18, Sept. 15, 1922, Aug. 10, 17, 1923. See *Modern News*, Aug. 18, 1922 for the election statistics.

74. See, "Strike and the Still," 403–25; and Alexander, *Ku Klux Klan in the Southwest*, 19, 27, and his "White Robed Reformers," "White Robes in Politics," and "Defeat, Decline, Disintegration."

75. The state government did not maintain voting returns on the township level for this period. Some returns were gleaned from newspapers, but they were only sporadically reported on the township level.

6. Delta Blues

1. W. D. Ezell, "Narrative Report of County Extension Workers," 1926, 15, H. S. Hinson, "Narrative Report of County Extension Workers," 1928, 8, and A. Raybon Sullivant, "Narrative Report of County Extension Workers," 1929, 12, Federal Extension.

2. *Marked Tree Gazette*, Aug. 8, 1916.

3. *Modern News*, Dec. 7, 1928, April 12, 26, May 10, June 7, Nov. 8, 1929, April 18, 1930. There seems to have been no well-organized opposition to the passage of the stock law on the ridge. For a similar situation among the yeomen of upcountry Georgia in the late nineteenth century, see Steven Hahn, *Roots of Southern Populism*, 267.

4. *Modern News*, Dec. 7, 1928, March 22, 29, May 10, June 7, and Nov. 11, 1929; Whitaker v. Mitchell, in *Arkansas Reports*, vol. 179, *Cases Determined in the Supreme Court of Arkansas*, 993–97.

5. Connor, "National Farm Organizations," 32.

6. Gile, "Organization and Management," 12; Alexander, *Arkansas Plantation*, 16.

7. Smith, "Farm Real Estate Trends," 409; Alexander, *Arkansas Plantation*, 14, 16–17; Fite, *American Farmers*, 46.

8. Fite, *American Farmers*, 34–48; Christensen, "Agricultural Pressure," 33–34; Connor, "National Farm Organizations," 32.

9. Gile, "Status of Cooperative Cotton Marketing," 5, 14.

10. Real Property, 1930.

11. "Historical Appraisal of Extension Work in Poinsett County," Extension Service; *Modern News*, April 1, 8, May 13, Aug. 19, Sept. 9, 1921.

12. Sargent, "Narrative Report of County Extension Workers," 1922, 5, 6, 20, 24, 26, Federal Extension; *Modern News*, July 21, 1922.

13. Sargent, "Narrative Report of County Extension Workers," 1922, 39–40, Federal Extension.

14. Real Property, 1920, 1930; Personal Property, 1920; *Modern News*, June 9, 1922.

15. Ezell, "Narrative Report of County Extension Workers," 1926, 15, Alexine Ledford, "Narrative Report of County Extension Workers," 1926, 1, 8, 19, Federal Extension.

16. Ezell, "Narrative Report of County Extension Workers," 1927, 6, 7, Federal Extension; "Notes of Meeting of Red Cross Representatives with Mr.

Hoover and Mr. Fieser, Marion Hotel, Arkansas," June 6, 1927, p. 3, Red Cross, box 224, DR 224.01; *Marked Tree Tribune*, May 6, 1927.

17. "Notes of Meeting of Red Cross Representatives with Mr. Hoover and Mr. Fieser, Marion Hotel, Arkansas," June 6, 1927, 3, Red Cross, box 224, DR 224.01; *Marked Tree Tribune*, Dec. 23, 1927; Woodruff, *As Rare as Rain*, 56–58.

18. Quoted in Daniel, *Deep'n As It Come*, 102–3.

19. "Narrative Report of Nursing Activities," July 30, 1927, 1, Red Cross, box 735, DR 224.08; *Marked Tree Tribune*, June 3, 1927.

20. *Marked Tree Tribune*, Dec. 21, 1928.

21. Daniel, *Deep'n As It Come*, 92–108; A. L. Schafer, "Memorandum: Return of Refugees," n.d., Red Cross, box 774, DR 224.6.

22. Hinson, "Narrative Report of County Extension Workers," 1928, 10–11, Federal Extension.

23. See a document attached to Hinson's regular annual report but dated Nov. 2, 1928 entitled "Report of County Agent of Poinsett County from June 6, 1928, to Oct. 29, 1928, a period of four months and twenty-four days," 1–3, Federal Extension.

24. Sullivant, "Narrative Report of County Extension Workers," Nov. 30, 1929, 3–4, ibid.

25. *Modern News*, May 13, 1938.

26. Sullivant, "Narrative Report of County Extension Workers," 1929, 2–3, Federal Extension.

27. Gile, "Development of Agricultural Credit Corporations," 3.

28. Randall, "Landlord-Tenant Problem," 3.

29. Sullivant, "Narrative Report of County Extension Workers," 1929, 6, 8, 9, Federal Extension.

30. *Modern News*, Dec. 7, 1928, April 12, 26, May 10, June 7, Nov. 8, 1929, April 18, 1930.

31. Sullivant, "Narrative Report of County Extension Workers," 1929, 10–11, Federal Extension.

32. Ibid., 13.

33. "Report on Drouth Situation," Aug. 20, 1930, 2, Red Cross, box 772, DR 401.11.

34. Ibid.

35. *Marked Tree Tribune*, Dec. 23, 1927; DeWitt Smith to M. R. Reddy, Sept. 26, 1930, 1, and article from *Labor* clipped to this memo, Red Cross, box 778, DR 401.653.

36. John L. Buxton to William M. Baxter, Jr., March 20, 1931, to H. E. Downey, n.d., Red Cross, box 778, DR 401.653. See also Charles E. Sullengers (of the Osceola Chapter of the American Red Cross in nearby Mississippi County) to Albert Evans, Feb. 6, 1931, 1, Red Cross, ibid.

37. Buxton to Baxter, March 30, 1931, ibid.

38. E. M. Perry to Mr. Crampton, Feb. 5, 1931, ibid.

39. Buxton to H. E. Downey, n.d., ibid.

40. The ability of a local chapter to distribute relief, whether through direct relief or works programs, was critical. In one community in the central Arkansas delta, the local chapter was unable to get the proper paperwork, and a riot occurred. Woodruff, *As Rare as Rain*, 56–58.

41. Sullivant, "Narrative Report of County Extension Workers," 1931, 3–4, Federal Extension.

42. "A Short History of the Arkansas Farm Credit Company," n.d., 1–2, and "Arkansas Farm Credit Company: Loan Regulations," n.d., 1–3, Red Cross, box 742, DR 224.6511; *Modern News*, May 8, 1931; Sullivant, "Narrative Report of County Extension Workers," Dec. 1, 1931, 16, Federal Extension.

7. The New Deal and the Old Plantation

1. *Marked Tree Tribune*, Oct. 23, 1936.
2. Ibid., Sept. 17, Oct. 1, 1936.
3. Mitchell, *Mean Things Happening*, 35–36.
4. Benedict and Stine, *Agricultural Commodity Programs*; Woofter, *Landlord and Tenant*; Daniel, *Breaking the Land*; Mertz, *New Deal Policy*; Conrad, *Forgotten Farmers*; Kirby, *Rural World's Lost*.
5. *Modern News*, March 15, 1935.
6. *Marked Tree Tribune*, Jan. 10, 1935.
7. Portis interview.
8. Minutes of Meeting, Aug. 1936, bk. 4, 86.
9. *Sixteenth Census of Agriculture, 1940*, 22; *Marked Tree Tribune*, Feb. 20, March 5, 1936; *Modern News*, April 24, 1936.
10. R. L. McGill, "Narrative Report of County Extension Workers," 1934, 2, Federal Extension; Leuchtenburg, *Franklin D. Roosevelt*, 49, 75–76; Burns, *Roosevelt*, 188; R. G. Tugwell to T. Roy Reid, June 27, 1933, AAA, Cotton.
11. A. Raybon Sullivant, "Narrative Report of County Extension Workers," 1933, 4–10, Federal Extension.
12. McGill, "Narrative Report of County Extension Workers," 1934, 2, ibid.
13. Sullivant, "Narrative Report of County Extension Workers," 1933, 8, ibid.
14. Ibid., 7; J. E. McKell, "Narrative Report of District Extension Worker," 1931, 1, 5, ibid.; *Modern News*, June 30, 1933.
15. Real Property, 1930, 1940.
16. "Arkansas Rice Growers' Co-operative Association Holds Its Annual Meeting," 20, Rice Millers Papers.
17. Prairie land was worth about one-third of the value of delta land. Real Property, 1930, 1940.
18. E. A. Miller to All Extension Service District Agents, Aug. 1, 1933, AAA, Cotton.

19. Memorandum from Jack L. Levy, Assistant Attorney, Litigation Section, Sept. 15, 1934, AAA, quoting p. 26 of "Instructions and Regulations Pertaining to the Cotton Act of April 21, 1934," located in this AAA file. Lee Wilson & Company's problem with the Mississippi County committee evolved out of a long-term political battle between contending planter factions in that county.

20. *Marked Tree Tribune*, May 9, 1935.

21. *Modern News*, June 30, 1933, Jan. 12, 1934, Jan. 31, 1935.

22. The newspapers ceased printing the names of the county and township committeemen after 1935, and they were not given in the AAA records or the annual reports of the county extension agent.

23. W. I. Myers to C. C. Davis, Jan. 13, 1934, Davis to Myers, Jan. 19, 1934, FCA.

24. *Marked Tree Tribune*, Jan. 17, Nov. 21, 1935.

25. Ibid., Jan. 1, 1935; *Modern News*, Jan. 18, 1935.

26. *Marked Tree Tribune*, Jan. 10, 1935.

27. Maddox was county judge between 1923 and 1925, and A. B. Caplinger was county judge between 1925 and 1929. Court Records, 1923–1929.

28. McGill, "Narrative Report of County Extension Workers," 1934, 5, Federal Extension.

29. Ibid; McGill, "Narrative Report of County Extension Workers," 1935, 7, ibid.; *Marked Tree Tribune*, May 16, 1935.

30. The Supreme Court ruling in early 1936 that the processing tax was unconstitutional had little effect on the crop reduction program in Poinsett County. Administrators in Washington quickly transformed the processing tax into a soil conserving payment. By the end of February the substitute measure had been legislated and signed by the president. By June the state extension agent in Arkansas had received complete instructions for implementation of the new program from the AAA. C. A. Cobb to C. C. Randall, June 16, 1936, AAA.

31. Benedict and Stine, *Agricultural Commodity Programs*, 12, 17; *Marked Tree Tribune*, Oct. 14, 1937. Poinsett County farmers, in fact, planted 77,472 acres in 1937 compared to 57,220 acres in 1936. McGill, "Narrative Report of County Extension Workers," 1936, 1–2, 7, and "Narrative Report of County Extension Workers," 1937, 7, Federal Extension; *Marked Tree Tribune*, April 15, 1937.

32. Benedict and Stine, *Agricultural Commodity Programs*; *Modern News*, Aug. 13, 1937.

33. *Marked Tree Tribune*, March 18, 1937, March 17, 1938; *Modern News*, Oct. 3, 1939.

34. McGill, "Narrative Report of County Extension Workers," 1934, 5, Federal Extension; *Sixteenth Census, Agriculture, 1940*, 64; McGill, "Narrative Report of County Extension Workers," 1937, 5, Federal Extension.

35. McGill, "Narrative Report of County Extension Workers," 1935, 10; "Narrative Report of County Extension Workers," 1936; "Narrative Report of County Extension Workers," 1937, 8–9, Federal Extension.

36. Ross Mauney, "Narrative Report of County Extension Workers," 1938, 14–15, Federal Extension.

37. The figures for the state are almost as dramatic. In 1930 cotton farmers in Arkansas produced 1,398,475 bales on 3,446,485 acres, valued at $4,292,572, and in 1940 1,351,209 bales on 2,056,775 acres, valued at $3,205,995. *Fifteenth Census, Agriculture, 1930*, 1172; *Sixteenth Census, Agriculture, 1940*, 62.

38. *Modern News*, Oct. 3, 1939.

39. The cultivation of soybeans is not labor intensive. Daniel, *Breaking the Land*, 242.

40. Blake, "Farm Tenancy in Arkansas," 46; Osgood and White, "Land Tenure in Arkansas," 29–31. Blake finds that the increase in the use of wage hands was caused partly by partial mechanization and by planter attempts to circumvent New Deal regulations regarding sharing crop reduction payments. Osgood and White link the decrease in sharecropping and the corresponding increase in wage labor to the rise in mechanization.

41. Street, *New Revolution in the Cotton Economy*, 162. Street is careful to describe the partial mechanization that took place in the cotton fields before the invention of a practical cotton picker, when tractors were used in breaking the land, seedbed preparation, and planting and cultivating, but they were not used for the harvest. McNeely and Barton show that between "1932 and 1938 the number of tractors per 10,000 acres of cropland increased from 12.5 to 29.6 in Mississippi County"; Osgood and White show that "from 1938 to 1944 the number of tractors per 10,000 acres of cropland increased from 27.3 to 60" in the same county. McNeely and Barton, "Land Tenure in Arkansas," 17; Osgood and White, "Land Tenure in Arkansas," 30–31.

42. Daniel, *Breaking the Land*, 56. According to Daniel, both growers and millers had "reacted to the disorganization of the rice industry at the turn of the century by organizing. After steady expansion from 1884 to 1910, prairie growers saturated the U.S. rice market, and, lacking foreign markets of any size, the industry required more planning. Both organizations attempted to discover foreign markets, increase domestic demand and lobby for a protective tariff."

43. *Modern News*, Jan. 27, 1933.

44. "Rice—As Agriculture's Guinea Pig," 9, Rice Millers Papers.

45. Daniel, *Breaking the Land*, 61.

46. "Rice—As Agriculture's Guinea Pig," 1, Rice Millers Papers; Louis Hogue to Charles G. Miller, Nov. 7, 1934, AAA, Rice.

47. The rice millers had begun to form an organization in 1900 and became the Rice Millers Association in 1914. In response, rice growers formed the Southern Rice Growers Association in 1910. It was 1921 before Arkansas rice growers joined. Daniel, *Breaking the Land*, 53–55.

48. McGill, "Narrative Report of County Extension Workers," 1934, 6–7, Federal Extension; S. M. Garwood, Production Credit Commissioner, to Charles G. Miller, Chief, Rice Section, Jan. 17, 1935, AAA, Rice.

49. McGill, "Narrative Report of County Extension Workers," 1934, 6–7, Federal Extension.

50. "Arkansas News," n.d., 13, Rice Millers Papers.

51. The complainant is quoted, but his name is not revealed in, Charles B. Howe to C. C. Randall, March 8, 1935, AAA, Rice.

52. Mrs. Lizzie Scholze's letter is referred to in Charles B. Howe to C. C. Randall, March 7, 1935, AAA, Rice.

53. I. W. Duggan to Charles C. Ruffman, Sept. 14, 1938, to Arthur Phillips, Oct. 12, 1938, to J. B. Daniels, April 21, 1939, to J. W. Ruesewald, May 25, 1939, ibid.

54. McGill, "Narrative Report of County Extension Workers," 1936, 20, Federal Extension.

55. *Marked Tree Tribune*, Sept. 15, 1938.

56. Benedict and Stine, *Agricultural Commodity Programs*, 233–34.

57. Ibid., 383–84; L. A. Dhonau, State Compliance Supervisor, to W. O. Fraser, Assistant Chief, Corn-Hog Section, June 30, 1936, H. A. Wallace to Hattie Caraway, March 27, 1940, AAA.

58. Sullivant, "Narrative Report of County Extension Workers," 1933, 9, 11–12, Federal Extension.

59. Mauney, "Narrative Report of County Extension Workers," 1938, 7, 21, ibid.

8. Meeting on the "Turn Row"

1. The epigraph is from *The Story of a Union That Would Not Die*, 7, pamphlet from H. L. Mitchell's personal collection, photo copy in possession of author. The England farmers hardly constituted rioters although the press typically referred to the "England riot"; See Woodruff, *As Rare as Rain*, 57–59. *Kansas City and Topeka Plaindealer*, Sept. 18, 1933, 6; *Pine Bluff Daily Graphic*, Aug. 4, 1934, 1; *Chicago Defender*, Jan. 14, 1933, 14; *Pittsburgh Courier*, Feb. 4, 1933, 10; Kelley, *Hammer and Hoe*, 39.

2. *Crisis*, Jan. 1934, 20; Aptheker, *Correspondence of W. E. B. Du Bois*, 1: 475–76; Rudwick, *W. E. B. Du Bois*, 272–83; Marable, *W. E. B. Du Bois*, 139–40; Broderick, *W. E. B. Du Bois*, 269–73; *Kansas City and Topeka Plaindealer*, March 2, June 22, 1934; *Chicago Defender*, April 21, June 16, 1934; *Pittsburgh Courier*, June 16, July 7, Aug. 11, 1934; Lewis, *W. E. B. Du Bois*, 466–500. See also, Logan, *W. E. B. Du Bois*, Kellogg, *NAACP*; and Aptheker, *W. E. B. Du Bois*. For STFU organizer Ward Rodgers's essay, see the *Crisis*, Aug. 1935, 248. Black newspaper editors did recognize and occasionally report on the STFU: *Pittsburgh Courier*, Feb. 23, 1935; *Cleveland Call and Post*, March 2, April 6, Oct. 1, 1936; *Kansas City and Topeka Plaindealer*, March 15, 1935, March 20, 1936, March 30, Oct. 15, 1937.

3. NAACP Minutes, reel 2; NAACP Miscellaneous Correspondence; Fuller, "Cotton Kingdom"; White, "Negro and the Communists"; Daniel, *Shadow of Slavery*; Martin, *Angelo Herndon Case*; Foner, *American Communism and Black*

Americans; Foner, *American Socialism and Black Americans*; Haywood, *Black Bolshevik*; Klehr, *Heyday of American Communism*; and Howe and Coser, *American Communist Party*, 204–16, 356; *Cleveland Call and Post*, April 27, March 2, 1935.

4. *Cleveland Call and Post*, March 2, 1935; Hays, Memorandum for Mr. Porter, 1934, Hays Papers; *Modern News*, Sept. 23, 1910; Bettis to E. B. Whitaker, Feb. 4, 1935, 2–4, AAA.

5. McGill, "Narrative Report of County Extension Workers," 1934, 5, Federal Extension; Hays, Memorandum to Mr. Porter, 3, Hays Papers; *Modern News*, March 23, 1934; Mertz, *New Deal Policy*; Grubbs, *Cry from the Cotton*; Conrad, *Forgotten Farmers*; Kirby, *Rural Worlds Lost*, 61; Daniel, *Breaking the Land*, 104–5; and Saloutos, *American Farmer and the New Deal*, 87–97; Mitchell, *Mean Things Happening*, 102.

6. Mitchell, *Mean Things Happening*, 17–18, 31; transcript of Mitchell interview of East, East Papers; Mitchell interview; MS Census, Population, 1910, 1920; Real Property, 1910, 1920, 1930; and Personal Property, 1920, 1930, 1934. Black newspaper editors commented on the discrimination faced by blacks in employment in various works programs: *Pittsburgh Courier*, Sept. 7, 1933; *Chicago Defender*, May 12, 1934; *Cleveland Call and Post*, Aug. 18, 1933, July 6, 13, 1934; *Kansas City and Topeka Plaindealer*, Jan. 20, Aug. 18, 1933, July 6, 13, 1934, Aug. 21, 1936.

7. Mitchell, *Mean Things Happening*, 36, 38; Transcript of Mitchell interview of East, 4, East Papers. See also Dunbar, *Against the Grain*, 124.

8. Donovan, *Governors of Arkansas*, 180–81; Gertrud Gates to Winthrop Lane, June 18, 1934, 2, FERA, box 460. For the controversy within the state FERA, see Whayne, "Reshaping the Rural South," 359–63. See also Walker, *Civil Works Administration*; Carothers, *Chronology of the Federal Emergency Relief Administration*; Williams, *Federal Aid for Relief*; Whiting, *Final Statistical Report*; Hopkins, *Spending to Save*; Armstrong, *We Too Are the People*; and Daniel, *Breaking the Land*, 72–90.

9. Gates to Lane, June 18, 1934, 2, and Arkansas FERA Report, Sept. 1933, FERA, box 460; *Modern News*, Jan 19, Feb. 9, 1934; W. T. Brown to Civil Works Administration, Dec. 6, 1933; John H. Carlew to Civil Works Administration, Dec. 8, 1933, CWA, box 33.

10. *Modern News*, Feb. 9, 1934; Daniel, *Breaking the Land*, 72–90.

11. *Marked Tree Gazette*, Aug. 14, 1908; *Marked Tree Tribune*, March 16, 1934. For socialism in the Southwest, see Green, *Grass-Roots Socialism*; and McWhiney, "Louisiana Socialists," 315–36.

12. Mitchell, *Mean Things Happening*, 40; *Marked Tree Tribune*, Feb. 16, 1934. For the STFU's early association with socialism, see Dunbar, *Against the Grain*, 83–87.

13. Mitchell, *Mean Things Happening*, 61, 199; *Marked Tree Tribune*, Feb. 18, 1927; East interview.

14. Mitchell interview with East, 4, 8, East Papers; Emrich Papers, account books, 1930; *Modern News*, Nov. 21, 28, 1930; Real Property, 1910, 1920, 1930.

15. Mitchell, *Mean Things Happening*, 46–47; Mitchell interview of East, 6, East Papers.

16. Arkansas Historical Records Survey, WPA, 1941, Churches. The printed census does not provide population figures by race on the township level, so it is impossible to estimate the number of seats in churches according to race.

17. Grubbs, *Cry from the Cotton*, 64–66. See also Denisoff, *Great Day Coming* and *Sing a Song*; Frank "Negro Revolutionary Music"; Wiley, "Songs of the Gastonia Textile Strike"; Margaret Larkin, "The Story of Ella May," *New Masses* 5 (Nov. 1929): 3–4; Lawrence Gellert, "Negro Songs of Protest," *New Masses* 6 (April 1931): 6–8; Harold Preece, "Folk Music of the South," *New South* 1 (March 1938): 14; Schatz, "Songs of the Negro Worker," 6–8; *A View of the 40th Anniversary Meeting*, Southern Tenant Farmers' Union, pamphlet in possession of the author.

18. Kelley, *Hammer and Hoe*, 105. See also Haywood, *Black Bolshevik*; Foner and Shapiro, *American Communism and Black Americans*; and Wright, *American Hunger*.

19. See Wilmore, *Black Religion and Black Radicalism*; Pope, *Millhands and Preachers*; and Flynt, *Dixie's Forgotten People*, 98–100.

20. Hubbell, "Always a Simple Feast," 198; Wilmore, *Black Religion and Black Radicalism*; Pope, *Millhands and Preachers*; Fraser et al., *Web of Southern Social Relations*; Kelley, *Hammer and Hoe*, 107; Burton, "Coming of Age of Southern Males," 206–7; Jones, *Labor of Love*, 102.

21. Ownby, *Subduing Satan*; Mitchell, *Mean Things Happening*, 117.

22. Foner, *Women and the American Labor Movement*, 297–98; Mann, "Slavery, Sharecropping, and Sexual Inequality," 792–93; Miller and Neth, "Farm Women in the Political Arena," 357–80.

23. Davis, *Trials of the Earth*, 115; Payne, "'What Ain't I Been Doing?'" 129.

24. Neth, "Building the Base," 340; Foner, *Women and the American Labor Movement*, 297. See also Osterud, "Gender and the Transition to Capitalism," 29; Riney-Kehrberg, "Separation and Sorrow"; Hall, *Labor of Love*, 105; Holt, "A Time to Plant"; Bercaw, "The Politics of the Household During the Transition from Slavery to Freedom in the Mississippi Delta"; Lerner, *Black Women in White America*; and Davis, *Good Wives*.

25. Whayne, "Significance of Race, Class, and Family in the Battle for Prohibition," 130–49; Ownby, *Subduing Satan*; Ayers, *Promise of the New South*.

26. Jones, "'Tore Up and a-Movin'," 17.

27. Harry Crews "Rough South," televised interview; Toni Morison, *Bluest Eye*, audio tape.

28. Weeks, *From Madison to Marked Tree*, 49.

29. Jones, "'Tore Up and a-Movin'"; Janiewski, "Sisters Under the Skin"; Sharpless, "Southern Women and the Land," 38.

30. Stith interview.

31. Mitchell, *Mean Things Happening*, 52; Hays, Memorandum for Mr. Porter, Hays Papers; Mitchell, *Mean Things Happening*, 54.

32. White, "'Massacring Whites,'" 715–16; Rogers, "The Elaine Race Riots"; Desmarais, "Military Intelligence Reports"; Butts and James, "The Underlying Causes of the Elaine Riot"; Woodruff, "African-American Struggles for Citizenship."

33. Mitchell, *Mean Things Happening*, 47–48.

34. Mertz, *New Deal Policy*; Grubbs, *Cry from the Cotton*; Conrad, *Forgotten Farmers*; Mitchell, *Mean Things Happening*; and Daniel, *Breaking the Land*, 98–99, 230–31; Kirby, *Rural Worlds Lost*, 152.

35. Tyronza Township (school district 22) Personal Property, 1920, 1934. All the tables deflate the 1920 dollars to what they would be worth in 1934 in order to draw a more accurate picture of the decline in real dollars, using 1926 as the base year. *Historical Statistics of the United States*, 103–104, 116. Thanks are due to Paul Paskoff of Louisiana State University for assistance in deflating the 1920 values and to Jeffrey J. Ryan and William Miller for help in arranging the table.

36. Crowley's Ridge farmers believed that the delta planters "voted" their tenants and often paid their poll taxes for them in order to do so. Roberts interview.

37. *Fourteenth Census, Population, 1920*; *Sixteenth Census, Population, 1940*.

38. Grubbs, *Cry from the Cotton*, 19–26, 40–41, 45–46.

39. Daniel, *Breaking the Land*; Kirby, *Rural Worlds Lost*; Mertz, *New Deal Policy*, especially 47–48; Grubbs, *Cry From the Cotton*, especially 42–43, 46–47, 53–58; Conrad, *Forgotten Farmers*.

40. Hays, memorandum for Mr. Porter, Hays Papers.

41. C. C. Davis, Administrator, AAA to "District Agents and Others Who are to Assist with Landlord-Tenant Problem," May 5, 1934, 10, STFU Papers, reel 1.

42. Ibid., 5–6.

43. STFU Clippings File, Cross County Historical Society; Mitchell, *Mean Things Happening*, 129.

44. Cross County Court Records, True Bills, 1936; Mitchell, *Mean Things Happening*, 73.

45. STFU Clippings File, Cross County Historical Society.

46. Mertz, *New Deal Policy*, 29; Daniel, *Breaking the Land*, 168–83; Agricultural Censuses, 1935–1969.

47. Naison, "Black Agrarian Radicalism," 56–57.

48. Mitchell, *Mean Things Happening*, 60; *Modern News*, Jan. 25, 1935; Grubbs, *Cry From the Cotton*, 70–71; *Marked Tree Tribune*, Jan. 24, 1935. Rodgers was convicted in the Marked Tree court, but on appeal to the Circuit Court the charges were dismissed. *Modern News*, Jan. 25, 1935; *Marked Tree Tribune*, March 5, 1936.

49. Howard Kester, "Acts of Tyranny and Terror," STFU Papers, reel 2; Mitchell interview of East, 8, East Papers; for other reports of violence, see Bunche, "Share-Croppers in the United States," 214; for biographical information on Harry Ritter, see *Marked Tree Tribune*, May 27, 1954.

50. Brief for Appellants, Arkansas Supreme Court Case no. 3821, STFU Papers, reel 1.

51. Mitchell, *Mean Things Happening*, 68.

52. Stith interview; Grubbs, *Cry from the Cotton*, 66–69; *Modern News*, Feb. 8, 1935; STFU Clippings File, Cross County Historical Society.

53. Grubbs, *Cry from the Cotton*, 62; *Marked Tree Tribune*, June 13, 1935; and Mitchell, *Mean Things Happening*, 80.

54. Bettis to Whitaker, Feb. 4, 1935, 3, FERA State Files, box 400; Mitchell interview of East, 5, East Papers; Grubbs, *Cry from the Cotton*, 176; Dunbar, *Against the Grain*; E. B. McKinney to Wiley Harris, July 29, Aug. 11, 1938, reel 8, STFU Papers.

55. The STFU became a part of the CIO's United Cannery, Agricultural, Packing, and Allied Workers of America (UCAPAWA). See Mitchell, *Mean Things Happening*; Grubbs, *Cry from the Cotton*, 176–81; and Kirby, *Rural World's Lost*, 267–71. For blacks and unionism, see Foner, *Organized Labor and the Black Worker*; and Cayton and Mitchell, *Black Workers and the New Unions*.

56. *Kansas City and Topeka Plaindealer*, July 6, 13, 1934.

57. *Marked Tree Tribune*, Feb. 14, 1935.

58. *Modern News*, July 2, 1937, June 24, 1938.

59. Ibid., Sept. 5, Oct. 3, 1935, March 18, April 15, 1937; *Modern News*, Feb. 4, 11, Oct. 7, 1938.

60. *Marked Tree Tribune*, July 25, 1935; *Modern News*, Nov. 8, 1935. Dyess colony was a rural rehabilitation community established in W. R. Dyess's home county.

61. *Modern News*, April 29, 1938; *Marked Tree Tribune*, May 6, 1938; *Lepanto News Record*, May 5, 1938.

62. *Marked Tree Tribune*, May 5, July 28, 1938.

63. St. Francis Levee Board, *History of the District*, 83, 84; Minutes of the Meetings, Dec. 11, 1934; Board of Commissioners, "Report of Drainage District Seven," 2.

64. *Marked Tree Tribune*, Sept. 2, 1937; *Lepanto News Record*, Sept. 3, 1937; *Modern News*, Sept. 3, 1937.

65. Mertz, *New Deal Policy*, 107.

66. Less than one hundred families were placed in the two communities, and these modest numbers could not account for the survival of tenancy in Poinsett County. Tenancy also began to decline in other counties with resettlement communities, including Mississippi County, home to Dyess Colony, the largest resettlement community in Arkansas.

67. McGill, "Narrative Report of County Extension Workers," 1935, 11, Federal Extension; Holley, *Uncle Sam's Farmers*; Baldwin, *Poverty & Politics*; "Campbell Farms" undated proposal, FSA, box 128.

68. Memorandum from C. B. Baldwin to Dr. Gray, Sept. 28, 1935, RA, box 45; "Poinsett County Project, RR-AK–13," Sept. 1, 1935, McGill to T. Roy Reid, Dec. 13, 1935, Ray B. Johnston to T. Roy Reid, April 19, 1937, FSA, box 128.

69. Johnston to Reid, April 19, 1937, P. F. Aylesworth to E. R. Henson, Feb. 26, 1936, FSA, box 128.
70. "Poinsett Farms Project, S-AK–118," May 20, 1938, Lewis E. Long to E. R. Henson, Sept. 23, 1937, ibid.
71. Alvin Inman to J. O. Walker, March 4, 1938, ibid., box 129; "Poinsett Farms Project S-AK–118," May 20, 1938, ibid, box 128.
72. J. C. Walker to A. C. Hervey, May 3, 1937, ibid., box 129; Johnston to Reid, April 19, 1937, ibid., box 128.
73. Inman to President Roosevelt, Feb. 21, 1938, Walker to Inman, March 17, 1938, ibid., box 129.
74. Bill Medley and L. V. Salter of the Campbell Farm and E. L. Dubose and Howard Moore of the Northern Ohio Farm were the only Poinsett County clients located in the FSA alphabetical file of clients who had purchased farms in the RA/FSA projects (FSA, box 328). Hervey interview; Kirkendall *Social Scientists and Farm Politics in the Age of Roosevelt*, 195–254; Baldwin, *Poverty and Politics*.

9. No Way for a Man To Live

1. *Marked Tree Tribune*, June 26, 1952.
2. Thatcher interview. See also Daniel, *Breaking the Land*; Kirby, *Rural Worlds Lost*; and Cobb, *Most Southern Place on Earth*; Schulman, *From Cotton Belt to Sunbelt*; Cobb, *Industrializing Southern Society*; Goldfield, *Promised Land*.
3. Lepanto and Tyronza have been particularly unsuccessful in attracting industries; Marked Tree has attracted some industries but has had great difficulty keeping them. Truman, because of its proximity to Jonesboro and possibly because of the Singer Sewing Machine Company located there, has been most successful. Its population has increased more than any other town in the county: Truman, plus 2,194 (37 percent); Marked Tree, plus 330 (10.3 percent); Lepanto, plus 163 (8.8 percent). Tyronza has actually declined in population: Tyronza, minus 55 (10.8 percent). Harrisburg has increased by 433 (22.4 percent). Weiner has increased by 71 (9.9 percent). *Nineteenth Census, Population, 1940*, 47, 108.
4. Rasmussen, *History of the Emergency Farm Labor Supply Program*; *Marked Tree Tribune*, Feb. 26, 1942; Voss, "Prisoner of War Camps," 14.
5. *Marked Tree Tribune*, Aug. 20, Sept. 10, 22, 1942, June 17, July 29, Aug. 19, 1943. See also Woodruff, "Pick or Fight."
6. *Marked Tree Tribune*, July 29, 1943.
7. "Digest of Materials Proposed for Presentation to the Senate Post-War Committee Hearing: Prospects for Reabsorption of Population and Workers in Agricultu[r]e at the End of the War," May 8, 1944, 2, Agriculture, box 2.
8. *Marked Tree Tribune*, May 11, 1944. This newspaper article describes the association of farmers as already in existence but does not indicate when it was formed. In addition to L. V. Ritter, it included R. L. McGill, R. H. Taylor,

G. E. Davis, Herrick Norcross, D. F. Portis, J. G. Stuckey, J. H. Prestidge, and Dr. G. O. Campbell.

9. Ibid., March 23, April 27, 1944. It was later decided that they did not need as many as 75 guards.

10. Boyer, "Narrative Report of County Agent," 1944, 47–50, Federal Extension; *Marked Tree Tribune*, Sept. 7, 1944.

11. *Marked Tree Tribune*, June 15, 1944, Feb. 22, 1945. See also Shea, "German Prisoner of War"; Robin, *Barbed-Wire College*; Clark, *Farm Work and Friendship*; Moore, *Faustball Tunnel*; Faulk, *Group Captives*; Powell, *Splinters of a Nation*; Gansberg, *Stalag, U.S.A.*; and Koop, *Stark Decency*.

12. For the black experience in securing equal opportunity in war industries, see Reed, *Seedtime for the Modern Civil Rights Movement*.

13. Cockcroft, *Outlaws in the Promised Land*; Corwin, *Immigrants*; Garcia, *Operation Wetback*; Garcia y Griego, *Importation of Mexican Contract Laborers to the United States*; Galarza, *Merchants of Labor* and *Strangers in Our Fields*; Geenens, "Economic Adjustments to the Termination of the Bracero Program"; Guerin-Gonzales, *Mexican Workers and American Dreams*; Haas, *Bracero in Orange County*; Kiser and Kiser, *Mexican Workers*; Monto, *Roots of Mexican Labor Migration*; Taylor, *Mexican Labor*; Rasmussen, *History of the Emergency Farm Labor Supply Program*; Reisler, *By the Sweat of Their Brow*; and Valdes, *Al Norte*.

14. Byron Mitchell to J. Otis Garber, Nov. 1, 1948, 2, Labor, file 1945–51, box 6.

15. W. L. Wright, "Narrative Report of County Agent," 1946, 86, Federal Extension.

16. Ibid., 87. According to the farm agent, rice farmers were using 25 combines during the 1946 season; in 1945 they used only 6 combines. In addition, the rice farmers pooled their machinery, on a voluntary basis, so that every rice farmer in the Poinsett County prairie had access to a combine.

17. *Marked Tree Tribune*, Aug. 15, 1946; Wright, "Narrative Report of County Agent," 1946, 86, Federal Extension.

18. *Memphis Press Scimitar*, Sept. 2, Oct. 1, 1942, Nov. 15, 1944, Mississippi Valley Collection clippings file.

19. Ibid., Nov. 15, 1944. See also Wilcox, *Farmer in the Second World War*.

20. *Marked Tree Tribune*, July 16, 1942. Out of 4,136 farms, 365 (8.8 percent) reported tractors in 1939; out of 4,599 farms, 793 (17.2 percent) reported tractors in 1944.

21. *Marked Tree Tribune*, June 10, 1943, April 6, 1944.

22. Carl F. Taeusch to Ray C. Smith, Sept. 15, 1942, Agriculture, box 1, entry 212. See also Binswanger, *Agricultural Mechanization*; Hirschhorn, *Beyond Mechanization*; Price, *Political Economy of Mechanization*; Berardi and Geisler, *Social Consequences and Challenges of New Agricultural Technologies*; Hallam, *Size, Structure, and the Changing Face of American Agriculture*; Davis, *Economics of Farm Mechanization*; Bertrand, et al., *Factors Associated with Agricultural Mechanization*; Gregor, *Industrialization of U.S. Agriculture*; Raper, *Machines in the Cotton Fields*; and Street, *New Revolution*. For mechanization outside the

United States, see Stavis, *Politics of Agricultural Mechanization in China*; Richards and Martin, *Migration, Mechanization, and Agricultural Labor Markets in Egypt*; Butler, *Farm Mechanization in the Soviet Union*; and Sargen, *Tractorization*. For mechanization and small farmers, see Williams, *Strategy for Survival*; and OEEC, *Mechanization of Small Farmers in European Countries*.

23. Stith, "Statement to the President's Commission on Migratory Labor," Aug. 31, Sept. 1, 1950, 1, Truman Papers.

24. Commission to the President, March 1, 1951, B-file, ibid.

25. Stith, "Statement to the President's Commission on Migratory Labor," Aug. 31, Sept. 1, 1950, 1, ibid.; Merce Whayne interview.

26. List of those giving testimony to the commission, B-file, Truman Papers.

27. "Planters Are Feuding over Mexican Pickers," *Memphis Press Scimitar*, Sept. 20, 30, 1948, Mississippi Valley Collection clippings file.

28. Thatcher interview.

29. "A Proposal to Bring Up to Parity the Existing Facilities of Extension Service With Negroes in the Southern States," Jan. 1943, 1–5, box 17, Campbell Papers; Whayne, "Segregated Farm Program in Poinsett County"; Woodruff, "Mississippi Delta Planters."

30. *Marked Tree Tribune*, May 16, 1946. The county ceased utilizing black agents after the 1954 census of agriculture revealed a dramatic decrease in black farmers. C. A. Vines to Elvira Heard, Feb. 8, 1957, Extension Service.

31. Lena H. Eddington, Negro Home Demonstration Agent, "Narrative Report of County Home Demonstration Agent," 16, Federal Extension; U.S., *Selective Service in Wartime*, xv, 9; *Marked Tree Tribune*, July 11, 1942, Jan. 1, 1945; T. R. Betton, Negro Movable School Agent, "Annual Report of Movable Demonstration School Work," for Dec. 1, 1945, to June 30, 1946, 3, Federal Extension.

32. L. J. Jackson, Negro County Agent, "Narrative Report of County Agent," 1946, 3, 5, 13, Federal Extension.

33. J. C. Barnett, District Agent, Supervisor of Negro Agents, "Narrative Report of Extension Work for Negroes, Arkansas," 20, ibid.; Orville L. Freeman, Department of Agriculture, Office of the Secretary, to the President of the United States, June 17, 1965, 7–10, box 358, ACES; W. B. Hill, sworn affidavit made to Curtis M. Crowe, Special Agent, office of the Inspector General, U.S. Department of Agriculture, box 358, Jan. 7, 1966, 1, Agriculture.

34. Between 1940 and 1965 white farmers declined by 62.7 percent while black farmers declined by 90.7 percent. Between 1940 and 1955, when the agricultural census included separate categories for tenants as opposed to sharecroppers, black tenants declined from 79 to 15 and white tenants declined from 1,366 to 992; black sharecroppers declined from 563 to 302 while white sharecroppers declined from 1,170 to 1,056.

35. In 1940 black tenants and sharecroppers farmed an average of 16.0 acres while white tenants and sharecroppers farmed an average of 51.9 acres. By 1965 the few black tenants and sharecroppers who remained operated an average of 15.7 acres while white tenants and sharecroppers operated an aver-

age of 211.6 acres. The size of the farm units operated by white tenants and sharecroppers suggests that they had particiated, at least to some extent, in the mechanical revolution in the postwar period. For black migration away from southern farms in the postwar period, see Goldfield, *Black, White and Southern*.

36. SNCC Papers, microfilm series (reels 5, 17, 19, 37, and 53 contain material relating to the "Arkansas Project"); ACHR (see especially boxes 1–8); Faubus Papers (boxes 538–45 pertain to race relations outside of Little Rock). For SNCC activity elsewhere in the South, see Waskow, *From Race Riot to Sit-In*, 227–28; Carson, *In Struggle*; Sellers, *River of No Return*; Zinn, *SNCC*; Dittmer, *Local People*.

37. Letter from T. R. Betton, Agricultural Agent for Negro Work, n.d., 2 (first page missing), Extension Service; Mitchell, *Mean Things Happening*, 160.

38. *Marked Tree Tribune*, June 1, July 20, 1950, Oct. 25, 1951.

39. Ibid., Oct. 10, 1957. David Chappell finds that many southerners shared these sentiments (Chappell, *Inside Agitators*). Editorials dealing with the issues of desegregation and voting rights appeared in the *Marked Tree Tribune*, Sept. 26, Oct. 10, 1957, Jan. 1, Feb. 16, 1961, April 5, 1962, Feb. 2, Sept. 19, 1963, April 23, May 28, Aug. 13, Sept. 3, 1964, May 20, Dec. 1, 1965.

40. Ibid., Aug. 13, 1964.

41. Ibid., Aug. 11, 1965.

42. Ibid., May 20, 1965.

43. According to Jean Thatcher, Ritter owned 50 percent of the St. Francis Valley Company. Thatcher interview.

44. Ibid.; *Marked Tree Gazette*, Dec. 28, 1950.

45. Thatcher interview.

46. See Kirby, *Rural Worlds Lost*.

47. Daniel, "Going among Strangers," and *Breaking the Land*, 167.

48. Crews, *Childhood*, 128–29.

BIBLIOGRAPHY

Manuscript Collections

Arkansas History Commission, Little Rock
    Mayor's Court, Harrisburg. Microfilm
    Palmer, Edwin, Papers. Microfilm
    Treiber, Jacob, Papers
    Works Progress Administration Place Files, Arkansas
Arkansas State Bank Department
    Liquidation Record
Auburn University Archives
    Agricultural Cooperative Extension Service Records
Cross County Historical Society, Cross County Courthouse, Wynne, Ark.
    Cross County Historical Society clippings file
Mississippi Valley Collection, Memphis State University, Memphis
    Memphis Press Scimitar clippings file
Special Collections Division, University of Arkansas Library, Fayetteville
    Abbott Family Papers
    Arkansas Historical Records Survey, Church Records
    University of Arkansas (System), Cooperative Extension
        Service Records
    Arkansas Council on Human Relations
    Davis, Jeff, Papers
    East, Henry Clay, Papers
    Emrich, John A., Papers
    Faubus, Orval Eugene, Papers
    Hays, Lawrence Brooks, Papers
    Robinson, Joseph Taylor, Papers
    Sanborn Map Company. Sanborn Fire Insurance Maps, Marked Tree, Ark., 1908.
    Thornburgh, George, Collection, Temperance Scrapbooks

Truman Library, Independence, Missouri
   Truman, Harry S., Papers
Tuskegee University Archives
   Campbell, Thomas M., Papers
University of Arkansas Libraries, Fayetteville
   Roosevelt, Theodore, Papers. Microfilm ser. 1, reel 53
   Southern Tenant Farmers Union Papers. University Microfilms, reels 1–13, 1934–40
   Student Non-Violent Coordinating Committee Papers, 1959–72. Microfilm ser., reels 5, 17, 19, 30, 31, 37, 53
University of Arkansas, Little Rock, Archives and Special Collections
   XV Papers
University of Southwest Louisiana, LaFayette
   Rice Millers Association Papers

County Records

Court Records. Poinsett County Courthouse, Harrisburg, Ark., 1886–1965
Court Records. Cross County Courthouse, Wynne, Ark., 1890–1915
Minutes of the Meetings of the Board of Commissioners of Drainage District 7. Office of Drainage District 7, Marked Tree, Ark., 1920–65
Mortgage Records. Poinsett County Courthouse, Harrisburg, Ark., 1909, 1910
Personal Property Records. Poinsett County Courthouse, Harrisburg, Ark., 1920, 1933, 1934
Real Property Records. Poinsett County Courthouse, Harrisburg, Ark., 1880, 1900, 1910, 1920, 1930, 1931, 1940, 1950, 1960
Record of the Proceedings of the Council of the Incorporated Town of Marked Tree, Ark., City Courthouse, Marked Tree, 1897–1908

Federal Records

Agricultural Adjustment Administration, NA, RG 145
Agricultural Stabilization and Conservation Service, NA, RG 145
Bureau of Agricultural Economics, NA, RG 83
Civil Works Administration, NA, RG 69
Commodity Credit Corporation, NA, RG 145
Department of Agriculture, NA, RG 16
Department of Justice, NA, RG 60
Farm Credit Administration, NA, RG 145
Farmers Home Administration, NA, RG 96
Farm Security Administration, NA, RG 96
Federal Emergency Relief Administration, NA, RG 69

Federal Extension Service, Narrative Reports of Regional and County Extension Workers, Arkansas Extension Service, College of Agriculture, University of Arkansas and U.S. Department of Agriculture, Cooperating. Fort Worth, Tex.: Federal Regional Archives, NA, RG 33, 1922–65

Federal Manuscript Census, Poinsett County, Ark., Schedule of Population, NA, 1860, 1870, 1880, 1900, 1910, 1920

Federal Manuscript Census, Poinsett County, Ark., Schedule of Slaves, NA, 1860

Federal Manuscript Census, Poinsett County, Ark., Schedule of Agriculture, NA, 1880

General Land Office Records, Arkansas Sunk Lands, Interior Department, Appellate Case Files, NA, RG 67

Public Works Administration, NA, RG 135

Reconstruction Finance Corporation, NA, RG 234

Record Book, U.S. District Court, Eastern District of Arkansas, Helena, E39 Criminal, vol. Nov. 30, 1898–March 10, 1908, Federal Regional Archives, Fort Worth, Tex., RG 21

Red Cross Records, NA, RG 200

Resettlement Administration, NA, RG 96

Soil Conservation Service, NA, RG 114

U.S. Public Health Service, NA, RG 90

Works Progress Administration, NA, RG 69

Oral History Projects

Arkansas State University Oral History Project, State College, Ark.
Mitchell, H. L., Oral History Project, Columbia University, 1956–57.

Audio-Visual Sources

Crews, Harry. "The Rough South of Harry Crews." University of North Carolina Center for Public Television, 1992.
Morison, Toni. *The Blueist Eye*. Random House, 1994.

Interviews

Abbott, Dorothy. Dec. 12, 31, 1994
Arnold, Mary Ann. Marked Tree, Ark., Feb. 29, 1988
Cart, Raymond. Weiner, Ark., April 1, 1983
Dawson, Virgie. Marked Tree, Ark., March 20, 1983
Dewey, Craft. Memphis, Dec. 31, 1985
East, Clay. Tucson, Ariz., July 20, 1982

Emrich, J. A. Tyronza, Ark., March 25, 1983
Fendler, Oscar. Blytheville, Ark., Aug. 31, 1981
Freeman, Joanna. Weiner, Ark., March 24, 1983
Harvey, Addie. Telephone, Dec. 12, 1994
James, Jimmie. Wynne, Ark., Nov. 8, 1994
Marshall, Gertrud. Lepanto, Ark., Dec. 30, 1986
Mitchell, H. L. Montgomery, Ala., June 27, 1982
Portis, D. F. Lepanto, Ark., Dec. 27, 1985
Ruesewald, Fred. Weiner, Ark., April 1, 1983
Schloze, Otto. Waldenberg, Ark., April 1, 1983
Senteney, Chester. Weiner, Ark., Oct. 25, 1986
Stith, George. Gould, Ark., March 28, 1983
Thatcher, Jean. Marked Tree, Ark., July 10, 1992
Whayne, Aud Earl Sr. Lepanto, Ark., Dec. 12, 1982
Whayne, Merce. Lepanto, Ark., Dec. 31, 1994
Wilson, Michael. Wilson, Ark., Nov. 2, 1982
Ziegenhorn, Raymond. Jonesboro, Ark., April 2, 1983

Newspapers

Arkansas Democrat
Arkansas Gazette
Chicago Defender
Cleveland Call and Post
Crisis
Forrest City Times
Harrisburg Modern News
Helena World
Lee County Courier
Lepanto News Record
Marked Tree Gazette (name changed to *Marked Tree Tribune* in 1917)
Marked Tree Tribune
Memphis Commercial Appeal
Memphis Press Scimitar
Pine Bluff Daily Graphic
Pittsburgh Courier
Topeka Plaindealer (*Kansas City and Topeka Plaindealer*)
Wynne Star Progress
Tri-City Tribune

State Publications

Rose, U. M. *Digest of Arkansas Reports, 1820-[to date]*. Little Rock, 1982.
State of Arkansas. *Acts of Arkansas, 1909*. Little Rock, Ark., 1910.
———. *Acts of Arkansas, 1921*. Little Rock, Ark., 1921.

Annual Reports

Annual Report of the St. Francis River Levee Distict. West Memphis, Ark., 1895–1916.
Report of the Commissioner of the General Land Office. Washington, D.C., 1906–21.

U.S. Government and Agricultural Experiment Station Publications

Bertrand, Alvin L., J. L. Charlton, Harold A. Pedersen, R. L. Skrabanek, and James D. Tarver. "Factors Associated with Agricultural Mechanization in the Southwest Region." Arkansas *Agricultural Experiment Station Bulletin* 567 (1956): 3–33.
Brodie, D. A. "Diversified Farming in the Cotton Belt." *Yearbook of the United States Department of Agriculture: 1905* (1906): 207–12.
Capstick, Daniel F. "Economics of Mechanical Cotton Harvesting." Arkansas *Agricultural Experiment Station Bulletin* 662 (1960): 3–34.
——. "Cost of Operating Farm Tractors in Eastern Arkansas." Arkansas *Agricultural Experiment Station Bulletin* 652 (1962): 3–26.
Capstick, Daniel F., and William P. Nelson. "Cost of Owning and Operating Miscellaneous Farm Machinery in Eastern Arkansas." Arkansas *Agricultural Experiment Station Bulletin* 661 (1962): 3–20.
Charlton, J. L. "Social Aspects of Farm Ownership and Tenancy in the Arkansas Coastal Plain." *Southwest Regional Agricultural Experiment Station Bulletin* 545 (1951): 1–85.
Congressional Record. 67th Cong., 1st sess., House. Washington, D.C., 1921.
——. Report 298. Washington, D.C., 1921.
Gile, B. M. "The Development of Agricultural Credit Corporations in Arkansas with State Aid in 1931." Arkansas *Agricultural Experiment Station Bulletin* 281 (1932): 2–39.
——. "Organization and Management of Agricultural Credit Corporations in Arkansas." Arkansas *Agricultural Experiment Station Bulletin* 259 (1931): 2–55.
——. "The Status of Cooperative Cotton Marketing in Arkansas." Arkansas *Agricultural Experiment Station Bulletin* 245 (1929): 2–44.
Grinstead, Mary Jo., Bernal L. Green, and J. Martin Redfern. "Rural Development and Labor Adjustment in the Mississippi Delta and Ozarks of Arkansas." Arkansas *Agricultural Experiment Station Bulletin* 795 (1975): 3–29.
Hearings Held before the Committee on Public Lands of the House of Representatives on H.R. 19637. Washington, D.C., 1910.
Holly, William C., Ellen Winston, and T. J. Woofter, Jr. *The Plantation South.* Washington, D.C., 1940.
House Reports, vol. 2, 67th Cong., 1st sess., Report 178, Washington, D.C., 1921.

Industrial Commission. *Industrial Commission Report on Agriculture and Agricultural Labor.* Washington, 1901.

Journal of the Senate, 65th Cong., 3rd sess. Washington, D.C., 1919.

———. 67th Cong., 1st sess., Washington, D.C., 1921.

———. 2d sess. Washington, D.C., 1921.

Knapp, S. A. "Causes of Southern Rural Conditions and the Small Farm as an Important Remedy." *Yearbook of the United States Department of Agriculture: 1908* (1909): 311–20.

"Land Tenure in the Southwestern States: Summary of Significant Findings of the Regional Land Tenure Research Project." *Southwest Regional Agricultural Experiment Station Bulletin* 482 (1948): 1–31.

McNeeley, J. G., and Glen T. Barton. "Land Tenure in Arkansas: II. Change in Labor Organization on Cotton Farms." Arkansas *Agricultural Experiment Station Bulletin* 397 (1940): 3–26.

Osgood, Otis T., and John W. White. "Land Tenure in Arkansas: IV. Further Changes in Labor Used on Cotton Farms, 1939–1944." Arkansas *Agricultural Experiment Station Bulletin* 459 (1945): 2–31.

Rasmussen, Wayne D. *A History of the Emergency Farm Labor Supply Program, 1943–47.* Agricultural Monograph no. 13, U.S. Department of Agricultural Economics, Washington, D.C., Sept. 1951.

———. "Price-Support and Adjustment Programs from 1933 through 1978: A Short History." *Agriculture Information Bulletin* 424 (1979): 1–32.

Rasmussen, Wayne D., Gladys L. Baker, and James S. Ward. "A Short History of Agricultural Adjustment, 1933–75." *Agriculture Information Bulletin* 391 (1976): 1–21.

Senate Reports, vol. 2, 67th Cong., 2d sess., no. 908. Washington, D.C., 1922.

Slusher, M. W., and Harold Scoggins. "Cotton Production Practices in Arkansas." Arkansas *Agricultural Experiment Station Bulletin* 507 (1951): 3–91.

Solberg, Erling D. "Legal Aspects of Farm Tenancy in Arkansas." Arkansas *Agricultural Experiment Station Bulletin* 468 (1947): 3–84.

Southern, John H. "Farm Land Ownership in the Southwest." Arkansas *Agricultural Experiment Station Bulletin* 502 (1950): 3–47.

Southern, John H., Harold Scoggins, and John W. White. "Arkansas Land Prices in War and Peace." Arkansas *Agricultural Experiment Station Bulletin* 517 (1951): 3–27.

Southwestern Land Tenure Research Committee. "Tenure Improvement for a Better Southwest Agriculture." Arkansas *Agricultural Experiment Station Bulletin* 491 (1949): 3–38.

U.S. Department of Agriculture, Agricultural Adjustment Administration. *A Report of Administration of the Agricultural Adjustment Act, May 1933 to February 1934.* Washington, D.C., 1934.

———. *A Report of Administration of the Agricultural Adjustment Act, May 12, 1933, to December 31, 1935.* Washington, D.C., 1936.

———. *A Report of the Activities Carried On by the Agricultural Adjustment Administration.* Washington, D.C., 1939.

———. Commodity Credit Corporation. *A Report of the President of the Commodity Credit Corporation, 1940.* Washington, D.C., 1941.

U.S. Department of Commerce, *Historical Statistics of the United States: Colonial Times to 1957.* Washington, D.C., 1961.

U.S. Selective Service System, *Selective Service in Wartime: Second Report of the Director of Selective Service, 1941–42.* Washington, D.C., 1943.

White, John W. "Combinations of Enterprises on Plantations in the Lower Arkansas River Delta." Arkansas *Agricultural Experiment Station Bulletin* 449 (1944): 3–75.

U.S. Census

U.S. Department of Commerce. *Eighth Census of the United States: 1860. Population* and *Agriculture.* Washington, D.C., 1862.

———. *Ninth Census of the United States: 1870. Population* and *Agriculture.* Washington, D.C., 1872.

———. *Tenth Census of the United States: 1880. Population* and *Agriculture.* Washington, D.C., 1882.

———. *Twelfth Census of the United States: 1900. Population* and *Agriculture.* Washington, D.C., 1902.

———. *Thirteenth Census of the United States: 1910. Population* and *Agriculture.* Washington, D.C., 1912.

———. *Fourteenth Census of the United States: 1920. Population* and *Agriculture.* Washington, D.C., 1922.

———. *United States Census of Agriculture: 1925.* Washington, D.C., 1927.

———. *Fifteenth Census of the United States: 1930. Population* and *Agriculture.* Washington, D.C., 1932.

———. *United States Census of Agriculture: 1935.* Washington, D.C., 1936.

———. *Sixteenth Census of the United States: 1940. Population* and *Agriculture.* Washington, D.C., 1942.

———. *United States Census of Agriculture: 1945.* Washington, D.C., 1946.

———. *Seventeenth Census of the United States: 1950. Population* and *Agriculture.* Washington, D.C., 1952.

———. *United States Census of Agriculture: 1954.* Washington, D.C., 1956.

———. *Eighteenth Census of the United States: 1960. Population* and *Agriculture.* Washington, D.C., 1962.

———. *United States Census of Agriculture: 1964.* Washington, D.C., 1966.

———. *Nineteenth Census of the United States: 1970. Population* and *Agriculture.* Washington, D.C., 1972.

———. *United States Census of Agriculture: 1974.* Washington, D.C., 1977.

———. *Twentieth Census of the United States: 1980. Population* and *Agriculture.* Washington, D.C., 1982.

———. *United States Census of Agriculture: 1978.* Washington, D.C., 1981.

———. *Twenty-First Census of the United States: 1980. Population* and *Agriculture.* Washington, D.C., 1983.

———. *United States Census of Agriculture: 1982*. Washington, D.C., 1984.
———. *United States Census of Agriculture: 1987*. Washington, D.C., 1989.

Privately Printed Pamphlets

Bank of Harrisburg. *Pictorial History of the Harrisburg Area*. Harrisburg, Ark., n.d.
Bank of Weiner. *Second Annual Rice and Amusement Carnival*. Weiner, Ark., 1919.
Board of Commissioners. *Report of Drainage District No. Seven of Poinsett County*. N.p., n.d.
Carnival Committee, *Second Annual Rice and Amusement Carnival*, Sept. 19, 20, 21, 1929. Weiner, Ark., [1929].
Chowning, Robert W. *History of Cross County, Arkansas*. Wynne, Ark., 1955.
Cross County Historical Society, Inc. *Naming a Confederate County, Cross County, Arkansas, 1862–1873*. Wynne, Ark., 1973.
Dawson, Virgie Waskom. *One Hundred Years of Progress*. N.p., March 1983.
Drainage District No. 7. *Report of Drainage District No. 7*. Marked Tree, Ark., 1965.
Edrington, Mabel F. *History of Mississippi County, Arkansas*. N.p., 1962.
Jernigan, Gail, and Sue Chambers. *A History of Lepanto, Arkansas*. Lepanto, Ark., 1989.
Lynch, Leila B. *Then and Now: Thru the Years in Weiner, Arkansas 1866–1971*. N.p., 1971.
St. Louis Bank for Cooperatives. *Farmer Co-ops in Arkansas*, St. Louis, n.d.
Senteney, Grace. *Just Born or A Look through the Twentieth Century*. Weiner, Ark., n.d.
Weeks, James E. *From Madison to Marked Tree by the St. Francis River*. N.p., n.d.

Books

Alexander, Donald Crichton. *The Arkansas Plantation, 1920–1942*. New Haven, 1943.
Aptheker, Herbert, ed. *The Correspondence of W. E. B. Du Bois: Selections*. 2 vols. Amherst, Mass., 1973–76.
Armstrong, Louise Van Voorhis. *We Too Are the People*. Boston, 1938.
Arsenault, Raymond. *The Wild Ass of the Ozarks: Jeff Davis and the Social Bases of Southern Politics*. Philadelphia, 1984.
Ashmore, Harry S. *Arkansas: A Bicentennial History*. New York, 1978.
Ayers, Edward L. *The Promise of the New South: Life after Reconstruction*. New York and Oxford, 1992.
———. *Vengeance and Justice: Crime and Punishment in the 19th Century*. New York, 1984.

Bailey, Joseph Cannon. *Seaman A. Knapp: Schoolmaster of American Agriculture.* New York, 1971.
Baker, James T. *Brooks Hays.* Macon, Ga., 1989.
Baldwin, Sidney. *Poverty and Politics: The Rise and Decline of the Farm Security Administration.* Chapel Hill, N.C., 1968.
Bendix, Reinhard. *Nation-Building and Citizenship: Studies of Our Changing Social Order.* Berkeley, Los Angeles, and London, 1964.
Benedict, Murray R., and Oscar C. Stine. *The Agricultural Commodity Programs.* New York, 1956.
Berardi, Gigi M., and Charles C. Geisler. *The Social Consequences and Challenges of New Agricultural Technologies.* Boulder, Colo., 1984.
Bertrand, Alvin Lee, et al. *Factors Associated with Agricultural Mechanization in the Southwest Region.* Fayetteville, Ark., 1956.
Billings, Dwight B. *Planters and the Making of a "New South": Class, Politics, and Development in North Carolina, 1865–1900.* Chapel Hill, N.C., 1979.
Binswanger, Hans P. *Agricultural Mechanization: A Comparative Historical Perspective.* Washington, D.C., 1984.
Boyer, Paul S. *Urban Masses and Moral Order in America, 1820–1920.* Cambridge, Mass., 1978.
Brandfon, Robert Leon. *Cotton Kingdom of the New South: A History of the Yazoo-Mississippi Delta from Reconstruction to the Twentieth Century.* Cambridge, Mass., 1967.
Broderick, Francis L. *W. E. B. Du Bois: Negro Leader in a Time of Crisis.* Stanford, Calif., 1959.
Brown, Richard Maxwell. *Historical Studies of American Violence and Vigilantism.* New York, 1975.
Brown, Roy M. *The North Carolina Chain Gang: A Study of County Convict Road Work.* Chapel Hill, N.C., 1927.
Brundage, W. Fitzhugh. *Lynching in the New South: Georgia and Virginia, 1880–1930.* Urbana, Ill., 1993.
Buenker, John D. *Urban Liberalism and Progressive Reform.* New York, 1973.
Burns, James MacGregor. *Roosevelt: The Lion and the Fox.* New York, 1956.
Butler, Karl Douglas. *Farm Mechanization in the Soviet Union.* Washington, D.C., 1959.
Carothers, Doris. *Chronology of the Federal Emergency Relief Administration, May 12, 1933, to December 31, 1935.* Washington, D.C., 1937.
Carson, Clayborne. *In Struggle: SNCC and the Black Awakening of the 1960s.* Cambridge, Mass., 1981.
Carstensen, Vernon, ed. *The Public Lands: Studies in the History of the Public Domain.* Madison, Wis., 1962.
Casdorph, Paul D. *Republicans, Negroes, and Progressives in the South, 1912–1916.* University, Ala., 1981.
Cayton, Howard R., and George S. Mitchell. *Black Workers and the New Unions.* Chapel Hill, N.C., 1939.

Clark, Norman H. *The Dry Years: Prohibition and Social Change in Washington.* Seattle, 1965.
Clark, Penny. *Farm Work and Friendship: The German Prisoner of War Camp at Lake Wabaunese.* Emporia, Kan., 1988.
Cobb, James C. *Industrialization and Southern Society, 1877–1984.* Lexington, Ky., 1984.
———. *The Most Southern Place on Earth: The Mississippi Delta and the Roots of Regional Identity.* New York and Oxford, 1992.
Cobb, James C., and Michael V. Namorato, eds. *The New Deal and the South.* Jackson, Miss., 1984.
Cockcroft, James D. *Outlaws in the Promised Land: Mexican Immigrant Workers and America's Future.* New York, 1986.
Cohen, William. *At Freedom's Edge: Black Mobility and the Southern White Quest for Racial Control, 1861–1915.* Baton Rouge, La., and London, 1991.
Conrad, David Eugene. *The Forgotten Farmers: The Story of Sharecroppers in the New Deal.* Urbana, Ill., 1965.
Conkin, Paul K. *Tomorrow a New World: The New Deal Community Program.* New York, 1959.
Cortner, Richard C. *A Mob Intent on Death: The NAACP and the Arkansas Riot Cases.* Middletown, Conn., 1988.
Corwin, Arthur F. *Immigrants—and Immigrants: Perspectives on Mexican Labor Migration to the United States.* Westport, Conn., 1978.
Cox, Thomas R. *Mills and Markets: A History of the Pacific Coast Lumbering Industry to 1900.* Seattle, 1974.
Crews, Harry. *A Childhood, the Biography of a Place.* New York, 1978.
Crozier, Ethelred W. *The White Caps: History of an Organization in Sevier County.* Sevierville, Tenn., 1963.
Cutler, James Elbert. *Lynch-Law: An Investigation into the History of Lynching in the United States.* Montclair, N.J., 1969.
The Cyclopaedia of Temperance and Prohibition. New York, 1891.
Danbon, David B. *The Resisted Revolution: Urban America and the Industrialization of Agriculture, 1900–1930.* Ames, Iowa, 1979.
Daniel, Pete. *Breaking the Land: The Transformation of Cotton, Tobacco, and Rice Cultures since 1880.* Chicago, 1985.
———. *Deep'n As It Come: The 1927 Mississippi River Flood.* Oxford, 1977.
———. *Shadow of Slavery: Peonage in the South, 1901–1969.* Chicago, 1972.
———. *Standing at the Crossroads: Southern Rural Life since 1900.* New York, 1986.
Dannenbaum, Jed. *Drink and Disorder: Temperance Reform in Cincinnati from the Washingtonian Revival to the WCTU.* Chicago, 1984.
Darnton, Robert. *The Great Cat Massacre and Other Episodes in French Cultural History.* New York, 1984.
Davis, Elizabeth Gould. *The Economics of Farm Mechanization in the United States, 1950–1960.* Washington, D.C., 1960.
Davis, Helen Dick, ed. *Trials of the Earth: The Autobiography of Mary Hamilton.* Jackson, Miss., 1992.

Davis, Natalie Zemon. *Society and Culture in Early Modern France*. Stanford, Calif., 1975.
Dawley, Alan. *Class and Community: The Industrial Revolution in Lynn*. Cambridge, Mass., 1976.
DeCanio, Stephen J. *Agriculture in the Postbellum South: The Economics of Production and Supply*. Cambridge, Mass., 1974.
Denisoff, R. Serge. *Great Day Coming: Folk Music and the American Left*. Urbana, Ill., 1971.
———. *Sing a Song of Social Significance*. Bowling Green, Ohio, 1983.
Dittmer, John. *Black Georgia in the Progressive Era, 1900–1920*. Urbana, Ill., 1980.
———. *Local People: The Struggle for Civil Rights in Mississippi*. Urbana, Ill., 1994.
Donovan, Timothy, Willard B. Gatewood, Jr., and Jeannie M. Whayne. *The Governors of Arkansas: Essays in Political Biography*. Fayetteville, Ark., 1995.
Duis, Perry R. *The Saloon: Public Drinking in Chicago and Boston, 1880–1920*. Chicago, 1983.
Dunbar, Anthony P. *Against the Grain: Southern Radicals and Prophets, 1929–1959*. Charlottesville, Va., 1981.
Efferson, John Norman. *The Production and Marketing of Rice*. New Orleans, 1952.
Ellis, Edward Robb. *A Nation in Torment: The Great Depression, 1929–1939*. New York, 1970.
Escott, Paul D. *Slavery Remembered: A Record of Twentieth-Century Slave Narratives*. Chapel Hill, N.C., 1979.
Etheridge, Elizabeth W. *The Butterfly Caste: A Social History of Pellagra in the South*. Westport, Conn., 1972.
Faulk, Henry. *Group Captives: The Re-education of German Prisoners of War in Britain, 1945–1948*. London, 1977.
Feldman, Herman. *Prohibition: Its Economic and Industrial Aspects*. New York, 1927.
Fite, Gilbert. *American Farmers: The New Minority*. Bloomington, Ind., 1981.
———. *Cotton Fields No More: Southern Agriculture, 1865–1980*. Lexington, Ky., 1984.
Fligstein, Neil. *Going North: Migration of Blacks and Whites from the South, 1900–1950*. New York, 1981.
Flynt, J. Wayne. *Dixie's Forgotten People: The South's Poor Whites*. Bloomington, Ind., 1979.
Foner, Eric. *Reconstruction: America's Unfinished Revolution, 1863–1877*. New York, 1988.
Foner, Philip S. *American Socialism and Black Americans: From the Age of Jackson to World War II*. Westport, Conn., and London, 1977.
———. *Organized Labor and the Black Worker, 1619–1973*. New York and Washington, D.C., 1974.
———. *Women and the American Labor Movement*. Vol. 2. New York, 1980.
Foner, Philip S., and Herbert Shapiro, eds. *American Communism and Black Americans: A Documentary History, 1930–1934*. Philadelphia, 1991.

Ford, Arthur M. *Political Economics of Rural Poverty in the South.* Cambridge, Mass., 1973.
Fox-Genovese, Elizabeth, and Eugene D. Genovese. *Fruits of Merchant Capitalism: Slavery and Bourgeois Property in the Rise and Expansion of Capitalism.* New York, 1983.
Fraser, Walter J., Jr., R. Frank Saunders, Jr., and Jon L. Wakelyn. *The Web of Southern Social Relations: Women, Family, and Education.* Athens, Ga., 1985.
Freidel, Frank. *FDR and the South.* Baton Rouge, La., 1965.
Galarza, Ernesto. *Merchants of Labor: The Mexican Bracero Study, an Account of the Managed Migration of Mexican Farm Workers in California, 1942–1960.* Charlotte, Calif., 1964.
———. *Strangers in Our Fields.* Washington, D.C., 1956.
Gansberg, Judith M. *Stalag, U.S.A.: The Remarkable Story of German POWs in America.* New York, 1977.
Garcia, Juan Ramon. *Operation Wetback: The Mass Deportation of Mexican Undocumented Workers in 1954.* Westport, Conn., 1980.
Garcia y Griego, Manuel. *The Importation of Mexican Contract Laborers to the United States, 1942–1964: Antecedents, Operations, and Legacy.* San Diego, Calif., 1981.
Gaventa, John. *Power and Powerlessness: Quiescence and Rebellion in an Appalachian Valley.* Urbana, Ill., 1980.
Gemmell, Gordon, and Carl K. Eicher. *A Framework for Research on the Economics of Farm Mechanization in Developing Countries.* East Lansing, Mich., 1973.
Genovese, Eugene D. *Roll Jordan Roll: The World the Slaveholders Made.* New York, 1971.
Gill, Flora. *Economics and the Black Exodus: An Analysis of Negro Emigration from the Southern United States, 1910–1970.* New York, 1979.
Goldfield, David R. *Black, White, and Southern: Race Relations and Southern Culture, 1940 to the Present.* Baton Rouge, La., and London, 1990.
———. *Promised Land: The South since 1945.* Arlington Heights, Ill., 1987.
Goodspeed Publishing Company. *Biographical and Historical Memoirs of Northeast Arkansas.* Chicago, 1889.
Grantham, Dewey W. *Southern Progressivism: The Reconciliation of Progress and Tradition.* Knoxville, Tenn., 1983.
Graves, John William. *Town and Country: Race Relations in an Urban-Rural Context, Arkansas, 1865–1905.* Fayetteville, Ark., 1990.
Green, James R. *Grass-Roots Socialism: Radical Movements in the Southwest, 1895–1943.* Baton Rouge, La., 1978.
Gregor, Howard F. *Industrialization of U.S. Agriculture: An Interpretive Atlas.* Boulder, Colo., 1982.
Grubbs, Donald H. *Cry from the Cotton: The Southern Tenant Farmers Union and the New Deal.* Chapel Hill, N.C., 1971.
Guerin-Gonzales, Camille. *Mexican Workers and American Dreams: Immigration, Repatriation, and California Farm Labor, 1900–1939.* New Brunswick, N.J., 1994.

Gusfield, Joseph. *Symbolic Crusade: Status Politics and the American Temperance Movement.* Urbana, Ill., 1963.

Haas, Lisbeth. *The Bracero in Orange County: A Work Force for Economic Transition.* La Jolla, Calif., 1981.

Hacker, Louis M. *The Course of American Economic Growth and Development.* New York, 1970.

Hagood, Margaret Jarman. *Mothers of the South: Portraiture of the White Tenant Farm Woman.* New York, 1939.

Hahn, Steven. *The Roots of Southern Populism: Yeoman Farmers and the Transformation of the Georgia Upcountry, 1850–1890.* New York, 1983.

Hall, Jacquelyn Dowd. *Revolt against Chivalry: Jessie Daniel Ames and the Women's Campaign against Lynching.* New York, 1979.

Hallem, Arne, ed. *Size, Structure, and the Changing Face of American Agriculture.* Boulder, Colo., 1993.

Harrell, David Edwin. *A Social History of the Disciples of Christ.* Nashville, Tenn., 1966–73.

Hawley, Ellis W. *The New Deal and the Problem of Monopoly, 1933–1939.* Princeton, N.J., 1965.

Hays, Brooks. *Politics Is My Parish: An Autobiography of Brooks Hays.* Baton Rouge, La., and London, 1981.

Haywood, Harry. *Black Bolshevik: Autobiography of an Afro-American Communist.* Chicago, 1978.

Henri, Florette. *Black Migration: Movement North, 1900–1920.* Garden City, N.Y., 1975.

Higgs, Robert. *Competition and Coercion: Blacks in the American Economy, 1865–1914.* Chicago, 1977.

Hirschhorn, Larry. *Beyond Mechanization: Work and Technology in a Postindustrial Age.* Cambridge, Mass., 1984.

Hofstadter, Richard. *The Age of Reform: From Bryan to F.D.R.* New York, 1956.

Holley, Donald. *Uncle Sam's Farmers: The New Deal Communities in the Lower Mississippi Valley.* Urbana, Ill., 1975.

Holley, William C., Ellen Winston, and T. J. Woofter, Jr. *The Plantation South, 1934–1937.* Washington, D.C., 1940.

Hopkins, Harry Lloyd. *Spending to Save: The Complete Story of Relief.* New York, 1936.

Howe, Irving, and Lewis Coser with Julius Jacobson. *The American Communist Party: A Critical History (1919–1957).* Boston, 1957.

Hurt, R. Douglas. *American Farm Tools: From Hand-Power to Steam-Power.* Manhattan, Kan., 1982.

———. *The Dust Bowl: An Agricultural and Social History.* Chicago, 1981.

Isaac, Paul E. *Prohibition and Politics: Turbulent Decades in Tennessee, 1885–1920.* Knoxville, Tenn., 1965.

Jensen, Vernon H. *Lumber and Labor.* New York, 1945.

Johnson, Charles S. *Shadow of the Plantation.* Chicago, 1934.

Johnson, Daniel M., and Rex R. Campbell. *Black Migration in America: A Social Demographic History.* Durham, N.C., 1981.

Jones, Jacqueline. *Labor of Love, Labor of Sorrow: Black Women, Work, and Family from Slavery to the Present.* New York, 1985.

Kelley, Robin D. G. *Hammer and Hoe: Alabama Communists during the Great Depression.* Chapel Hill, N.C., and London, 1990.

Kellogg, Charles Flint. *NAACP: A History of the National Association for the Advancement of Colored People.* Vol. 1. Baltimore, 1967.

Kennedy, Louise V. *The Negro Peasant Turns Cityward: Effects of Recent Migrations to Northern Centers.* New York, 1930.

Kirby, Jack Temple. *Darkness at the Dawning: Race and Reform in the Progressive South.* Philadelphia, Pa., 1972.

———. *Rural Worlds Lost: The American South, 1920–1960.* Baton Rouge, La., 1987.

Kirkendall, Richard S. *Social Scientists and Farm Policies in the Age of Roosevelt.* Columbia, Mo., 1966.

Kiser, George C., and Martha Woody Kiser. *Mexican Workers in the United States: Historical and Political Perspectives.* Albuquerque, N.Mex., 1979.

Klehr, Harvey. *The Heyday of American Communism: The Depression Decade.* New York, 1984.

Koop, Allen V. *Stark Decency: German Prisoners of War in a New England Village.* Hanover, N.H., 1988.

Kousser, J. Morgan. *The Shaping of Southern Politics: Suffrage Restrictions and the Establishment of the One-Party South, 1880–1910.* New Haven, 1974.

Lamb, Robert Byron. *The Mule in Southern Agriculture.* Berkeley, Calif., 1963.

LeMaster, Carolyn Gray. *A Corner of the Tapestry: A History of the Jewish Experience in Arkansas, 1820s–1990s.* Fayetteville, Ark., 1994.

Lerner, Gerda, ed. *Black Women in White America: A Documentary History.* New York., 1973.

Leuchtenburg, William E. *Franklin D. Roosevelt and the New Deal.* New York, 1963.

Lewis, David Levering. *W. E. B. Du Bois: Biography of a Race, 1868–1919.* New York, 1993.

Link, William A. *The Paradox of Southern Progressivism.* Chapel Hill, N.C., 1992.

Litwack, Leon F., *Been in the Storm So Long: The Aftermath of Slavery.* New York, 1979.

Logan, Rayford W., ed. *W. E. B. Du Bois: A Profile.* New York, 1971.

McConnell, Grant. *The Decline of Agrarian Democracy.* Berkeley, Calif., 1953.

McMath, Robert C. *A History of the Southern Farmers' Alliance.* Chapel Hill, N.C., 1975.

Mandle, Jay R. *The Roots of Black Poverty: The Southern Plantation Economy after the Civil War.* Durham, N.C., 1978.

Marable, Manning. *W. E. B. Du Bois: Black Radical Democrat.* Boston, 1986.

Martin, Charles H. *The Angelo Herndon Case and Southern Justice.* Baton Rouge, La., 1976.

Meier, August. *Negro Thought in America, 1880–1915: Racial Ideologies in the Age of Booker T. Washington.* Ann Arbor, Mich., 1963.

Meier, August, and Elliott M. Rudwick. *From Plantation to Ghetto: An Interpretative History of American Negroes*. New York, 1966.
Mertz, Paul E. *New Deal Policy and Southern Rural Poverty*. Baton Rouge, La., 1978.
Mitchell, H. L. *Mean Things Happening: The Life and Times of H. L. Mitchell, Co-Founder of the Southern Tenant Farmers Union*. Montclair, N.J., 1979.
Monto, Alexander. *The Roots of Mexican Labor Migration*. Westport, Conn., and London, 1994.
Moore, Barrington Jr. *Injustice: The Social Bases of Obedience and Revolt*. White Plains, N.Y., 1978.
Moore, John Hammond. *The Faustball Tunnel: German POWs in America and Their Great Escape*. New York, 1978.
NAACP. *Thirty Years of Lynching in the United States, 1889–1918*. New York, 1969.
National Rural Center. *Production Efficiency and Technology for Small Farms*. Washington, D.C., 1980.
Niswonger, Richard. *Arkansas Democratic Politics, 1896–1920*. Fayetteville, Ark., 1990.
Nolan, Patrick B. *Vigilantes on the Middle Border: A Study of Self-Appointed Law Enforcement in the States of the Upper Mississippi from 1840 to 1880*. New York and London, 1987.
North, Douglass Cecil. *The Economic Growth of the United States, 1790–1860*. Englewood Cliffs, N.J., 1961.
Orfield, Matthias W. *Federal Land Grant to the States with Special Reference to Minnesota*. Minneapolis, 1915.
Organization for European Economic Co-Operation. *The Mechanization of Small Farms in European Countries*. Paris, 1950.
Ottoson, Howard W., ed. *Land Use Policy and Problems in the United States*. Lincoln, Nebr., 1963.
Ownby, Ted. *Subduing Satan: Religion, Recreation, and Manhood in the Rural South*. Chapel Hill, N.C., 1990.
Perkins, Van L. *Crisis in Agriculture: The Agricultural Adjustment Administration and the New Deal, 1933*. Berkeley, Calif., 1969.
Polanyi, Karl. *The Great Transformation: The Political and Economic Origins of Our Time*. Boston, 1957.
Polk's Arkansas State Gazetteer and Business Directory. Vol. 5, 1906–7. Memphis, 1906. Copy in Special Collections Division, University of Arkansas Library, Fayetteville.
Polk's Arkansas State Gazetteer and Business Directory. Vol. 6, 1912–13. Memphis, 1912. Copy in Special Collections Division, University of Arkansas Library, Fayetteville.
Pope, Liston. *Millhands and Preachers: A Study of Gastonia*. New Haven, 1942.
Powell, Kent. *Splinters of a Nation: German Prisoners of War in Utah*. Salt Lake City, 1989.
Powell, Lawrence N. *New Masters: Northern Planters during the Civil War and Reconstruction*. New Haven, 1980.

Price, Barry L. *The Political Economy of Mechanization in U.S. Agriculture*. Boulder, Colo., 1983.
Ransom, Roger, and Richard Sutch. *One Kind of Freedom: The Economic Consequences of Emancipation*. Cambridge, Mass., 1977.
Raper, Arthur F. *Machines in the Cotton Fields: Mechanization Comes to the Southern Cotton Farm*. Atlanta, 1946.
——. *Preface to Peasantry: A Tale of Two Black Belt Counties*. Chapel Hill, N.C., 1936.
——. *The Tragedy of Lynching*. Chapel Hill, N.C., 1933.
Rawick, George P., ed. *The American Slave: A Composite Autobiography, Arkansas Narratives*. Westport, Conn., 1972.
Reed, Merl E. *Seedtime for the Modern Civil Rights Movement: The President's Committee on Fair Employment Practice, 1941–1946*. Baton Rouge, La., and London, 1991.
Reidy, Joseph P. *From Slavery to Agrarian Capitalism in the Cotton Plantation South: Central Georgia, 1800–1880*. Chapel Hill, N.C., 1992.
Reisler, Mark. *By the Sweat of Their Brow: Mexican Immigrant Labor in the United States, 1900–1940*. Westport, Conn., 1976.
Richards, Alan, and Philip L. Martin, eds. *Migration, Mechanization, and Agricultural Labor Markets in Egypt*. Boulder, Colo., 1983.
Roark, James L. *Masters without Slaves: Southern Planters in the Civil War and Reconstruction*. New York, 1977.
Robin, Ron Theodore. *The Barbed-Wire College: Reeducating German POWs in the United States during World War II*. Princeton, N.J., 1995.
Roe, Daphne A. *A Plague of Corn: The Social History of Pellagra*. Ithaca, N.Y., 1973.
Rorabaugh, W. J. *The Alcoholic Republic: An American Tradition*. New York, 1979.
Rosengarten, Theodore. *All God's Dangers: The Life of Nate Shaw*. New York, 1974.
Royce, Edward. *The Origins of Southern Sharecropping*. Philadelphia, 1993.
Rudwick, Elliott M. *W. E. B. Du Bois: A Study in Minority Group Leadership*. Philadelphia, 1960.
St. Francis River Levee Board. *History of the St. Francis River Levee District*. West Memphis, Ark., 1946.
Salmond, John A. *The Civilian Conservation Corps, 1933–1942*. Durham, N.C., 1967.
Saloutos, Theodore. *The American Farmer and the New Deal*. Ames, Iowa, 1982.
Sargen, Nicholas Peter. *"Tractorization" in the United States and Its Relevance for the Developing Countries*. New York, 1979.
Schlesinger, Arthur M., Jr. *The Age of Roosevelt: The Crisis of the Old Order, 1919–1933*. Boston, 1957.
Schulman, Bruce J. *From Cotton Belt to Sunbelt: Federal Policy, Economic Development, and the Transformation of the South, 1938–1980*. New York and Oxford, 1991.

Scott, James. *Weapons of the Weak: Everyday Forms of Peasant Resistance.* New Haven, 1985.
Scott, Roy V. *The Reluctant Farmer: The Rise of Agricultural Extension to 1914.* Urbana, Ill., 1970.
Scott, Roy V., and J. G. Shoalmire. *The Public Career of Cully A. Cobb: A Study in Agricultural Leadership.* Jackson, Miss., 1973.
Sellers, Cleveland, with Robert Terrell. *The River of No Return: The Autobiography of a Black Militant and the Life and Death of SNCC.* Jackson, Miss., 1990.
Shapiro, Herbert. *White Violence and Black Response: From Reconstruction to Montgomery.* Amherst, Mass., 1988.
Shay, Frank. *Judge Lynch, His First Hundred Years.* New York, 1938.
Shover, John. *First Majority—Last Minority: The Transformation of Rural Life in America.* Dekalb, Ill., 1976.
Sinclair, Andrew. *Prohibition: The Era of Excess.* Boston, 1962.
Skocpol, Theda. *States and Social Revolutions: A Comparative Analysis of France, Russia, and China.* Cambridge, Mass., 1979.
Smith, Kenneth L. *Sawmill: The Story of Cutting the Last Great Virgin Forest East of the Rockies.* Fayetteville, Ark., 1986.
Stavis, Benedict. *The Politics of Agricultural Mechanization in China.* Ithaca, N.Y., 1978.
Sternsher, Bernard. *The Negro in Depression and War: Prelude to Revolution, 1930–1945.* Chicago, 1969.
Strausberg, Stephen F. *A Century of Research: Centennial History of the Arkansas Agricultural Experiment Station.* Fayetteville, Ark. 1989.
Street, James. *New Revolution in the Cotton Economy: Mechanization and Its Consequences.* Chapel Hill, N.C., 1957
Taylor, Paul S. *Mexican Labor in the United States.* Vols. 1–2. Berkeley, Calif., 1968.
Terris, Milton. *Goldberger on Pellagra.* Baton Rouge, La., 1964.
Thomas, Norman. *The Plight of the Share-Cropper.* New York, 1934.
Thompson, E. P. *The Making of the English Working Class.* New York, 1966.
Thompson, George. *Arkansas and Reconstruction.* Port Washington, N.Y., 1976.
Timberlake, James H. *Prohibition and the Progressive Movement, 1900–1920.* Cambridge, Mass., 1963.
Tindall, George Brown. *The Emergence of the New South, 1913–1945.* Baton Rouge, La., 1967.
Tokes, Charlotte. *Labor and Lumber.* New York, 1931.
Tolnay, Stewart E., and E. M. Beck. *A Festival of Violence: An Analysis of Southern Lynchings, 1882–1930.* Urbana, Ill., and Chicago, 1995.
Tucker, David M. *Arkansas: A People and Their Reputation.* Memphis, 1985.
Ulrich, Laurel. *Good Wives: Image and Reality in the Lives of Women in Northern New England, 1650–1750.* New York, 1982.
Valdes, Dennis Nodin. *Al Norte: Agricultural Workers in the Great Lakes Region, 1917–1970.* Austin, Tex., 1981.

Vance, Rupert B. *Human Factors of Cotton Culture: A Study in the Social Geography of the American South*. Chapel Hill, N.C., 1929.

Vanderwood, Paul J. *Night Riders of Reelfoot Lake*. Memphis, 1969.

Waldrep, Christopher. *Night Riders: Defending Community in the Black Patch, 1890–1915*. Durham, N.C., and London, 1993.

Walker, Forrest A. *The Civil Works Administration: An Experiment in Federal Work Relief, 1933–1934*. New York, 1979.

Waskow, Arthur I. *From Race Riot to Sit-In, 1919 and the 1960s: A Study in the Connections between Conflict and Violence*. Garden City, N.Y., 1966.

West, Elliott. *The Saloon on the Rocky Mountain Mining Frontier*. Lincoln, Nebr., and London, 1979.

White, Walter. *Rope and Faggott: A Biography of Judge Lynch*. New York, 1929.

Whiting, Theodore E. *Final Statistical Report of the Federal Emergency Relief Administration*. Washington, D.C., 1942.

Wilcox, Walter W. *The Farmer in the Second World War*. New York, 1973.

Williams, C. Fred, ed. *A Documentary History of Arkansas*. Fayetteville, Ark., 1984.

Williams, Edward Ainsworth. *Federal Aid for Relief*. New York, 1939.

Williams, Thomas T., ed. *Strategy for Survival of Small Farmers—International Implications*. Tuskegee, Ala., 1984.

Williamson, Joel. *The Crucible of Race: Black-White Relations in the American South since Emancipation*. New York, 1984.

Wilmore, Gayraud S. *Black Religion and Black Radicalism: An Interpretation of the Religious History of Afro-American People*. Maryknoll, N.Y., 1983.

Wilson, Charles Reagan. *Baptized in Blood: The Religion of the Lost Cause*. Athens, Ga., 1980.

Wood, Peter. *Black Majority: Negroes in Colonial South Carolina from 1670 through the Stono Rebellion*. New York and London, 1974.

Woodman, Harold D. *New South—New Law: The Legal Foundations of Credit and Labor Relations in the Postbellum Agricultural South*. Baton Rouge, La., and London, 1995.

Woodruff, Nan Elizabeth. *As Rare as Rain: Federal Relief in the Great Southern Drought of 1930–1931*. Chicago, 1985.

Woodward, C. Vann. *Origins of the New South, 1877–1913*. Baton Rouge, La., 1951.

———. *The Strange Career of Jim Crow*. New York, 1955.

Woofter, T. J., Jr. *Landlord and Tenant on the Cotton Plantation*. New York, 1936.

Wright, Gavin. *Old South, New South: Revolutions in the Southern Economy since the Civil War*. New York, 1986.

Wright, Richard. *American Hunger*. New York, 1977.

Zagrando, Robert. *The NAACP Crusade against Lynching, 1909–1950*. Philadelphia, 1980.

Zinn, Howard. *SNCC: The New Abolitionists*. Boston, 1965.

Articles

Alexander, Charles C. "Defeat, Decline, Disintegration: The Ku Klux Klan in Arkansas, 1924 and After." *Arkansas Historical Quarterly* 22 (1963): 311-31.
———. "White-Robed Reformers: The Ku Klux Klan Comes to Arkansas, 1921-1922." *Arkansas Historical Quarterly* 22 (1963): 8-23.
———. "White Robes in Politics: The Ku Klux Klan in Arkansas, 1922-1924." *Arkansas Historical Quarterly* 22 (1963): 195-214.
Alston, Lee J. "The Wright Interpretation of Southern U.S. Economic Development: A Review Essay of *Old South, New South* by Gavin Wright." *Agricultural History* 61 (1987): 52-67.
Atack, Jeremy. "The Agricultural Ladder Revisited: A New Look at an Old Question with Some Data for 1860." *Agricultural History* 63 (1989): 1-25.
Auerbach, Jerold S. "Southern Tenant Farmers: Socialist Critics of the New Deal." *Arkansas Historical Quarterly* 27 (1968): 113-31.
Balch, Jim. "The Story of Richwood Township." *Arkansas Historical Quarterly* 16 (1957): 366-82.
Barton, Glen T., and J. G. McNeely. "Recent Changes in the Status of Laborers and Tenants on Arkansas Plantations." *Journal of Land and Public Utility Economics* 15 (1939): 235-47.
Boehm, Randolph H. "Mary Grace Quackenbos and the Federal Campaign against Peonage: The Case of Sunnyside Plantation." *Arkansas Historical Quarterly* 50 (1991): 40-59.
Bogue, Margaret Beattie. "The Swamp Land Act and Wet Land Utilization in Illinois, 1850-1890." *Agricultural History* 25 (1951): 169-80.
Brooks, Roy L. "The Use of the Civil Rights Acts of 1866 and 1871 to Redress Employment Discrimination." *Cornell Law Review* 62 (1977): 258-88.
Brown, Richard Maxwell. "The American Vigilante Tradition." Pp. 144-205 in *Violence in America: Historical and Comparative Perspectives*, ed. Hugh Davis Graham and Ted Robert Gurr. New York, 1969.
Brundage, David. "The Producing Classes and the Saloon: Denver in the 1880s." *Labor History* 26 (1985): 29-52.
Bunche, Ralph J. "Share-Croppers in the United States." *Plebs* 39 (Sept. 1937): 212-15.
Burton, Orville Vernon. "The Effects of the Civil War and Reconstruction on the Coming of Age of Southern Males, Edgefield County, South Carolina." Pp. 204-24 in *The Web of Southern Social Relations, Women, Family, and Education*, ed. Walter J. Fraser, Jr., R. Frank Saunders, Jr., and Jon L. Wakelyn. Athens, Ga., 1985.
Butts, J. W., and Dorothy James. "The Underlying Causes of the Elaine Riot of 1919." *Arkansas Historical Quarterly* 20 (1961): 95-105.
Campbell, Randolph B. "Population Persistence and Social Change in Nineteenth-Century Texas: Harrison County, 1850-1880." *Journal of Southern History* 48 (1982): 185-204.

Cavert, William L. "The Technological Revolution in Agriculture, 1910–1955." *Agricultural History* 30 (1956): 18–27.
Christensen, Alice M. "Agricultural Pressure and Governmental Response in the United States, 1919–1929." *Agricultural History* 11 (1937): 33–42.
Clements, William M., and Larry D. Ball. "'This Was the Beginning of Clearing of Land': The Development and Use of the East Arkansas Stump Saw." *Arkansas Historical Quarterly* 55 (1986): 41–52.
Colbert, Douglas L. "Challenging the Challenge: Thirteenth Amendment as a Prohibition against the Racial Use of Pereemptory Challenges." *Cornell Law Review* 76 (1990): 1–75.
Connor, James R. "National Farm Organizations and United States Tariff Policy in the 1920's." *Agricultural History* 32 (1958): 32–43.
Cotterill, R. S. "The National Land System in the South, 1803–1812." *Mississippi Valley Historical Review* 16 (1930): 495–506.
Craig, Lee A. "Constrained Resource Allocation and the Investment in the Education of Black Americans: The 1890 Land-Grant Colleges." *Agricultural History* 65 (1991): 73–84.
Crawford, Sidney. "Arkansas Suffrage Qualifications." *Arkansas Historical Quarterly* 2 (1943): 331–39.
Crosby, Earl W. "The Struggle for Existence: The Institutionalization of the Black County Agent System." *Agricultural History* 60 (1986): 123–136.
Dabney, A. L. "1912 Flood on the Lower Mississippi." *Engineering News* 67 (June 13, 1912): 1121–27.
Dalfiume, Richard M. "The 'Forgotten Years' of the Negro Revolution." *Journal of American History* 55 (1968): 90–106.
Danbom, David B. "The Agricultural Experiment Station and Professionalization: Scientists' Goals for Agriculture." *Agricultural History* 60 (1986): 246–255.
Daniel, Pete. "Going among Strangers: Southern Reactions to World War II." *Journal of American History* 77 (1990): 886–911.
Daniels, Richard S. "Blind Tigers and Blind Justice: The Arkansas Raid on Island 37, Tennessee." *Arkansas Historical Quarterly* 38 (1979): 259–70.
Degler, Carl N. "A Century of the Klans: A Review Article." *Journal of Southern History* 31 (1965): 435–43.
Desmarais, Ralph H. "Military Intelligence Reports on Arkansas Riots, 1919–1920." *Arkansas Historical Quarterly* 33 (1974): 175–94.
Dethloff, Henry C. "Rice Revolution in the Southwest, 1880–1900." *Arkansas Historical Quarterly* 29 (1970): 66–75.
Dew, Lee A. "The Blytheville Case and the Regulation of Arkansas Cotton Shipments." *Arkansas Historical Quarterly* 38 (1979): 116–30.
———. "The J.L.C. & E.R.R. and the Opening of the 'Sunk Lands' in Northeast Arkansas." *Arkansas Historical Quarterly* 27 (1968): 22–39.
Dillard, Tom. "To the Back of the Elephant: Racial Conflict in the Arkansas Republican Party." *Arkansas Historical Quarterly* 33 (1974): 3–15.

Durrill, Wayne K. "Producing Poverty: Local Government and Economic Development in a New South County, 1874–1884." *Journal of American History* 71 (1985): 764–81.

Dyson, Lowell K. "Radical Farm Organizations and Periodicals in America, 1920–1960." *Agricultural History* 45 (1971): 111–20.

Elkins, F. Clark. "The Agricultural Wheel and Consolidation." *Arkansas Historical Quarterly* 29 (1970): 152–75.

———. "Arkansas Farmers Organize for Action." *Arkansas Historical Quarterly* 13 (1954): 231–48.

Farmer, Rod. "Direct Democracy in Arkansas, 1910–1918." *Arkansas Historical Quarterly* 40 (1981): 99–118.

Fields, Barbara Jean. "Ideology and Race in American History." Pp. 143–178 in *Region, Race and Reconstruction: Essays in Honor of C. Vann Woodward*, ed. J. Morgan Kousser and James M. McPherson. New York and Oxford, 1982.

Fite, Gilbert C. "The Agricultural Trap in the South." *Agricultural History* 60 (1986): 38–50.

———. "Farmer Opinion and the Agricultural Adjustment Act, 1933." *Mississippi Valley Historical Review* 49 (1962): 656–73.

Foley, Neil F. "Chicanos and the Culture of Cotton in Central Texas, 1880–1990: Reshaping Class Relations in the South." Pp. 111–26 in *Community Empowerment and Chicano Scholarship*, ed. Mary Romero and Cordelia Candelaria. Berkeley, Calif., 1990.

Forsythe, James L. "World Cotton Technology since World War II." *Agricultural History* 54 (1980): 208–22.

Foti, Thomas. "The Rivers Gifts and Curses." Pp. 30–57 in *Arkansas Delta: Land of Paradox*, ed. Jeannie Whayne and Willard B. Gatewood, Jr. Fayetteville, Ark., 1993.

Frank, Richard. "Negro Revolutionary Music." *New Masses* 11 (May 15, 1934): 29–30.

Fries, Robert F. "The Mississippi River Logging Company and the Struggle for the Free Navigation of Logs, 1865–1900." *Mississippi Valley Historical Review* 35 (1948): 429–48.

Fuller, Henry. "The Cotton Kingdom, 1931." *New Republic* 69 (Dec. 16, 1931):129–34.

Ganoe, John T. "The Origin of a National Reclamation Policy." *Mississippi Valley Historical Review* 18 (1931): 34–52.

Gates, Paul W. "Research in the History of the Public Lands." *Agricultural History* 48 (1974): 31–50.

———. "Tenants of the Log Cabin." *Mississippi Valley Historical Review* 49 (1962): 3–31.

Gatewood, Willard B., Jr. "Sunnyside: The Evolution of an Arkansas Plantation, 1848–1945." *Arkansas Historical Quarterly* 50 (1991): 5–29.

Gellert, Lawrence. "Negro Songs of Protest." *New Masses* 6 (April 1931): 6–8.

Gilbert, Jess, and Steve Brown. "Alternative Land Reform Proposals in the 1930s: The Nashville Agrarians and the Southern Tenant Farmers' Union." *Agricultural History* 55 (1981): 351–69.

Graves, John William. "Negro Disfranchisement in Arkansas." *Arkansas Historical Quarterly* 26 (1967): 199–225.

Hahn, Steven. "African-American Life in the Nineteenth-Century South: A Review Essay." *Arkansas Historical Quarterly* 50 (1991): 352–73.

Hamilton, David E. "Herbert Hoover and the Great Drought of 1930." *Journal of American History* 68 (1981): 852–75.

Harbaugh, William H. "The Limits of Voluntarism: Farmers, County Agents, and the Conservation Movement." Pp. 123–55 in *Essays in Honor of Arthur S. Link*, ed. John M. Cooper, Jr., and Charles New. Arlington, Ill., 1991.

Harris, Carl V. "Reforms in Government Control of Negroes in Birmingham, Alabama, 1890–1920." *Journal of Southern History* 38 (1972): 567–600.

Harrison, Robert W. "Clearing Land in the Mississippi Alluvial Valley." *Arkansas Historical Quarterly* 13 (1954): 352–71.

———. "Early State Flood-Control Legislation in the Mississippi Alluvial Valley." *Journal of Mississippi History* 23 (1961): 104–26.

———. "The Formative Years of the Yazoo Mississippi Delta Levee District." *Journal of Mississippi History* 13 (1951): 236–48.

Harrison, Robert W., and Walter M. Kollmorgen. "Land Reclamation in Arkansas under the Swamp Land Grant of 1850." *Arkansas Historical Quarterly* 6 (1947): 369–418.

———. "Socio-Economic History of Cypress Creek Drainage District and Related Districts of Southeast Arkansas." *Arkansas Historical Quarterly* 7 (1948): 20–52.

Heaney, Gerald W. "Busing, Timetables, Goals, and Ratios: Touchstones of Equal Opportunity." *Minnesota Law Review* 69 (1984–85): 735–820.

———. "Jacob Trieber: Lawyer, Politician, Judge." *University of Arkansas at Little Rock Law Review* 8 (1985–86): 421–78.

Helms, Douglas. "Eroding the Color Line: The Soil Conservation Service and the Civil Rights Act of 1964." *Agricultural History* 65 (1991): 35–53.

Hicks, Floyd W., and C. Roger Lambert. "Food for the Hungry: Federal Food Programs in Arkansas, 1933–1942." *Arkansas Historical Quarterly* 37 (1978): 23–43.

Higgs, Robert. "The Boll Weevil, the Cotton Economy, and Black Migration, 1910–1930." *Agricultural History* 50 (1976): 335–50.

Hobson, Edythe Simpson. "Twenty-seven Days on the Levee, 1927." *Arkansas Historical Quarterly* 39 (1980): 210–29.

Hohner, Robert A. "The Prohibitionists: Who Were They?" *South Atlantic Quarterly* 68 (1969): 491–505.

Holley, Donald. "The Negro in the New Deal Resettlement Program." *Agricultural History* 45 (1971): 179–94.

———. "The Second Great Emancipation: The Rust Cotton Picker and How It Changed Arkansas." *Arkansas Historical Quarterly* 52 (1993): 44–77.

Holmes, William F. "Labor Agents and the Georgia Exodus, 1899–1900." *South Atlantic Quarterly* 79 (1980): 436–48.
———. "Moonshiners and Whitecaps in Alabama, 1893." *Alabama Review* 34 (Jan. 1981): 31–49.
———. "Moonshining and Collective Violence: Georgia, 1889–1895." *Journal of American History* 67 (1980): 588–611.
———. "Whitecapping: Agrarian Violence in Mississippi, 1902–1906." *Journal of Southern History* 35 (1969): 165–85.
———. "Whitecapping in Georgia: Carroll and Houston Counties, 1893." *Georgia Historical Quarterly* 64 (1980): 388–404.
———. "Whitecapping in Mississippi: Agrarian Violence in the Populist Era." *Mid-America* 55 (1973): 134–48.
Horton, Donald C., and E. Fenton Shepard. "Federal Aid to Agriculture since World War I." *Agricultural History* 19 (1945): 114–19.
Hubbell, Kenneth. "Always a Simple Feast: Social Life in the Arkansas Delta." Pp. 184–207 in *Arkansas Delta: Land of Paradox*. ed. Jeannie Whayne and Willard B. Gatewood, Jr. Fayetteville, Ark., 1993.
Humphries, Frederick S. "1890 Land-Grant Institutions: Their Struggle for Survival and Equality." *Agricultural History* 65 (1991): 3–11.
Irwin, James R. "Farmers and Laborers: A Note on Black Occupations in the Postbellum South." *Agricultural History* 64 (1990): 53–60.
James, Felix. "The Tuskegee Institute Movable School, 1906–1923." *Agricultural History* 45 (1971): 201–9.
Janiewski, Dolores. "Sisters under Their Skins: Southern Working Women, 1880–1950." Pp. 13–36 in *Sex, Race, and the Role of Women in the South*, ed. Joanne V. Hawks and Sheila L. Skemp. Jackson, Miss., 1983.
Jenkins, Robert L. "The Black Land-Grant Colleges in Their Formative Years, 1890–1920." *Agricultural History* 65 (1991): 63–72.
Jones, Allen. "Improving Rural Life for Blacks: The Tuskegee Negro Farmers' Conference, 1892–1915." *Agricultural History* 65 (1991): 105–14.
———. "The Role of Tuskegee Institute in the Education of Black Farmers." *Journal of Negro History* 60 (1975): 252–67.
———. "The South's First Black Farm Agents." *Agricultural History* 50 (1976): 636–44.
Jones, Jacqueline. "'Tore Up and a-Movin': Perspectives on the Work of Black and Poor White Women in the Rural South, 1865–1940." Pp. 15–34 in *Women and Farming: Changing Roles, Changing Structures*, ed. Wava G. Haney and Jane B. Knowles. Boulder, Colo., 1988.
Kelley, Robin D. G. "'We Are Not What We Seem': Rethinking Black Working Class Opposition in the Jim Crow South." *Journal of American History* 79 (1993): 75–112.
Kerr, Norwood Allen. "Institutionalizing the New Agenda: The State Agricultural Experiment Stations, 1977–1981." *Agricultural History* 62 (1988): 279–95.
Kirby, Jack Temple. "Black and White in the Rural South, 1915–1954." *Agricultural History* 58 (1984): 411–22.

———. "The Southern Exodus, 1910–1960: A Primer for Historians." *Journal of Southern History* 49 (1983): 585–600.
———. "The Transformation of Southern Plantations." *Agricultural History* 57 (1983): 257–76.
Kirkendall, Richard S. "The Agricultural Colleges: Between Tradition and Modernization." *Agricultural History* 60 (1986): 3–21.
———. "The New Deal as Watershed: The Recent Literature." *Journal of American History* 54 (1968): 839–52.
Koeniger, A. Cash. "The New Deal and the States: Roosevelt versus the Byrd Organization in Virginia." *Journal of American History* 68 (1982): 876–96.
Kunze, Joel P. "The Purnell Act and Agricultural Economics." *Agricultural History* 62 (1988): 131–49.
Lackey, Daniel Boone. "Cutting and Floating Red Cedar Logs in North Arkansas." *Arkansas Historical Quarterly* 19 (1960): 361–70.
Lambert, Roger. "Hoover and the Red Cross in the Arkansas Drought of 1930." *Arkansas Historical Quarterly* 29 (1970): 3–19.
Larkin, Margaret. "The Story of Ella May." *New Masses* 5 (Nov. 1929): 3–4.
Larsen, Grace H., and Henry E. Erdman. "Aaron Sapiro: Genius of Farm Cooperative Promotion." *Mississippi Valley Historical Review* 49 (1962): 242–68.
Ledbetter, Calvin R., Jr. "Jeff Davis and the Politics of Combat." *Arkansas Historical Quarterly* 33 (1974): 15–37.
———. "The Long Struggle to End Convict Leasing in Arkansas." *Arkansas Historical Quarterly* 52 (1993): 1–27.
Lewis, Todd E. "Mob Justice in the 'American Congo': 'Judge Lynch' in Arkansas during the Decade after World War I." *Arkansas Historical Quarterly* 53 (1993): 156–84.
Lowitt, Richard. "Henry A. Wallace and the 1935 Purge in the Department of Agriculture." *Agricultural History* 53 (1979): 607–21.
Lucas, Marietta Ann. "Bracero Labor in Northeast Arkansas." *Craighead County Historical Quarterly* 6 (Summer 1968): 19–25.
McDean, Harry C. "Professionalism, Policy, and Farm Economists in the Early Bureau of Agricultural Economics." *Agricultural History* 57 (1983): 64–82.
McWhiney, Grady. "Louisiana Socialists in the Early Twentieth Century: A Study of Rustic Radicalism." *Journal of Southern History* 20 (1954): 315–36.
Mann, Susan A. "Slavery, Sharecropping, and Sexual Inequality." *Signs* 14 (1989): 774–98.
Marcus, Alan I. "The Ivory Silo: Farmer-Agricultural College Tensions in the 1870s and 1880s." *Agricultural History* 60 (1986): 22–36.
Martin, Robert E. "The Referendum Process in the Agricultural Adjustment Programs of the United States." *Agricultural History* 25 (1951): 34–46.
May, Henry F. "Shifting Perspectives on the 1920's." *Mississippi Valley Historical Review* 43 (1956): 405–27.
Mayberry, B. D. "The Tuskegee Movable School: A Unique Contribution to National and International Agriculture and Rural Development." *Agricultural History* 65 (1991): 85–104.

Milani, Ernesto R. "Peonage at Sunnyside and the Reaction of the Italian Government." *Arkansas Historical Quarterly* 50 (1991): 30–39.
Miller, Lorna Clancy, and Mary Neth. "Farm Women in the Political Arena." Pp. 357–80 in *Women and Farming: Changing Roles, Changing Structures*, ed. Wava G. Haney and Jane B. Knowles. Boulder, Colo., and London, 1988.
Mitchell, H. L. "The Founding and Early History of the Southern Tenant Farmers Union." *Arkansas Historical Quarterly* 32 (1973): 342–69.
Moneyhon, Carl H. "The Impact of the Civil War in Arkansas: The Mississippi River Plantation Counties." *Arkansas Historical Quarterly* 51 (1992): 105–18.
Moore, Gary E. "The Involvement of Experiment Stations in Secondary Agricultural Education, 1887–1917." *Agricultural History* 62 (1988): 164–76.
Moss, Jeffrey W., and Cynthia B. Lass. "A History of Farmers' Institutes." *Agricultural History* 62 (1988): 150–63.
Murray, Gail S. "Forty Years Ago: The Great Depression Comes to Arkansas." *Arkansas Historical Quarterly* 29 (1970): 291–312.
Naison, Mark D. "Black Agrarian Radicalism in the Great Depression: The Threads of a Lost Tradition." *Journal of Ethnic Studies* 1 (1973): 47–65.
Nelson, Lawrence J. "Welfare Capitalism on a Mississippi Plantation in the Great Depression." *Journal of Southern History* 50 (1984): 225–50.
Neth, Mary. "Building the Base: Farm Women, the Rural Community, and Farm Organizations in the Midwest, 1900–1940." Pp. 339–56 in *Women and Farming: Changing Roles, Changing Structures*, ed. Wava G. Haney and Jane B. Knowles. Boulder, Colo., and London, 1988.
Nipp, Robert E. "The Negro in the New Deal Resettlement Program: A Comment." *Agricultural History* 45 (1971): 195–200.
Osterud, Nancy Grey. "Gender and the Transition to Capitalism in Rural American." *Agricultural History* 67 (1993): 14–29.
Otto, John Solomon. "The Migration of the Southern Plain Folk: An Interdisciplinary Synthesis." *Journal of Southern History* 51 (1985): 183–200.
Paisley, Clifton. "The Political Wheelers and Arkansas' Election of 1888." *Arkansas Historical Quarterly* 25 (1966): 3–21.
Patton, Adell, Jr. "The 'Back-to-Africa' Movement in Arkansas." *Arkansas Historical Quarterly* 52 (1992): 164–77.
Payne, Elizabeth Anne. "'What Ain't I Been Doing?': Historical Reflections on Women and the Arkansas Delta." Pp. 128–49 in *Arkansas Delta: Land of Paradox*, ed. Jeannie Whayne and Willard B. Gatewood, Jr. Fayetteville, Ark., 1993.
Peterson, Arthur G. "Governmental Policy Relating to Farm Machinery in World War I." *Agricultural History* 17 (1943): 31–40.
Pickins, William. "The American Congo: The Burning of Henry Lowry." *Nation* 112 (March 23, 1921): 426–28.
Pisani, Donald J. "Reclamation and Social Engineering in the Progressive Era." *Agricultural History* 57 (1983): 46–63.
Pittman, Dan W. "The Founding of Dyess Colony." *Arkansas Historical Quarterly* 29 (1970): 313–26.

Pratt, William C. "Women and the Farm Revolt of the 1930s." *Agricultural History* 67 (Spring 1993): 214–23.
Preece, Harold. "Folk Music of the South." *New South* 1 (March 1938): 14.
Prunty, Merle, Jr. "The Renaissance of the Southern Plantation." *Geographical Review* 45 (1955): 459–91.
Rabinowitz, Howard N. "From Exclusion to Segregation: Southern Race Relations, 1865–1890." *Journal of American History* 63 (1976): 325–50.
Rasmussen, Wayne D. "The 1890 Land-Grant Colleges and Universities: A Centennial Overview." *Agricultural History* 65 (1991): 168–72.
Rector, William G. "From Woods to Sawmill: Transportation Problems in Logging." *Agricultural History* 23 (1949): 239–44.
"Red Cedar Shingles and Shakes: The Labor Story." *Journal of Forest History* 19 (1975): 112–27.
Reeves, William D. "PWA and Competitive Administration in the New Deal." *Journal of American History* 50 (1973): 357–72.
Reid, Joseph D., Jr. "Sharecropping as an Understandable Market Response: The Post-Bellum South." *Journal of Economic History* 33 (1973): 106–30.
Riegel, Robert E. "Trans-Mississippi Railroads during the Fifties." *Mississippi Valley Historical Review* 10 (1923): 153–72.
Riney-Kehrberg, Pamela. "Separation and Sorrow: A Farm Woman's Life, 1935–1941." *Agricultural History* 67 (1993): 185–96.
Ritter, Anna. "Marked Tree from 1883–1936." *Craighead County Historical Quarterly* 5 (Winter 1967): 24–27.
——. "Marked Tree from 1883–1936: Pt. 2." *Craighead County Historical Quarterly* 5 (Spring 1967): 24–28.
——. "Marked Tree from 1883–1936: Pt. 3." *Craighead County Historical Quarterly* 5 (Summer 1967): 26–29.
Robinson, Armstead L. "Beyond the Realm of Social Consensus: New Meanings of Reconstruction for African History." *Journal of American History* 68 (1981): 276–97.
Rogers, O. A., Jr. "The Elaine Race Riots of 1919." *Arkansas Historical Quarterly* 19 (1960): 142–50.
Rosenberg, Charles E. "Science, Technology, and Economic Growth: The Case of the Agricultural Experiment Station Scientist, 1875–1914." *Agricultural History* 45 (1971): 1–20.
Rosencrantz, Florence L. "The Rice Industry in Arkansas." *Arkansas Historical Quarterly* 5 (1946): 123–37.
Ross, Earle D. "Retardation in Farm Technology before the Power Age." *Agricultural History* 30 (1956): 11–18.
——. "Roosevelt and Agriculture." *Mississippi Valley Historical Review* 14 (1927): 287–310.
Rothstein, Morton. "The New South and the International Economy." *Agricultural History* 57 (1983): 373–84.
Salmond, John A. "The Civilian Conservation Corps and the Negro." *Journal of American History* 52 (1965): 75–88.

Saloutos, Theodore. "The Agricultural Wheel in Arkansas." *Arkansas Historical Quarterly* 2 (1943): 127–40.
———. "New Deal Agricultural Policy: An Evaluation." *Journal of American History* 61 (1974): 394–416.
Sautter, Udo. "Government and Unemployment: The Use of Public Works before the New Deal." *Journal of Southern History* 73 (1986): 59–86.
Schatz, Philip. "Songs of the Negro Worker." *New Masses* 5 (May 1930): 6–8.
Schlesinger, Andrew Bancroft, "Las Gorras Blancas, 1889–1891." *Journal of Mexican American History* 1 (1971): 87–143.
Schor, Joel. "The Black Presence in the U.S. Cooperative Extension Service since 1945: An American Quest for Service and Equity." *Agricultural History* 60 (1986): 137–53.
Scott, Roy V. "American Railroads and the Promotion of Forestry." *Journal of Forest History* 23 (1979): 72–81.
Seals, R. Grant. "The Formation of Agricultural and Rural Development Policy with Emphasis on African Americans: 2. The Hatch-George and Smith Lever Acts." *Agricultural History* 65 (1991): 12–34.
Sharpless, Rebecca. "Southern Women and the Land." *Agricultural History* 67 (1993): 30–42.
Shea, William L., ed. "A German Prisoner of War in the South: The Memoir of Edwin Pelz." *Arkansas Historical Quarterly* 44 (1985): 42–55.
Shideler, James H. "Herbert Hoover and the Federal Farm Board Project, 1921–1925." *Mississippi Valley Historical Review* 42 (1956): 710–29.
Shlomowitz, Ralph. "The Origins of Southern Sharecropping." *Agricultural History* 53 (1979): 557–75.
———. "Plantations and Smallholdings: Comparative Perspectives From the World Cotton and Sugar Cane Economies, 1865–1939." *Agricultural History* 58 (1984): 1–16.
Shofner, Jerrell H. "Negro Laborers and the Forest Industries in Reconstruction Florida." *Journal of Forest History* 19 (1975): 180–91.
Shover, John L. "The Communist Party and the Midwest Farm Crisis of 1933." *Journal of American History* 51 (1964): 248–64.
Sillers, Walter, Sr. "Flood Control in Bolivar County, 1838–1924." *Journal of Mississippi History* 9 (1947): 3–20.
Simonson, S. E. "Origin of Drainage Projects in Mississippi County." *Arkansas Historical Quarterly* 5 (1946): 263–73.
———. "The St. Francis Levee and High Waters on the Mississippi River." *Arkansas Historical Quarterly* 6 (1947): 419–29.
Slichter, Gertrude Almy. "Franklin D. Roosevelt and the Farm Problem, 1929–1932." *Mississippi Valley Historical Review* 43 (1956): 238–58.
Smith, C. Calvin. "Black Organization and White Suppression: The Elaine Race Riot, 1919." *Craighead County Historical Quarterly* 12 (Summer 1974): 2–11.
———. "The Politics of Evasion: Arkansas' Reaction to *Smith v. Allwright*, 1944." *Journal of Negro History* 67 (1982): 40–51.

Smith, John. "Farm Real Estate Trends in Arkansas." *Arkansas Historical Quarterly* 10 (1951): 409–14.

Smith, John L. "Reminiscences of Farming and Business in the Depression, 1929–1933." *Arkansas Historical Quarterly* 45 (1986): 321–29.

Swindle, J. L. "Lepanto: Land of Many Waters." *Craighead County Historical Quarterly* 3 (Spring 1965): 35–39.

Thelen, David P. "Social Tensions and the Origins of Progressivism." *Journal of American History* 56 (1969): 321–41.

Towns, Stuart. "Joseph T. Robinson and Arkansas Politics: 1912–1913." *Arkansas Historical Quarterly* 24 (1965): 291–307.

Venkataramani, M. S. "Norman Thomas, Arkansas Sharecroppers and the Roosevelt Agricultural Policies, 1933–1937." *Arkansas Historical Quarterly* 24 (1965): 3–28.

Voss, Larry D. "The Prisoner of War Camps and the Use of Its Labor in Northeast Arkansas." *Craighead County Historical Quarterly* 9 (Summer 1969): 11–14.

Walz, Robert B. "Arkansas Slaveholdings and Slaveholders in 1850." *Arkansas Historical Quarterly* 12 (1952): 38–74.

Wennersten, John R. "The Travail of Black Land-Grant Schools in the South, 1890–1917." *Agricultural History* 65 (1991): 54–62.

West, William Elliott. "Life on the Urban Frontier: A Look at Social Life in the Rocky Mountain Mining Camps." *Speculator* 1 (1984): 4–15.

Whayne, Jeannie M. "The Creation of a Plantation System in the Arkansas Delta in the Twentieth Century." *Agricultural History* 66 (1992): 63–84.

———. "The Power of the Plantation Model: The Sunk Lands Controversy." *Forest and Conservation History* (1993): 56–67.

———. "The Segregated Farm Program in Poinsett County, Arkansas." *Mississippi Quarterly* 45 (1992): 421–38.

———. "The Significance of Race, Class, and Family in the Struggle over Prohibition in Small Town Arkansas." *Locus* (1995): 129–49.

White, Walter. "'Massacring Whites' in Arkansas." *Nation* 109 (Dec. 6, 1919): 715–16.

———. "The Negro and the Communists." *Harpers* 164 (Dec. 1931): 62–72.

Wiener, Jonathan. "Class Structure and Economic Development in the American South, 1865–1955." *American Historical Review* 84 (1979): 970–1006.

Wik, Reynold M. "Henry Ford and the Agricultural Depression of 1920–1923." *Agricultural History* 29 (1955): 15–21.

Wiley, Stephen R. "Songs of the Gastonia Textile Strike of 1929: Models of Southern Working-Class Woman's Militancy." *North Carolina Folklore Journal* 30 (Fall-Winter 1982): 87–98.

Williams, C. Fred. "Frustration amidst Hope: The Land Grant Mission of Arkansas AM&N College, 1873–1972." *Agricultural History* 65 (1991): 115–30.

Williams, Thomas T., and Handy Williamson, Jr. "Teaching, Research, and Extension Programs at Historically Black (1890) Land-Grant Institutions." *Agricultural History* 62 (1988): 244–57.

Winters, Donald L. "The Agricultural Ladder in Southern Agriculture: Tennessee, 1850–1870." *Agricultural History* 61 (1987): 36–52.
———. "Tenant Farming in Iowa, 1860–1900: A Study of the Terms of Rental Leases." *Agricultural History* 48 (1974): 130–50.
Wolf, Eric R., and Sidney W. Mintz. "Haciendas and Plantations in Middle America and the Antilles." *Social and Economic Studies* 6 (1957): 380–412.
Wood, Stephen E. "The Development of Arkansas Railroads: The Great Railroad Boom." *Arkansas Historical Quarterly* 7 (1948): 155–93.
Woodman, Harold D. "Comments." *American Historical Review* 84 (1979): 997–1001.
———. "How New Was the New South?" *Agricultural History* 58 (1984): 529–45.
———. "New Perspectives on Southern Economic Development: A Comment." *Agricultural History* 49 (1975): 374–80.
———. "Postbellum Social Change and Its Effects on Marketing the South's Cotton Crop." *Agricultural History* 56 (1982): 215–30.
———. "Post Civil War Agriculture and the Law." *Journal of Southern History* 45 (1979): 319–37.
———. "Sequel to Slavery: The New History Views the Postbellum South." *Journal of Southern History* 43 (1977): 523–54.
Woodruff, Nan Elizabeth. "African-American Struggles for Citizenship in the Arkansas and Mississippi Deltas in the Age of Jim Crow." *Radical History Review* 55 (1993): 33–51.
———. "Mississippi Delta Planters and Debates over Mechanization, Labor, and Civil Rights in the 1940s." *Journal of Southern History* 60 (1994): 263–84.
———. "Pick or Fight: The Emergency Farm Labor Program in the Arkansas and Mississippi Deltas during World War II." *Agricultural History* 64 (1990): 74–85.
Worley, Ted R. "The Control of the Real Estate Bank of the State of Arkansas, 1836–1855." *Mississippi Valley Historical Review* 37 (1950): 403–26.
Wyatt-Brown, Bertram. "Leroy Percy and Sunnyside: Planter Mentality and Italian Peonage in the Mississippi Delta." *Arkansas Historical Quarterly* 50 (1991): 60–84.
Wye, Christopher G. "The New Deal and the Negro Community: Toward a Broader Conceptualization." *Journal of American History* 49 (1972): 621–39.

Dissertations and Master's Theses

Bercaw, Nancy. "The Politics of the Household during the Transition from Slavery to Freedom in the Mississippi Delta." Ph.D. diss., Univ. of Pennsylvania, 1996.
Blake, Earl. "Farm Tenancy in Arkansas." M.A. thesis, Univ. of Arkansas, Fayetteville, 1939.
Drummond, Boyce Alexander, Jr. "Arkansas Politics: A Study of a One-Party System." Ph.D. diss., Univ. of Chicago, 1957.

Ellenburg, Martha Ann. "Reconstruction in Arkansas." Ph.D. diss., Univ. of Missouri, 1967.

Fain, James Harris. "Political Disfranchisement of the Negro in Arkansas." M.A. thesis., Univ. of Arkansas, 1961.

Finley, Randy. "The Freedman's Bureau in Arkansas." Ph.D. diss., Univ. of Ark., 1992.

Geenens, Ronald Burke. "Economic Adjustments to the Termination of the Bracero Program." Ph.D. diss., Univ. of Arkansas, 1967.

Harris, Michael Mehlman. "The Resettlement Adminstration and the Problems of Tenant Farmers in Arkansas, 1935–1936." Ph.D. diss., New York Univ., 1970.

Holley, James Donald. "The New Deal and Farm Tenancy: Rural Resettlement in Arkansas, Louisiana, and Mississippi." Ph.D. diss., Louisiana State Univ., 1969.

Holt, Sharan Ann. "A Time to Plant: The Economic Lives of Freedpeople in Granville County North Carolina, 1865–1900." Ph.D. diss., Univ. of Pennyslvania, 1991.

Hunt, George Murrell. "A History of the Prohibition Movement in Arkansas." M.A. thesis, Univ. of Arkansas, 1933.

Kjeldsen, Richard G. "Importance of Agricultural Exports to the Arkansas Delta Economy." M.A. thesis, Univ. of Arkansas, 1967.

Lewis, Elsie M. "From Nationalism to Disunion: A Study of the Secession Movement in Arkansas, 1850–1861." Ph.D. diss., Univ. of Chicago, 1947.

Moyers, David. "Arkansas Progressivism: The Legislative Record." Ph.D. diss., Univ. of Arkansas, 1986.

Russell, Marvin Frank. "Republican Party of Arkansas, 1874–1913." Ph.D. diss., Univ. of Arkansas, 1985.

Segraves, Joe T. "Arkansas Politics, 1874–1918." Ph.D. diss., Univ. of Kentucky, 1973.

Taylor, Orville W. "Negro Slavery in Arkansas." Ph.D. diss., Duke Univ., 1956.

Walz, Robert Bradshaw. "Migration into Arkansas, 1834–1880." Ph.D. diss., Univ. of Texas at Austin, 1958.

Whayne, Jeannie M. "Reshaping the Rural South: Land, Labor, and Federal Policy in Poinsett County, Arkansas, 1900–1940." Ph.D. diss., University of California, San Diego, 1989.

Wheeler, John McDaniel. "The People's Party in Arkansas, 1891–1896." Ph.D. diss., Tulane Univ., 1975.

White, James Harold. "Trends in Arkansas Agricultural Production and Income from 1934 to 1951." M.A. thesis, Univ. of Arkansas, 1953.

White, Lonnie Joe. "Arkansas Territorial Politics, 1819–1936." Ph.D. diss., Univ. of Texas at Austin, 1961.

INDEX

A. F. Oliver Company, Memphis, 44–45
A.F. and A.M., 153
Abbott, Della, and blacks, 86, 99; wife to Eliga "Kidd" Abbott, 100
Abbott, Eliga ("Kidd"): and family (photograph), 82; background and personal and real wealth, 100–101; prosperity short-lived, 103; and drainage taxes, 104
Adamsville, Tenn., 115
Adkins, W. A., 76
Agricultural Adjustment Administration (AAA): and complaints about, 164; and the cotton program, 158, 159; and the crop reduction program, 158; and Cully Cobb, 188; and emergency crop loans, 166; and the FCA, 165; and Jerome Frank, 187; and landlord-tenant conflict, 204–5; and need to court southern planters, 203–4, 206; new regulations in response to STFU, 205–6; and planter control of local programs, 216; and programs for ridge farmers, 173–74; and the rice program, 159, 171–73; and the STFU, 208; on success of, 167–68; and township committeemen, 161

Agricultural Advisory Board, 166
Agricultural Credit Associations, 155
Ainsworth, William, 117
Alabama, 71, 185
Albright, Charles, 147
Alexander, Douglas, 105–6, 110
Alexandria, Mo., 117
Allen, Thomas H., 59
American Farm Bureau Federation, 140
American Federation of Labor, 226
American Legion Post: and Elaine Race Riot, 76; for blacks, 230
Anderson's Chapel, 38
Anthony, Henry, 61
antienticement legislation, 75
Anti-Saloon League: abandons local option strategy, 33–34; and the statewide prohibition statute, 35–36
Arkansas Act 12, 52
Arkansas City, Ark., 100
Arkansas Cotton Growers Association, 141
Arkansas County, 47
Arkansas Farm Credit Association, 155
Arkansas General Assembly, 110; and STFU, 186
Arkansas Rice Growers Association, 163

Arkansas Rural Rehabilitation Corporation, 214
Arkansas Supreme Court: and Poinsett County election dispute, 134; and the stock law, 139
artesan wells, lack of, 63

backwoods culture, 73–75
Baird, J. C., Jr., 227
banks: Bank of Marked Tree and Drainage District One, 107–8; Bank of Tyronza, 150, 192; closure, 158; Harrisburg, 155; and Ernest Ritter, 35
Baptist Church, Marked Tree, 38
Barnes, Robert B., 202
Beal, Emmet, 62
Bennett, T. J., 161
Bettis, T. A.: as FERA director sympathetic to planters, 186, 188; describes Britt McKinney, 209; as RA supervisor, 214
Betton, F. R., 209, 230
Betton, T. R., 230
Biggs, B. P., 165
Black, Dr. J. R., 134
blacks: black doctor leaves Marked Tree, 29; decline in number of farmers, 229; enmeshed in plantation system, 6; farm agent, 229; forms of resistance, 56–57; and economic decline, 200–202; and Elaine, 205; and the extension service, 228; home demonstration agent, 228, 229; and homesteading, 86, 99; and increase in population of, 67; and land tenure, 99; not attracted to Marked Tree, 29; as plantation labor, 73; population increase, 99; and prohibition, 36; as unskilled or menial laborers, 29
Blanton, C. A. ("Dad"), 131, 132
Blytheville, Ark., 176
Bolivar Township: identified, 12; and the railroad, 17

Bolivar Township East of Big Bay, 117, 118
boll weevil, 75, 145
Bone dry law, 37
bootleggers, 21, 73, 195
Boston, Ed, 191
Bott Brothers, 117
Brackenseik brothers, 187
Bradford, W. D., 135, 136
Brandfon, Robert, 4
Bratton, O. S., and Elaine Race Riot, 76
Brigance, T. C., 149
Brinkley, Ark., 62, 173
Brough, Charles (Governor), 76
Brown, Captain J. H., 49
Brown, W. T., 189
Brown v. Board of Education, 230
Brunner, John, 147
Buck, David Seely, 97
Bull Moose Republicans, 28, 35, 129
Bureau of Agricultural Economics, 216
Burtinett, Claud, 57
Burton, E. P., 215, 225
businessmen-planters: become a plantation elite, 2; emerge in 20th-century Poinsett County, 15; face challenges, 3; of Marked Tree, 23; and modernization of plantation, 4; new to plantation system, 5; plantation system, 39
Buxton, John E., 152, 153
Byrn, E. T., 161

Cairo, Ill., 147
Campbell, Dr. G. O., 214–15
Campbell Farms, 214–15
Campbellite Church, 48
Caplinger, A. B., 166
Caraway, Congressman T. H., 108, 109
Carlew, John, 189
Carlson, A., 225
Carpenter, C. T., 191, 198, 207

Catlett and Foley, 117
Chamber of Commerce: of Harrisburg, 150; of Lepanto, 155
Chapin, S. C., 169
Chapman, E. R.: origins, 24; wife refuses to settle in Marked Tree, 27
Chapman, W. B., 24
Chapman and Dewey: and the Agricultural Adjustment Administration, 167; ceases business (permanent), 232; closes mills (temporary), 152; comes to Poinsett County, 1; comes to Marked Tree, 18; continues in operation, 37; and cotton picker, 224; as creditor, 45; and doodlums, 27; drainage, 103, 105, 160; and Drainage District Seven, 112; and farm company, 114; and Greenwood Township, 98; and Harrisburg mills, 117; and labor supply, 24; land holdings, 24; and Lepanto, 97; and Marked Tree Town Council, 30; and Marked Tree POW camp, 222; purchases Harding Mill site, 18, size of operation, 26; and STFU, 207; and sunk lands, 91, 92, 98, 102; and transient millworkers, 22
Cheatham, A. W., 86
Chickaninny Farm, 62
children, numbers increase in Marked Tree, 38
Christian Church, 48–49
churches: black, 193, 194; importance in life of tenants and sharecroppers, 218; and Marked Tree, 27, 38; and millworkers' attitude toward, 28; and role in delta communities, 194; and STFU, 193; white, 193; women dominate membership, 194
CIO (Congress of Industrial Organizations), 210
Civil Rights Act of 1866, 50

Civil Rights Act of 1964: and whitecapping, 52; and end of black extension, 229
Civil rights movement, 229–30
Civil War, 70; challenge to Harrisburg elite from delta during, 14; sharecropping arises after, 5
Civil Works Administration: controversy over jobs, 188, 189; and farm laborers, 189–90; and planter influence over, 210; and tenants and sharecroppers, 187
Clampet, William R., 50, 70–71
Clark, W. T., 174
Clarke, James P.: and the Oldfield bill, 95; and the sunk lands, 90
Claunch, John B., landholdings, 38
Clements, S. A., 163
Cleveland Call and Post, 185–86
Cobb, Cully, and the cotton adjustment program, 204; and the CWA, 188; and Roosevelt's New Deal coalition, 203–4
Cobb, James, 3, 4
Coker, J. W., 165
Cole, Benjamin (Judge), 130
Collins, L. J., 126
commissary system, 1, 39, 54–55, 203
Commission on Migratory Labor, 226
Conaster, James, 96–97
Conaster, John, 96–97
Conaster, Ruben, 96–97, 100
convict labor, 31, 45, 63
Cook, W. L., 161
Cooke, Levi, 95
Corps of Engineers: and drainage, 108, 109; and the St. Francis Levee project, 212
cotton: and AAA, 158; acres grown in 1937, 167; associations, 141; and the boll weevil, 75; businessmen-planters' lack of familiarity with, 5; and chattel mortgages, 39; children chopping (photo-

cotton (*cont.*)
graph), 177; and the crop lien, 44–45; and the crop reduction program, 157, 159, 161–62, 165, 167, 169–70, 168; destruction of, 120–21; and drought of 1930, 152; expansion of, 115, 221; and factors, 41, 140; and landlord-tenant conflict, 204; and mechanical picker, 224–25, 231; and Memphis factors, 40; and plantation agriculture, 2; and planters' Mexican labor program, 224–25; and Poinsett County, 66; and the prairie, 121, 233; prices drop, 120, 140; prices high, 116; and the ridge, 41, 119; southern growers and farm legislation, 140; in the swamp, 9–10; yields, 150, 168

county agent (*see also* Extension Service): and advice to whites only, 229; and black home demonstration agent, 228, 229; and county agricultural committee, 162; county plan of work committee, 161, 162; and credit needs, 155; and crop reduction program, 160–61; and delta planters, 151, 156, 168, 221; expansion of black agents, 227–28; W. D. Ezell, 138, 146, 149; and H. W. Hinson, 138, 149; and labor needs, 221; R. L. McGill, 161, 167, 186–87; orientation of, 144; and prairie farmers, 151; and ridge farmers, 151; E. F. B. Sargent, 145; soil-building efforts, 168–69; A. Raybon Sullivant, 138, 149, 150–51, 154, 155, 161, 174; tenure in Poinsett County, 144

County Democratic Central Committee, 126, 127–28, 129, 130, 131, 132

county government, struggle for control over, 5, 130
courthouse, burning of, 111; controversy over, 123–24, 133; described by Edwin Palmer, 10
Craft, Henry, 227
Craig, O. T., 58, 59
Craighead County, 13, 42, 88, 89, 117, 122
Crawford, Isaac, 62
Crisis, 184–85
Crittenden County, 13, 42, 89, 142, 197, 198, 207, 208, 230
crop lien, 2, 39, 43–45, 121, 203
crop reduction program, 158–59, 164, 184, 186, 199, 203, 206, 215
Cross, David, 14, 15–16
Cross County, 13, 14, 15, 48–53, 69, 71–72, 89, 102, 122, 142, 205, 207–9, 230
Crowley's Ridge: challenge from delta businessmen-planters, 15; county seat located on, 2; and David Cross, 14; description of, 10–12; and Edwin Palmer's trip, 10; falls to Union forces during Civil War, 14; and flood of 1927, 147; map, 11; and Missouri drainage project, 108; and New Deal programs, 173–74; and plow-up campaign committee, 162; and Poinsett County drainage project, 110; and politics, 124, 128; and railroad, 16; and swamplands, 13, 15; and soil infertility, 174; and stock law, 138–39; and whitecappers, 71

dairy farming, ridge, 173, 174
Daniel, Pete, 3, 4, 59–60, 171
Davis, Jeff (Governor), 94, 125
Davis, Oliver, Mill, 18
Dawson, C. A.: as creditor, 45; as Ernest Ritter's business partner,

Dawson, C. A. (*cont.*)
    45; on farm labor committee,
    221; as general manager of E.
    Ritter and Company, 167
Dawson, W. C., 97
Dean, J. L., 150
Delta and Pine Land Company, 8,
    65–66
Democratic politics, 35, 124,
    127–28, 129, 191, 204
Dent, S. P., 118–20, 145
Department of Agriculture, 55, 216,
    225, 226–27
Department of the Interior (*see also*
    General Land Office): and
    homesteaders, 86–87, 88; prohibition on cutting timber,
    92–93; reversal of policy on
    sunk lands, 91; support for riparian claimants, 94; and suspension of final proofs, 92–93; and
    trespassers on sunk lands, 90
desegregation, 230
Dewey, Curtis, 142, 160, 167
Dewey, H. C., 112
Dewey, W. C., 24
DeWitt, Ark., 172
Dibble, C. H., and evictions, 209;
    photograph of evictees, 178; and
    STFU, 205
Dickens, Felix, 214–15
Dilatush, M. T., 155
Disciples of Christ, 48
disfranchisement, 35
diversification, 119, 120, 145
Dobson Township, 131
doodlum, 27, 39
drainage: enterprises, 1, 14; and influence of World War I boom, 3;
    and labor, 63; and land prices,
    116; and ridge farmers, 13; and
    taxes, 103, 142–43, 151
Drainage District One, 106–7, 111
Drainage District Seven, 110, 114,
    118, 132, 134–35, 159–60,
    189, 212

Drainage District Six, 108, 111
Drainage District Two, 105–6, 110
drainage districts: enabling legislation, 104; sources of opposition,
    104–5
Driver, William J.: on bootleggers,
    21; and homesteaders, 101–2;
    and sunk lands claimants, 94; on
    whitecapping, 52
drought of 1930–31, 3, 151–52,
    155, 160
DuBard, J. D., 154
Du Bois, W. E. B., 184–85
Dudley, Wade, 62
Duren, E. M., 115
Dyerle, G. E., 168
Dyess, W. R., 188

E. Ritter and Company (*see also*
    Ernest Ritter): agricultural crisis,
    142; and cotton, 115; and C. A.
    Dawson, 167, 221; and Greenwood Township sunk lands,
    103; and M. W. Hazel, 107;
    homesteaders, 86; increase in
    landholdings, 142; and V. O. Isbell, 219; and Mexican laborers,
    227; and officers of, 45; and
    Ozark Trail meeting, 123; and
    postwar transformation,
    231–32; prosperity, 114; retirement plan and insurance, 233;
    and Harry Ritter, 207; and L. V.
    Ritter, 112, 149, 160; and Jean
    Thatcher, 220
Eagle Nursery, 119
Earle, Ark., 120
East, Alex, and furnishing tenants and
    sharecroppers after plow-up,
    193; and the STFU, 192
East, Henry ("Clay"): and the Civil
    Works Administration controversy, 187, 188; decides to organize a union, 191; and Alex East,
    192; describes Britt McKinney,
    209; intimidation of, 208; and

East, Henry ("Clay") (*cont.*)
  Norman Thomas, 190; and violence against STFU, 207
East, Maxine, 188
Eddington, Lena, 228
El Alamein, 222
Elaine, Ark., 135, 219
Elaine race riot, 75–77, 135, 199, 205
election: fraud, 113; residency requirement, 30
El Paso, Tex., 58
Embrey, Hannah: and drainage taxes, 105, 111; president of Homesteaders Union, 86
Emrich, J. A. (John): Agricultural Credit Association, 155; bank goes under, 162; and cotton, 115; and courthouse controversy, 124; as creditor, 45; on Farm Debt Adjustment Committee, 166; and association with Hiram Norcross, 192, 193; and the Ozark Trail, 123; on plan of work committee, 161; on plow-up campaign committeeman, 166; and post–World War I crisis, 142; and Harry Ritter, 207; and Louis Ritter, 45, 149; and STFU, 198; and the St. Francis Levee Board, 129; as township committeeman, 161; and the Tyronza Business Men's Club, 149–50
England, Ark., 184, 208
Ernest Pan, Miss., 100
Evans, Mose, 62
evictions: and Hiram Norcross, 199, 208; from Resettlement community, 215
Extension Service (*see also* county agent): 4-H Club, 149; and the civil rights movement, 230; and cooperative marketing, 145; and credit needs of farmers, 150; and crop reduction program, 160–61; and diversification, 150–51; and drought of 1930–31, 151; experiment station, 172; and farm labor clerk, 224–25; and landlord/tenant conflict, 204–5; and Mexican labor program, 224; orientation of, 118–20, 140, 144; and planters' labor needs, 221; and T. Roy Reid, 160; and stock law, 138, 139
Ezell, W. D., 138, 146, 149

Fairview plantation, 192
Fallis, R. R., 49
Farm Bloc, 140–41
Farm Bureau, 145
Farm Credit Administration, 165
Farm Debt Adjustment Association, 166
Farm Debt Adjustment Committee, of Poinsett County, 166
Farmers and Merchants Bank of Marked Tree, organized, 34
Farmers Union: destruction of warehouse, 120; and farm bloc, 140; and postwar planning, 224, 225; and support for county agent, 145
Farm Security Administration, 213–16
Federal Emergency Relief Administration (FERA): and Thurman A. Bettis, 186, 209; caseworkers, 188–89; and W. R. Dyess, 188; and Maxine East, 188; and Brooks Hays, 187, 204; and Ward Rodgers, 206; works program, 189
Federal Feed, Seed, and Food Loan, 155
Federal Land Bank of St. Louis, 159, 165, 166
Finn, H. L.: and Drainage District Seven, 111; and Homesteaders Union, 99; property, 100

Finney, E. C.: acting secretary of the interior, 94; and legislation disadvantageous to homesteaders, 101
First National Bank of Helena, 53
Fisher, Newt, 34, 115
Fisher, Ark., 173
floods, 9, 19; flood of 1912, 64; flood of 1927: 3, 144, 146–47, 149, 160
Flournoy, Thomas, 110
Foley, F. G., 117
Forrest City, Ark., 17, 42
Forrest City Commercial Club, 108, 109
Fourteenth Amendment, 53
Fox, Mayor, 207
Frank, Jerome, 187
freedmen, force compromise with planters, 6
French, C. R., 111
Frisco Railroad, 31
Fruit Growers Association, 119
Fuller, W. A., 19
Fuller Brothers Mill, 28
Futrell, Gov. Junius Marion: on FERA, 188; and STFU, 205

Gabriel, Annie, 148
Gant, J. B., 126
Gant, J. G., 110, 111
Garden Club, for blacks, 230
Garfield, James R., and sunk lands, 91
General Land Office (*see also* Department of the Interior): homestead certificates of entry, 109; and H. N. Pharr, 89–90; and riparian owners, 112; and suspension of final proofs, 92; and trespassers on sunk lands, 91–92
Georgia, 48, 185
German POWs, 222–23, 228
Gilmore, Ark., 197
Glenn, Earl, 173

Going, L. C.: cocounsel mentioned, 102; convict labor, 64; and the Going Amendment, 36; and Marked Tree Town Council, 36; political career briefly described, 125; and prohibition, 46; and Twenty-ninth Congressional District, 36; and Ernest Ritter, 129; and whitecappers, 51
Going Amendment, 36
Goldberger, Herman, 205
Goodin, Joe, 43, 165
Gould, Jay, 16
Gracy, Constable, 68
grandfather clause, 22, 35
Great Lakes, 26
Green, Steve, 57
Green County, 13, 88, 89, 122
Greenwood, Charlie B.: and Drainage District One, 107; homesteaded Lepanto town site, 97; mayor of Lepanto, 96; photograph of, 81; supports homesteaders, 96, 98–99
Greenwood, Fred: homesteader, 86; and support for homesteaders, 99
Greenwood Township: created, 45; and black farmers, 99; and Chapman and Dewey, 98; and Drainage District Seven, 111; and homesteaders, 100; and H. H. Howington, 163; opposition to other delta elites, 130; and Ernest Ritter land purchase in, 94, 95, 98, 103; and sunk lands, 96
Griffin, L. D., 131–32
Grismore-Hyman Mill, Lepanto, 97
Gulf of Mexico, 147

Hall, H. C., and sunk lands dispute, 93
Hall, Joe, 132–33, 135
Hall family, 72

Hamilton, Mary, *Trials of the Earth*, 18–19
Handcox, John, 193
Harding, President Warren G., 87, 94, 112
Harding Sawmill, purchased by Chapman and Dewey, 24
Harihan Bridge, 43
Harris, Benjamin, 14, 15, 126, 146
Harris, George, 19
Harris, T. D., 124, 130, 131, 132
Harrisburg: banks close, 158; and challenge from delta during Civil War, 14; and cotton warehouse, 120; and county agent, 118, 120, 149; county seat of Poinsett County, 10; courthouse clique, 13; courthouse controversy, 113, 124, 133, 134; and crop reduction committeemen, 165; and the delta, 46; Drainage District Seven, 111–12; economic growth, 116–17, 123; and FERA works program, 188; and forest industries, 117; and Moutan Rice Mill, 225; mills, 119; political candidates, 135; and political power, 233; and POW camp, 222–23; and prairie connection, 122; prominent citizens, 126; and roads, 42; and the stock law, 139; and the STFU, 186; and White Hall, 50
Harrisburg Chamber of Commerce, 150
Harrisburg Hotel, 10
Harrisburg State Bank, 110
Haverstick family, 74
Hays, Brooks, 187, 198, 204, 213
Hazel, Crockett J., 128
Hazel, M. W.: and County Democratic Central Committee, 126, 129, 132; as creditor, 45; and Drainage District One, 107; and Drainage District Seven, 112; as mayor, 30; on Marked Tree Town Council, 30; and prohibition, 35; and E. Ritter and Company, 35; and roads, 130; seeks state office, 125
Hazel, N. J., 107
Helena, Ark., 10, 76
Helena World, and whitecapping, 47–48, 52
Herndon, Angelo, 185
Higgs, Robert, 60
Hinson, H. W., 138, 149
Hitchcock, E. A., 91
Hodges, Reuben, 50, 71
Hodges v. U.S., 51, 52, 54, 70
Hog law, in Marked Tree, 31
Hogue, Louis, 161, 163, 166, 172, 222
homesteaders (*see also* Sunk Lands): and drainage districts, 104–5; dynamite passage, 86; perfect their claims, 100; and union, 2, 86, 99, 111, 191, 193, 217
Homesteaders Union, 86, 99, 111, 191, 193, 217
Hooten, J. C., 45, 132
Hopkins, Harry, 188
House Bill 6863, 101–2
House Committee on Public Lands, 94
House Resolution 96, 221
House of Commons, 208
Howington, H. H., 163, 165, 191
Hubbard, C. M., 64
Huddleston, J. A., 100
Hughes, T. B., 115
Hughes, Will, 57–58, 59

Illinois, 1, 17, 99, 100, 117
Illinois Farming Company, 114
Indiana, 17, 99
Inman, Alvin, 215
Intermediate Credit Bank, 155
Inter-river Improvement District, 108

Iowa, 18, 64
Isbell, V. O., 219, 220

Jackson, L. J., 229
Jackson, Marion, 97
Jackson County, 12
Jefferson County, 59
Jernigan, G. C., 165, 167, 173
Jim Crow, 2
Johnson, Fanny, 59
Johnson, George, 61, 63
Johnson, R. E. L., 93, 99
Johnson, Scott, 123
Joiner, J. H., 161, 162
Jones, J. Sam, 190
Jonesboro, Ark., 117-18, 145

Kansas City, Fort Scott, and Gulf Railroad, 16, 17, 18
Kansas City and Memphis Farm Company, 114
Kennedy, Herbert E., 219
Kennedy, Hubert, 219
Kentucky, 1, 71, 99
Kester, Howard, 207
Killough, O. N., 102
Kimbell, Irma, 230, 231
Kirby, Jack Temple, 3, 4
Kirkendall, Jim, 66, 67
Knapp, Seaman A., 144
Korean War, 226-27
Krier, H. J., 191, 227
Krier, John, 30, 37-38, 45, 130, 191
Ku Klux Klan, 53, 135-36, 208

Labor, 152, 154
labor agents, 61-62, 75
Ladies' Improvement Association, 39
land: buying and selling of, 115; clearing, 63; delta companies' proportion of, 116; holdings of whitecappers, 71-72; ownership rate, 39, 40, 41; on ridge individuals own greater proportion of, 116; rising prices of, 116; tenure, 40, 69, 103, 116, 142, 158, 220-21, 229, 231
Landers, A. F., 135
landlord and tenant, struggle, 203
landlord-merchant conflict, 68
Laughlin, Nurse, 148
Lawler, Oscar, 95
Lee, Henry, 55
Lee, J. T., 45, 96, 97, 98-99
Lee, Jennie, 208
Lee Wilson & Company (*see also* R. E. L. Wilson), 164
Lepanto: and the AAA program, 167; and Eliga Abbott, 103; and cotton, 115; creditors, 45; crop reduction committeemen, 165; and difficulty getting to courthouse, 124; and the eastern delta, 118; and the extension program, 168; and homesteaders, 110; and the Homesteaders Union, 86, 100, 191; land sales, 115; origins of, 97; and railroad, 116; and Red Cross, 148, 153; and roads, 42; and road to Memphis, 43; and sunk lands supporters, 96; Terrapin Derby, 157
Lepanto Chamber of Commerce, 155
Lepanto Homesteaders Union, 86, 100
Levy, Jack L., 164
Lewis, Talitha, 62
lien laws, 43-44
Lindsey, F. G., 113
liquor (*see also* prohibition): illegal traffic in, 21; and local option, 32-33
Little River, 17, 19, 22, 86, 103
Little River Township, 69, 72, 95, 98, 99, 111, 118, 131, 163
Little Rock, 14, 15, 95, 120, 230
livestock, 119-20, 142, 152, 173, 202, 220-21
Local option, 32-33

Lonoke County, 47, 56
Louisiana, 54
Lowry, Henry, 57–59, 60, 76–77
lumber industry (*see also* sawmills): and accidents, 26; attracts men to delta, 1; and black labor, 25; and boom and bust cycle, 37; and day labor, 25; and drainage, 105; and employment in, 118; and erratic nature of, 24–26; and excess labor, 27; and inaccessible swamplands, 1; laboring in, 19; mobility of workforce in, 25; northern owned, 5; origins of labor force, 26; and sunk lands, 92; and the swamps, 25; and underemployment, 25; and vulnerability to weather conditions, 27; workers in dependency relationship, 27; working conditions, 26

McCalla, John T., 165
McCracken, George, 173
McFessel, R. F., 107
McGee, C. B., 165
McGill, R. L., 161, 167, 186–87
McKinney, Wash, 50
McKinney, Rev. E. B. ("Britt"): and internal disputes within STFU, 209–10; photograph, 181; replaced by F. R. Betton, 209
McNary-Haugen bill, 141
Macon, Ga., 16
Maddox, H. P., 166
malaria, 9, 13, 63
Mandle, Jay, 59–60
Marian, Ark., 100
Marianna, Ark., 219
Marked Tree: access to the market, 41–42; access to Lepanto, 97; businessmen-planters, 2, 66; commerce and credit, 42; control of millworkers, 22; convict labor, 31, 63; cotton, 115, 177; county agent, 149; courthouse controversy, 113–14, 124, 133–34; creditors, 45; drainage, 106; Drainage District Seven, 111–12; E. Ritter and Company, 86; economic growth, 123; emerges from railroad camp, 12; employment figures, 37; factories, 31; FERA workers, 189; forest industries, 117; forest industry, 117; from sawmill settlement to plantation town, 37; frontier town, 22; M. W. Hazel, 125; hog law, 31; illegal traffic in liquor, 21; incorporated, 18; indirect route to Harrisburg, 42–43; intimidation of STFU, 207; institutional diversions, 27; H. J. Krier, 191; lack of women and children, 18; V. O. Isbell, 219; Britt McKinney, 209; malaria, 19; mayor's office, 129; W. B. Miller, 91, 94; Mississippi River bridge, 123; Dr. S. L. Mitchum, 68; mobile workforce, 24–25; newspaper, 130; occupational characteristics, 32; occupational figures, 37–38; plantation sector, 39; Poinsett County Taxpayers Association, 135; population of, 37; POW camp, 222, 223; predominance of males, 18; problems confronting early settlers, 19; Progressive era, 21; prohibition, 22, 37; prosperity, 118; railroad, 116; Red Cross, 147, 148, 152–53; Ernest Ritter, 119; L. V. Ritter, 162; Ward Rodgers, 207; roads, 42; saloons, 21, 28; St. Francis River, 108; the STFU, 197; B. F. Taylor, 163; Norman Thomas speaks, 190–91; transformation of, 37; transformation of workforce, 32; transient culture, 73

Marked Tree Bank and Trust Company, 35, 45
Marked Tree Farmers and Merchants Bank and Trust Company, 155
Marked Tree Gazette, 34–35, 86, 98, 99, 114, 124, 131, 132
Marked Tree Rotary Club, 149, 225
Marked Tree Town Council: composition of, 30; and control of millworkers, 30–31; and economic expansion, 31; and L. C. Going, 51; ordinances to control millworkers, 29–30; and prohibition, 33–34; and W. B. Miller, 107; and Ernest Ritter, 107
Marked Tree Tribune, 230
marketing cooperatives, and Poinsett County, 141
Martin, John, 115
Mayo, S. T. (Judge), and roads, 130–31
mechanization, 4, 170–71, 224–25, 231
Memphis, Tenn., 40, 43, 114, 123, 188, 205, 208
Memphis *Commercial Appeal*, 49
Methodist Episcopal Church, Marked Tree, 38
Mexican labor program, 224–25, 226, 227, 232
Mid-South Cotton Growers Cooperative Association, 154
Miller, E. A., 164, 187
Miller, W. B.: and the CDCC, 128; and Drainage District One, 107; fall from power, 102; on Marked Tree Town Council, 30; and the Oldfield bill, 95; and St. Francis Levee Board, 30, 91, 94, 102, 126; and sunk lands purchase, 91
millworkers: dominate Marked Tree's population, 23, 27; and Marked Tree ordinances, 22; and prohibition, 21–22; and saloons, 28

Mingo District, 108–9
Mink, C. W., 119
Mississippi, 1, 17, 48, 52, 54, 62, 64, 66, 71, 130, 184, 207, 227
Mississippi County, 10, 13, 29, 45, 52, 57–59, 80, 83, 86, 87–88, 89, 91, 94, 96, 99, 100, 110, 142, 152–53, 164, 207, 230
Mississippi River, 61, 108
Mississippi River bridge, 123
Missouri, 1, 3, 19, 99, 108–9
Missouri Pacific Railroad, 76
Mitchell, Harry L.: and the CWA controversy, 187, 188; decides to organize a union, 191; and Dibble plantation, 205; and internal disputes within STFU, 209; intimidation of, 207, 208; photograph of, 179; protests foreign labor program, 224; and socialism, 190
Mitchell, J. C., 130
Mitchell, J. K., 115
Mitchum, Dr. S. L., 68–69, 115
mobility, 59–61, 77, 196, 217–18
Modern News, 97, 119, 120, 124, 128
mortgages, 12, 39, 43–45
Mouton Rice Mill, 225
Movable School, 228–29
mules, 5, 220–21, 225
Music, Grant, 100

NAACP, 54, 184, 185, 210, 230
National Farm Labor Union, 226
National Grange, 140
National Relief Administration, 184
Nettleton, Ark., 124
New Deal (*see also specific programs*): challenge to, 4; define class relations, 8; and evictions, 184, 209; and farm laborers, 190; favored delta elite, 157, 160; impact on landless farmers, 202; and planters, 7, 212; tenants and sharecroppers, 199

New Madrid Earthquake, 87
New Madrid, Mo., 10
Newport, Ark., 12
Nichols, Jim, 100, 103, 104
nightriders, *see* whitecapping
Nodena Landing, 58
Norcross, Hiram, 8, 161, 163, 192, 198, 199, 204–6, 209
North Africa, 222
North Carolina, 12, 61, 99–100
Northern Ohio Cooperation Company, 214
Northern Ohio Farm, 214
Nunnally, Alvin, 198

Odd Fellows Hall, Tyronza, 188
Odd Fellows Lodge, 58
Oldfield bill, 94–96
Old Southwest, 3
Oliver Davis Sawmill, 19
Orrell, Geraldine Green, 148
Osceola, Ark., 10, 29
Osceola Times, 58, 86
Ouachita Mountains, 25
Owens, W. A., 145
Ozark Trail, 123

Palmer, Edwin, 9, 10, 12, 14, 15, 233
panic of 1907, 25
Paragould, Ark., 108
Parkin, Ark., 178
paternalism, 64–65, 193
Payne, W. R., 112
pellagra, 64
peonage, 55
Perry, E. M., 153, 154
Pharr, H. N., 89–90
Phillips, John P., 16
Phillips County, 47, 48, 78, 84, 85, 183
Pickens, William, 54, 59
Pierce, E. L., 147
Piggott, Ark., 119
Pine Bluff, Ark., 59

planter-tenant conflict, 57–59, 64–65
plow-up campaign, 161, 163, 164, 192–93
Poplar Bluff, Mo., 100, 115
Portis, D. F., Sr., 43
Portis, Dan, 167
Post War Planning Commission, 225–26
poultry farming, 145, 173
Powell, Mabern, 146
Powell, Robert L., 146
POW labor program, 221–24
Pratt, Charles, 76
Prestidge, J. H., 166
Production Credit Association, 165, 166
Progressive era, 21, 112
Progressive Farmers and Household Union of America, 75–76, 199
progressivism: and disfranchisement, 22; links delta and ridge elites, 2; and prohibition, 21–22, 36–37
prohibition: and blacks, 36; and the Going Amendment, 36; and legislative action, 36–37; links delta and ridge elites, 2; and local option approach, 32–33; Marked Tree merchants divided over, 28; and millworkers, 36; prevails, 3; and progressivism, 21–22, 36–37; statewide drive, 29; statute on 1912 ballot, 35, 36
Public Lands Committee, 95, 99
Public Works Administration, 210–12

race or racism, 56, 75, 69, 135, 196, 205, 230; impediment to solidarity of tenants and sharecroppers, 7–8; experiment with interracial cooperation abandoned, 8; fear of intermingling, 2; fluid nature of race relations, 2; intermingling in saloons, 28, 37; and

race or racism (*cont.*)
    racial composition of prairie wage laborers, 122; and STFU, 217; and whitecapping, 48–49
Radical Reconstruction, 15
railroads: attract men to delta, 1; availability of jobs, 62–63; black employees assassinated, 184; bonds, 15; camp at Little River and St. Francis River, 12; come to Poinsett County, 15–17; and Jay Gould, 16; injury to employee, 100; Kansas City to Memphis line, 17; Little Rock to Memphis line, 17, 42; between Marked Tree and Lepanto, 97; penetrates the swamp, 1; shortline to Truman, 118; work crew (photograph), 78; workers, 22
Rankin, Ill., 123
Ray, Lonnie, 43
Reconstruction Finance Corporation, 159–60, 212
Red Cross: aid to farmers, 155; and delta elite, 139–49, 156, 206; and drought of 1930–31, 151–52, 157; and flood of 1927, 147; and C. M. Hubbard, 64; local committees, 147; Marked Tree chapter and works program controversy, 151–53; resistance from local doctors, 148; shakes foundation of plantation system, 148–49; works programs, 189
Reddit, J. W., 145–46
Reddit, Carl, 145–46
Red River Bottom, 61–62
Reid, T. Roy, 160–61, 215
Republican party, 35, 111, 128–29, 140–41
Resettlement Administration, 166, 213–16
Reynolds, L. J., 161, 163

rice: and the AAA, 159, 171–72, 173, 174; acres in, 170; and the county agent, 118; in the delta, 233; farmers mechanize, 224; growers slow to cooperate with AAA program, 172; mill burns, 122; millers dissatisfied with AAA program, 171–72; prairie begins to produce, 39, 41; prairie production increases, 121
riparian: claimants secure favorable legislation, 103; injunctions against settlers, 93; claimants and legislation, 101–2; and petition to Department of Interior, 92; rights raised by Hoke Smith, 90
Ritter, Anna Hirschman: moves to Marked Tree, 18; moves to Memphis, 38–39; and women's organizations in Marked Tree, 39; and WCTU, 32
Ritter, Ernest: background, 64; and black sharecroppers, 64; builds housing subdivisions, 23, 39; opens cannery, 119; comes to Marked Tree, 18; contracts to build roads, 24; courthouse controversy, 133; as a creditor, 45; and delta development, 43; and Democrats, 102; dominates labor, 24; and drainage, 24, 88, 103, 107–8, 112; and the election of 1912, 35, 36; employment in Oliver Davis Mill, 18; family portrait, 80; fish, 23; Harry Ritter (son), 147; homesteads 160 acres, 18; and ice plant, 23; injury, 18; and L. V. Ritter (son), 149; and labor needs, 75; and landholdings, 23–24, 38; Marked Tree Town Council, 30, 37–38; Marked Tree Bank and Trust Company, 35, 45; Oldfield bill, 95, 96;

Ritter, Ernest (*cont.*)
Ozark Trail, 123; picture of store, 79; plantations, 23; postmaster, 23; as a Progressive, 28; and prohibition, 28, 34, 35, 36; as a Republican, 28, 35, 129; and roads, 42, 130; St. Francis Levee District, 92; sunk lands, 91, 92, 94–95, 96; WCTU, 32; wealth, 24; white tenants, 64; whitecapping, 52; wife and family move to Memphis, 38–39
Ritter, Harry, 147, 207
Ritter, L. V., 102–3, 112, 142, 149, 151, 154, 160, 161, 162, 166, 220, 222, 231
Ritter, Louis, 42, 45, 115, 149
roads: default on taxes, 142–44; difficult-to-maintain road between ridge and delta, 122–23; and difficulty getting to courthouse, 124; Harrisburg and Weiner road, 123; importance to Poinsett County delta, 43; linking prairie and ridge, 122–23; map, 41; Marked Tree–Harrisburg, 130; and Memphis, 43; and the plantation economy, 41–42; and political struggle, 130; and ridge connection to prairie, 42; and the role of Marked Tree merchants, 23, 41–42; taxes, 151
Robert, W. J., 161
Robinson, Joseph T., 94, 95–96
Rob Roy, Ark., 59
Rockefeller Foundation, 148
Rodgers, Ward A., 179, 206–7, 209
Rooks, J. W., 128
Roosevelt, Eleanor, 205
Roosevelt, Franklin, 158, 160, 203–4, 215
Roosevelt, Theodore, 87, 112
Royce, Edward, 6
Ryan, Steve, 233

St. Francis County, 14, 52, 94
St. Francis Land Company, 94–95
St. Francis Levee Board: appointments to, 125, 126; and J. A. Emrich, 129; and W. B. Miller, 30, 107
St. Francis Levee District: counties within, 89; founding and early history of, 125; importance of drainage efforts, 160; and initial drainage, 104; and New Deal works programs, 212; and nullification of sunk lands transactions, 91; and Oldfield bill, 95; promotes drainage, 89; and riparian rights, 90; and Ritter suit, 92; struggle for control over, 94, 102; and sunk lands, 88, 90, 91
St. Francis River, 10, 12, 17, 19, 22, 27, 108, 109, 130
St. Francis Valley Company, 232
St. Francis Valley Floodway project, 168, 212
St. Louis, Mo., 42, 193
St. Louis, Iron Mountain and Southern, comes to Harrisburg, 17
St. Louis Southwestern Railroad, comes to prairie, 16
St. Matthias Guild, 39
saloons, 10, 27, 28, 29, 33, 38
Sargent, E. F. B., 145
sawmills, 1, 18–19, 31–32, 70, 100, 122
Schonberger, Jake, 147
schools: built in Marked Tree, 22; segregated system, 23; taxes, 151
Schuster, Mr., 173
Scott, James, 56–57
Scottsboro boys, 185
Scott Township, 69, 70, 131
scrip, 114, 135
Searcy Township, 14
segregation, 2, 28, 196
Senteney, Chester, 163
Seymour, Albert, 43

Seymour, Layton, 43
sharecroppers or sharecropping: bring delta into cultivation, 5; come to Poinsett County, 2; and commissary system, 62; community of, 195; compared to millworkers, 27; as a compromise, 55–56; cotton crisis, 121; and the crop lien, 44; and the CWA, 188; and debt, 54–55; decline in number of, 229; definition of, 65; in the delta, 17; and destruction of, 218; and Dibble evictions, 209; and dietary deficiencies, 64; drop illusion of partnership with planters, 8; drought of 1930–31, 151, 152; Mose Evans, 62; and evictions, 208; face starvation, 158; and the furnishing system, 192–93; "going home" (photograph), 176; and illusion of partnership with planters, 6; and increase in, 142; in the law, 6, 65; Henry Lee, 55; and malaria, 63; meeting (photograph), 180; and mobility, 7; on the neoplantations, 220–21; and the New Deal, 199–200, 206; and Norman Thomas, 191; partnership with planters, 56; and pellagra, 64; perspective of status differs from planters, 7; on Poinsett County plantations, 114–15; position eroded, 227–28; working in the prairie, 121; and racism, 75; and the Red Cross, 148–49, 154; relationship with planters, 6; and resettlement administration communities, 213–14; restrictive laws, 7; and Ernest Ritter, 24; size of farmholding in Poinsett County, 8; STFU, 207; support for, 192; sympathy for, 227; the flood of 1927, 147; the extension service and blacks, 228; weaken delta challenge, 130; Harold Woodman describes, 6
Sharp, Grandisom M., 14–15
Shaw, Isaac, 198–99
Sheddan, Arthur, 57–58
Shelton, Dan, 70
Shelton, W. D., 34, 120, 135, 136
Sidle, William, 57
Singer Sewing Machine Company, 117, 135
Sloan, C. C., 34
Sloan, Homer, 115–16
Sloan, H. F., 161
Smith, C. H., 197–98
Smith, Hilliard C., 86–87
Smith, Hoke, 90
Smith, P. D., 120
Smith-Lever Act, 144
Socialist party, local ticket, 191
soil conservation payment, 215
South Carolina, 14
Southern Tenant Farmers' Union (STFU): AAA regulations in response to, 205–6; and Robert B. Barnes, 202; and churches, 193, 194; and cooperative farm in Mississippi, 207, and Dibble plantation, 205; and E. Ritter and Company, 232; first meeting, 198–99; founding of, 184, 185; and H. H. Howington, 163; and internal disputes, 209–10; as interracial union, 184, 197, 198–99, 217; and Norcross suit, 207–8; place in history, 5; and postwar planning, 225; protests foreign labor program, 224; reasons behind formation of, 8, 186, 191, 192, 193, 216; role of women, 194–95; and Roosevelt's New Deal coalition, 203–4; and C. H. Smith arrest, 197–98; and George Stith, 226; strikes, 207; violence against, 207

Southwestern Land and Timber Company, 105–6, 110
Southwestern Socialist Encampments, 190
soybeans, 151, 154, 168, 169, 170, 173, 221, 233
Stafford, Fred H., 209
Stafford, William, 70
State Democratic Central Committee (CDCC), 128
Staton, T. G., 30
Stith, George: and resistance, 60; and meeting on the turn row, 197; and Mexican labor, 227; and postwar labor needs, 226; and the separate water bucket ritual, 196–97; and STFU, 208
stock law, 23, 138–39, 149, 151
strikes, and STFU, 207
Stuck, Dorothy, 230
Stuckey, Sam, 43
Student Non-Violent Coordinating Committee, 230
Stultz, William A., 207, 209–10, 217
Stuttgart, Ark., 122, 172
Sullivant, A. Raybon: County Plan of Work Committee, 162; and Crowley's Ridge, 174; delta orientation, 149, 150–51; and the drought of 1930–31, 154; and plow-up campaign, 161; and Red Cross aid, 155; and the stock law, 138
sunk lands: creation of, 87; and lumbering prohibition, 92, 93; opened to homesteading, 88; preferential rights, 87, 93–94, 101–2, 112; and riparian rights, 90, 91; and J. A. Tellier, 95; trespassers, 90
Sunnyside schoolhouse, 191, 198, 199

Tallapoosa County, Ala., 184
taxes, default on, 3, 141–43, 151

Taylor, Dr. B. F., 161, 163
Taylor, Frederick, 226
Teapot Dome, 87, 112
Tellier, J. A., 95
tenants or tenancy: blacks and the extension service, 228; bring delta into cultivation, 5; cash tenancy, 142; compared to millworkers, 27; and cotton crisis, 121; crop lien, 44; CWA, 188; decline in number of, 206, 229; debt, 54–55; definition of, 65; in the delta, 17; destruction of, 218; deteriorating economic position of, 199–200; dietary deficiencies of, 64; and drought of 1930, 151, 152; and evictions, 208; furnishing system, 192–93; illusion of partnership with planters, 6, 8; increase of, 12; and the law, 6, 7, 65; within malaria zone, 63; from Mississippi, 66–67; mobility, 7; neoplantations, 220–21; New Deal regulations, 206; meet Norman Thomas, 191; and pellagra, 64; perspective of status differs from planters, 7; Poinsett County use of, 114–15; position eroded, 227–28; in the prairie, 121, 122; and the Red Cross, 148–49, 154; relationship with planters, 6; and resettlement administration communities, 213–14; and Ernest Ritter, 24, 75; size of farmholding in Poinsett County, 8; and STFU, 207; support for, 192, 227
Tennessee, 1, 17, 54, 99, 187, 222
Terrapin Derby, 157
Texas, 3, 49, 54
Thatcher, Jean, 220, 232, 233
Thirteenth Amendment, 51, 52
Thomas, Norman: photograph, 180; speaks in Marked Tree, 190–91

Thompson, C. M., 106, 110–11
Thompson, E. P., 7
Thompson, J. C., 190
Thompson, Joe, 34
Thompson, S. P.: and credit, 38; landholdings, 38, 115; prosperity of, 114; rise from saloonkeeper to merchant and banker, 34; runs for local office as Republican, 35
Thorn, H. B., 131, 186
Tookone, W., 124
tractors, 122, 170–71
Trieber, Judge Jacob: background, 53–54; on whitecappers, 71; and whitecapping cases, 51
Trotter Hotel, 134
Truman, Harry S., 226
Truman, Ark., 117, 118, 135, 147, 169, 214, 225
Truman Lions Club, 155
Tugwell, Rexford G., 160–61, 213
Tyronza, Ark., 42, 45, 115, 124, 129, 132, 145, 162, 186, 187, 191, 230
Tyronza Business Men's Club, 149, 150
Tyronza Central Railroad, 116
Tyronza Supply Company, 192, 193
Tyronza Township, 68, 69, 72, 99, 163, 193, 194, 195, 198, 199, 200–203, 204
Tyronza Unemployment League, 188
Tyrus, Burton, 61

Unemployed Citizens League, of Tyronza, 189
Union, Mo., 117
Union County, Ark., 184
Union Labor party, 208
U.S. Army, 86
U.S. Supreme Court: and race, 53; and STFU, 208; and sunk lands, 90, 91–92, 93, 101, 102; and whitecapping case, 151

U.S. v. Wilson, 93–94
University of Arkansas College of Agriculture, 220

wage labor, 41, 121–22, 142, 170, 206, 217, 221, 233
Walker, J. O., 215
Walker, W. T., 161, 163
Walker, Wiley, 66–67
Walz, S., 117
War Department, 108, 222
War Manpower Commission, 224
Warren, Sam, 100
Warren, W. W., 135, 145, 150
Warsaw, Ill., 117
Waskom, John G., 68
Watkins, C. J., 124
Weeks, James, 73–74
Weiner, Ark., 16, 122, 133–34, 172
Weiner Bank, 163
Weiner Rice Mill, 122
Weona, Ark., 118
Weona Land Company, 117–18
West Memphis, Ark., 212
Wharton, Charles W., 173
Whipple, William, 50–51
Whitaker, N. T., 110–11
White primary, 22
Whitecappers, 7–8, 51–52, 53, 69, 73
White Chapel, Ark., 48, 52
White Chapel Club, 48–49, 50, 71–72
White Hall, Ark., 50, 51, 70, 102, 162
White River, 78
white supremacy, 59, 75, 135
White, D. D., 123
White, J. A., 227
Wiener, Jonathan, 59–60
Willbeth plantation, 115–16
Williams, Burt, 198–99
Williams, L. D., 115
Willis, J. R., 126, 128, 130
Willis Township, 126–27, 136

Wilson, R. E. L.: and preferential rights to sunk lands, 102; and riparian rights to sunk lands, 91; and suit over sunk lands, 91; and sunk lands and lumbering, 92
Wilson, Ark., 164
Winn, Alexander M., 14–15
Wofford, C. O., 163, 173
Wolfe, Jess, 16
women: and churches, 39; numbers increase in Marked Tree, 38; organizations in Marked Tree, 39; role of in STFU, 194–95; and schools, 39; social and economic roles in southern rural culture, 195
Women's Christian Temperance Union, 28, 32, 35–36
Woodman, Harold, 6–7
Woodruff, Nan, 4
Woodruff County, 208, 209

Works Progress Administration, 63, 210–12
World War I: agricultural crisis following, 117, 118, 120, 122; boom, 3, 111
World War II, 118, 216, 219, 224, 228, 234; and impact on Resettlement communities, 215–16; postwar era rise of neoplantation, 5; stimulates massive exodus, 220; transformation of the plantation after, 3
Wright, Gavin, 3, 4, 7

XV Club, 53–54

Yates, W. G., 115
Yazoo Mississippi delta, 4
Yellow Town, 27

Zeigenhorn, R. A., 161, 163